MINEFILL 2020-2021

T0260369

PROCEEDINGS OF THE 13TH INTERNATIONAL SYMPOSIUM ON MINING WITH
BACKFILL, KATOWICE, POLAND, 25-28 MAY 2021

Minefill 2020-2021

Editors

Ferri Hassani
McGill University, Canada

Jan Palarski
Silesian University of Technology, Poland

Violetta Sokoła-Szewioła
Silesian University of Technology, Poland

Grzegorz Strozik
Silesian University of Technology, Poland

CRC Press
Taylor & Francis Group
Boca Raton   London   New York   Leiden

CRC Press is an imprint of the
Taylor & Francis Group, an **informa** business

A BALKEMA BOOK

CRC Press/Balkema is an imprint of the Taylor & Francis Group, an informa business

© 2021 Copyright: Authors

Typeset by Integra Software Services Pvt. Ltd., Pondicherry, India

The Open Access version of this book, available at www.tandfebooks.com, has been made available under a Creative Commons Attribution-Non Commercial-No Derivatives 4.0 license.

Published by: CRC Press/Balkema
        Schipholweg 107C, 2316XC Leiden, The Netherlands
        e-mail: Pub.NL@taylorandfrancis.com
        www.routledge.com – www.taylorandfrancis.com

ISBN: 978-1-032-07203-6 (Hbk)
ISBN: 978-1-032-07205-0 (Pbk)
ISBN: 978-1-003-20590-6 (eBook)
DOI: 10.1201/9781003205906
https://doi.org/10.1201/9781003205906

Minefill 2020-2021 – Hassani et al (eds)
© 2021 Taylor & Francis Group, London, ISBN 978-1-032-07203-6

Table of contents

Preface	ix
Committees	xi
Sponsors	xiii

Sponsor sessions

Efficient paste mix designs using new generation backfill admixtures – perception versus reality *F. Erismann & M. Hansson*	3
Paste fill stiffness investigation *J.D.V. Wickens & S. Wilson*	13
Pulsation dampening systems for piston pumps operating at backfill installations *T. Lutz & P. Peschken*	22
Paste quality control benchmarks *D. Stone*	35
Using paste backfill for stabilizing underground stopes at the Giant Mine remediation project *B. Ting, D. Stone, K. Ruptash & B. Fabien*	44
Investigation into the high transients experienced in Eleonore Mine's pastefill distribution system *M. McGuinness, K. Creber, J. Jacobs & B. Haley*	53

Paste backfill measurements and testing

The effect of curing conditions on the strength development of cemented backfill samples *B. Dennis & M. Helinski*	67
Experimental investigation on shear strength properties of interface between backfill and rock *G. Liu, W. Wu, L. Guo, X. Yang & Z. Zhang*	80
The concept of obtaining backfilling material using the dredging method *M. Gruszczyński, S. Czaban, & Sz. Zieliński*	90
The properties of the backfill mixtures based on own fine-grained waste *R. Pomykała & W. Kępys*	102
Cemented paste backfill failure envelope at low confining stress *M. Grabinsky & A. Pan*	108

Cemented paste backfill response to isotropic pressure 118
M. Grabinsky & M. Jafari

Evaluating cemented paste backfill plug strength and the potential for continuous pouring 127
M. Grabinsky, B. Thompson & W. Bawden

Binders, admixtures, and other chemicals to improve fill performance

The influence of the flocculant on the process of thickening and depositing of copper ore
flotation tailings 143
M. Gruszczyński, S. Czaban, R. Pratkowiecki, Z. Skrzypczak & P. Stefanek

Determining the required underground grout pack production profile for narrow tabular
mining operations 153
B. van der Spuy

Rheological yield stress measurement of paste fill: new technical approaches 169
M. Silva, M. Hansson & M. Costa e Silva

Backfill reticulation: pumping, piping, hydraulic analyses

Automated backfill diverter valves at Kirkland Lake Gold Fosterville Mine improve safety
and efficiency 185
R. Evans & G. Trinker

Filter revamping, the economic way to get old filters in tailings dewatering back on track 194
J. Hahn

Theoretical methods to improve the placement of paste backfill behind barricades 204
C. Lee

Laboratory testing on static and dynamic behaviours of backfill

A novel approach to assessing the early age strength of fibrecrete, using shear wave velocity 217
S. McGrath & M. Helinski

A design procedure for evaluation and prediction of in-situ cemented backfill performance 227
X. Wei & L. Guo

The uniaxial compressive strength of fiber-reinforced frozen backfill 234
H. Niu, F.P. Hassani, M.F. Kermani & M. He

Incorporating the monolithic nature of paste backfill into self-heating assessments 244
C. Lee, B. Timmis, D. Brown, V. Bertrand & M. Stewart

Minefill geomechanics, numerical modeling, mitigation of subsidence

DEM numerical modeling of longwall extraction of coal in "Mysłowice" colliery 261
G. Smolnik

Deformations of the mining area surface as a result of exploitation with sealing of caving
gobs 272
V. Sokoła-Szewioła, A. Mierzejowska & M. Poniewiera

Geomechanical safety aspects in hard rocks mining based on room-and-pillar and longwall mining systems   288
W. Pytel, B. Pałac-Walko & P. Mertuszka

Legal, safety, and environmental drivers for backfill

Filling underground voids to prevent water hazards in active and decommissioned hard coal mines   307
G. Strozik

Impact of the method of managing opencast excavations by filling with mining waste on the quality of leachate entering the surface water   317
S. Rzepecki, A. Grodzicka & K. Moraczewska-Majkut

Properties cemented backfill prepared on the basis selected coal combustion products (CCPs)   327
P. Pierzyna

Impact involving the sealing degree of caving goaf with fine-fraction hydraulic mixtures on the ventilation parameters of longwall headings   337
M. Popczyk & D. Musioł

Case studies

Geomechanical analysis of the rock mass stability in the area of the "Regis" shaft in the "Wieliczka" Salt Mine   349
G. Dyduch, P. Jarczyk & M. Jendryś

Case study: paste plant retrofit   360
L. Correia, B. Cothill, P. Antunes & S. James

Review of sandfill reticulation system at northern ontario mine   375
J. Landriault, D. Dewit & J. Yamine

Paste-waste design and implementation at newmont goldcorp's tanami operation   382
R.L. Veenstra & J.J. Grobler

Summary of improvements to the backfill system at DBS operations   398
A. Zajac & R.L. Veenstra

Case study – design and implementation of a high density cemented hydraulic fill system at Oz Mineral's Prominent Hill Mine   409
M. Helinski & J. Shaw

Green mining – use of hydraulic backfill in the Velenje coal mine   422
J. Kortnik

Author index   433

Preface

The series of International Symposium on Mining with Backfill (*Minefill*) explores both the theoretical and practical aspects of the application of mine fill, with many case studies from both underground and open-pit mines. *Minefill* attendees include mining practitioners, engineering students, operating and regulatory professionals, consultants, academics, researchers, and interested individuals and groups from within the wider community. There are presentations from people from around the world working in very diverse environments, with a range of local expertise showing their unique problems but common solutions.

Since the first Symposium held in Mount Isa in 1973, the series has had 12 editions. Since 1998 it has a form of a triennial symposium being held most frequently in Australia, Canada, South Africa, and the United States.

The Symposium topics evolve with the development of trends in mine fill technologies, overall mining problems, hazards, environmental standards, etc.

In the current edition of the Symposium *Minefill* 2020-2021, the following range of detailed themes has been proposed:

➢ Legal, safety, environmental and financial drivers for backfill
➢ Mine fill geomechanics, numerical modeling, the interaction between backfill and rock mass
➢ Bulkhead capacity and fill strategies
➢ Binders, admixtures, and other chemicals to improve fill performance
➢ Backfill reticulation: pumping, piping, hydraulic analyses
➢ Field fill mass blast and stress monitoring before and after exposure and data analysis
➢ Laboratory testing on static and dynamic behaviors of backfill
➢ Hazard and risk control practices in backfill application
➢ Application of new technology and new equipment in mining with backfill
➢ Influence of mining with backfill on the mitigation of subsidence related hazards
➢ Case studies.

This year, the Symposium is organized in the Silesia Coal Region in Poland, where hydraulic fill has been used for the first time in the world. Originally, the Symposium was planned for 2020, but due to the extraordinary epidemic situation related to COVID-19 (SARS-Cov-2 virus), the conference was moved to 2021, with the name of the Symposium being *Minefill 2020-2021*. The organization of the Symposium was adopted in a remote form on a specially dedicated platform.

Polish underground mines of salt, base metals, and coal have used sand, gravel, underground waste rock, tailing, smelter slag, and coal combustion by-products since the end of the 19th century to fill underground voids. Fills used in coal mines have evolved from early loosely dumped rock, ash, and slag through hydraulically placed sand fill (in the early 1880s), pneumatic (the year 1920), and throwing (the year 1947) stowing up to today's hydraulically transported, densified fills with fly ash, tailing and flue-gas desulphurisation by-products. The use of fly ash, slimes, and water mixture as compaction grout of roof fall materials in gob area has had a very great impact on mining practice in Polish coal mines (the year 1975). Additionally, in various locations, fly ash slurry has been injected into underground abandoned coal mine workings to provide structural support for surface damages reduction.

Mine backfill plays an essential role in the modern mining industry in the world and will continue in the upcoming years. Its main role is to improve stope safety by providing both local support (effect of backfill on the stability of the roof in the working area) and regional support (behaviour of the rock mass surrounding the mining excavation). Its other advantages include, amongst other things, the reduction of the quantity of mine waste to be disposed of on the surface.

Currently, the development of innovative fill materials and technologies that can be applied to unmanned and intelligent mining to reduce the negative impact of mining activities on the environment and communities is a key challenge faced by many engineering research efforts.

Minefill 2020-2021 Symposium is the best platform to learn the latest achievements and share knowledge and expertise with mining and environmental professionals across the world.

We hope that this Symposium will fill you with inspiration and will provide many fruitful opportunities for future cooperation between Participants.

We wish you a very productive symposium, a lively exchange of knowledge as well as many interesting discussions.

Gliwice, May 25-27, 2021.

The Chairman of Organizing Committee
Prof. SUT, Violetta Sokoła-Szewioła
PhD, DSc, Eng

The Chairman of Scientific Committee
Prof. Jan Palarski, PhD, DSc, Eng

Minefill 2020-2021 – Hassani et al (eds)
© *2021 Taylor & Francis Group, London, ISBN 978-1-032-07203-6*

Committees

Advisory Committee:

Professor Ferri Hassani, *McGill University, Canada*

Doctor David Stone, *President of MineFill Services, Washington, USA*

Professor Jan Palarski, *Silesian University of Technology, Poland*

Organizing Committee:

Chairman of the Symposium: Professor Violetta Sokoła-Szewioła, *PhD, DSc (Eng), Silesian University of Technology, Poland*

Co-chairman of the Symposium: Doctor Grzegorz Smolnik, *PhD (Eng), Silesian University of Technology, Poland*

Professor Franciszek Plewa, *Dean of the Faculty of Mining, Safety Engineering and Industrial Automation of the Silesian University of Technology, Poland*

Professor Jan Palarski, *Silesian University of Technology, Poland*

Professor Ferri Hassani, *McGill University, Canada*

Professor Grzegorz Strozik, *Silesian University of Technology, Poland*

Doctor Aneta Grodzicka, *Silesian University of Technology, Poland*

Doctor Patrycja Jarczyk, *Silesian University of Technology, Poland*

Doctor Dariusz musioł, *Silesian University of Technology, Poland*

Doctor Piotr Pierzyna, *Silesian University of Technology, Poland*

Doctor Marcin Popczyk, *President of the Foundation for the Faculty of Mining, Safety Engineering and Industrial Automation of the Silesian University of Technology, Poland*

Doctor Marta Matyjaszek, *Foundation for the Faculty of Mining, Safety Engineering and Industrial Automation of the Silesian University of Technology, Poland*

M.Sc. Małgorzata Wrzesień, *Silesian University of Technology, Poland*

Scientific Committee:

Chairman of Scientific Committee: Professor Jan Palarski, *PhD, DSc, (Eng), Silesian University of Technology, Poland*

Co-chairman of Scientific Committee: Professor Ferri Hassani, *PhD, FCIM,MIMMM, Webster Chair Professor, Department of Mining and Materials Engineering, McGill University, Canada*

Professor Zacharias Agioutantis, *University of Kentucky, United States*

Professor Marek Cała, *Agh University of Science & Technology, Poland*

Professor Elisabeth Clausen, *Rwth Aachen University, Germany*

Professor Stanisław Czaban, *Wrocław University of Environmental and Life Sciences, Poland*

Professor Christoph Dauber, *Technische Hochschule Georg Agricola, Germany*

Professor Lijie Guo, *Beijing General Research Institute of Mining and Metallurgy, China*

Professor Jože Kortnik, *University of Ljubljana, Slovenia*

Professor Jadwiga Maciaszek, *AGH University of Science and Technology, Poland*

Professor Václav Matoušek, *Czech Academy of Sciences, Prague, Czech Republic*

Professor Franciszek Plewa, *Silesian University of Technology, Poland*

Professor Radosław Pomykała, *Agh University of Science and Technology, Poland*

Professor Stanisław Prusek, *Central Mining Institute, Poland*
Professor Chongchong QI, *Central South University, Changsha, China*
Professor Antonio Bernardo Sánchez, *University of León, Spain*
Professor Serkan Saydam, *University of New South Wales, Australia*
Professor Nikolaus August Sifferlinger, *University of Leoben, Austria*
Professor Jerzy Sobota, *Wrocław University of Environmental and Life Science, Poland*
Professor Violetta Sokoła-Szewioła, *Silesian University of Technology, Poland*
Professor Grzegorz Strozik, *Silesian University of Technology, Poland*
Professor Pavel Vlasak, *Czech Academy of Sciences, Czech Republic*

Minefill 2020-2021 – Hassani et al (eds)
© 2021 Taylor & Francis Group, London, ISBN 978-1-032-07203-6

Sponsors

Platinum sponsor

Gold Sponsor

MINEFILL SERVICES

Silver Sponsor

BUILDING TRUST

Silver Sponsor

Silver Sponsor

Silver Sponsor

Silver Sponsor

Sponsor sessions

Minefill 2020-2021 – Hassani et al (eds)
© 2021 Taylor & Francis Group, London, ISBN 978-1-032-07203-6

Efficient paste mix designs using new generation backfill admixtures – perception versus reality

Fabian Erismann
Sika Tunneling and Mining, Switzerland

Martin Hansson
Sika Tunneling and Mining, Sweden

SUMMARY: Backfilling mined out stopes with cemented paste has become the standard in most modern, long-hole stoping- and cut- and fill operations globally. The implementation of large scale paste production plants and efficient underground paste reticulation systems contributed significantly to maximised ore extraction, higher levels of mine-scale, geotechnical stability, larger degree of automation and last but not least to a more conscious management of mine processing waste. However, paste backfilling comes at a cost.

Binder costs alone amount to 70-80% of total operating costs of paste plants depending on the strength requirements and characteristics of tailings used for backfilling and it is not uncommon that underground paste filling contributes up to 20% of the overall cost structure of an underground operation. With a view towards improving the paste mix designs and keeping cement consumption under control, new generation paste backfill admixtures have been developed over the past years. The mine backfill industry can be compared to the concrete industry 40 years ago, when high performance admixtures started to emerge for land mark projects, where exceptionally high concrete qualities and specific, fresh concrete properties were required (Aitcin and Wilson, 2015). These high range water reducers quickly transformed the construction industry and pushed concrete applications to new, previously unknown limits.

This paper intends to give an overview of the reality of admixtures in the mine backfill industry and their performance and justification in modern paste backfill plants. Data from different paste plants around the globe has been compiled for this study, covering different deposit types, strength requirements and paste mix designs. The study illustrates the powerful effect paste backfill admixtures can have on standard, paste mix designs and their potential to reduce binder consumption substantially by remaining within the required strength- and workability limits of the paste to improve the overall cost structure of backfill operations.

Keywords: paste fill, admixtures, binder, optimization, mining

1 INTRODUCTION

High range superplasticisers with the ability to significantly reduce water out of a given concrete mix, while maintaining the workability but strongly influence the strength- and durability characteristics of concrete, have been introduced to the global construction industry by the end of the 1990's (Ramachandran et al., 1998). Today, the fundamental reason to develop and utilise such superplasticisers in nearly every day mass produced concrete is the strong and lasting effect of a lower water content, or lower water to binder ratio to be more specific, on concrete. As an example: In order to

DOI: 10.1201/9781003205906-1

produce a 25 MPa concrete it is necessary to use around 300 Kg of cement without the use of an admixture. In contrast, when a 75 MPa concrete is produced, 450Kg of cement is needed plus some litres of superplasticizer to reduce the water to binder ratio. Hence, while using only 1.5 times as much cement, the strength of the concrete triples (Aitcin and Flatt 2016).

Similar to concrete, cemented paste is a cementitious system too, sharing similarities with concrete. The fact that reducing the water to binder ratio in concrete has a positive effect on strength development at a given cement content, is even more pronounced in water-saturated or a water-oversaturated system like a cemented paste, where the water to binder ratio is of an order of magnitude larger than in concrete. Reducing water in such systems has a very strong effect on the strength development of the paste fill (Erismann et al. 2016).

Today, this simple principle, which is illustrated in Figure 1 (left hand side), has only partly found its way into the mine paste backfill industry, where the use of such admixtures is not standardised and very often, no admixtures are used at all, in order to improve the cost performance of paste backfill plants (Erismann et al. 2017). Figure 1 (right hand side) illustrates that oversaturated paste mixes of water to binder ratios of 5 and higher have a poor correlation between the added amount of binder and strength development. This obviously represents a challenge in meeting strength requirements in wet mixes.

This paper should help to address this shortcoming by presenting three case studies from mines where admixtures are used and have a very positive effect on the quality and performance of the backfill system. The three mine sites are completely unrelated to each other from a geographical, mineralogical, structural- and ore forming process point of view. By choosing such different cases, the authors would like to emphasize how important a proper admixture selection is by looking at the mineralogical and physical footprint of these deposits and hence the generated tailings which are the base of the produced backfill mix.

By far the most common goal while using admixtures is to increase the solid content of the fill, which in turn should have a positive effect on strength development as illustrated in Figure 1. As yield stress is positively correlated with the solid content of the fill, e.g. the higher the solid content, the higher the yield stress (Silva 2017), increasing the solid content of the fill is expected to have an effect on the pumping pressure within the reticulation system of the mine. However, actual pressure data from the mines could show that this is not necessarily the case.

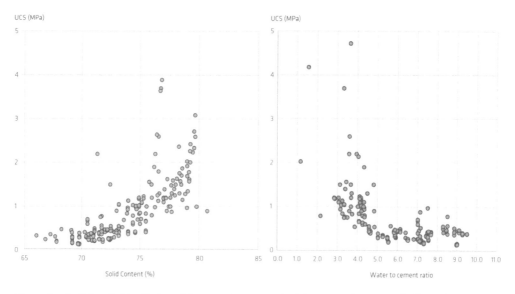

Figure 1. Left hand side: Paste solid content versus uniaxial compressive strength. Data from several paste plant projects. Right hand side: Relationship of water- to cement ratio and the uniaxial compressive strength.

In combination with admixtures, a pressure reduction can be achieved while strongly increasing the solid content at the same tame. This will be illustrated using the three case studies.

2 METHODOLOGY

For this paper, data from three different paste plants have been used over a period of three years. The mines where these paste plants are installed are located in North America, West Africa and Europe and are hosted by three different deposit types. The mine located in North America is a typical Carlin-type deposit. The mine in Europe is widely referred to as a SEDEX (Sedimentary Exhalative) type deposit and the one on West Africa a gold deposit of Orogenic type which is very common in this part of Africa (Partington and Williams 2000). Despite the differences in terms of ore-forming processes, alteration mineralogy, mineralization type and host lithology, these deposits also share similarities to a certain extent.

Particle size distribution as well as full phase mineralogical composition analyses were conducted at the Sika Technology laboratories in Zürich, Switzerland. Mineralogical analytics were done using scanning electron microscopy (SEM) to identify and confirm major and minor mineral phases and then quantify them using X-ray diffraction.

All mines utilise large tonnages of cemented paste fill and have different requirements in terms of early and final strength development as well as workability of the fill. Strength of the fills was tested according to the mine's specifications, usually testing the strengths after 7 days, 14 days, 28 days and 56 days. As common in many mines, workability is tested using standard concrete slump cones to measure the slump of the produced paste. Workability of the paste is related to yield stress and measuring the slump has proven to be a viable method to get an indication of the actual yield stress of the paste (Silva 2017). These workability limits often range in the 7-10 inch slump range (Erismann et al. 2017). Measuring the slump and final flow table spread is appropriate for an indication of the actual yield stress (Silva 2017), this test method remains the most common at mine sites. Special attention was also given to the behaviour of pumping pressure, both on surface and underground where possible and where the reticulation system is equipped to measure such data.

3 ADMIXTURE SELECTION

Selecting the right admixture for a certain paste is crucial to optimise the cost performance of a paste backfill system (Erismann et al. 2017). This can be illustrated based on the chosen ore deposit types described in Figure 2, where the simplified geological sections are shown. These deposits show the following characteristics:

A) Sedimentary Exhalative polymetallic (lead, zinc, silver, copper) deposit in Europe. These deposits are derived from an ancient, sub-seafloor, hydrothermal vent that deposited metals upon cooling and redox reactions of hydrothermal brines in contact with seawater on - or close to the seafloor (McKibben et al 1997). Usually, there is a distinct footwall alteration associated with these deposits. In this case, it is an intense potassic alteration, rich in Na_2O and K_2O with associated minerals such as Muscovite and Biotite.

B) Orogenic, shear-hosted gold lode deposits: This deposit type is a major host for gold globally. In this particular case the gold bearing shear zone is defined by a highly strained and brecciated, narrow zone that is heavily altered at either side of the shear including alteration products such carbonate, silica, albite, pyrite, chlorite and hematite.

C) Carlin-type, limestone/dolomite hosted gold deposit: This deposit type is characterised by the calcareous dominated host lithology and the distinct alteration features that can directly be linked to hydrothermal fluids responsible for the introduction of gold alongside alteration features such as carbonate dissolution, argillic alteration and silicification (Hausen and Kerr 1968, Radtke 1980, Bakken and Enaudi 1986).

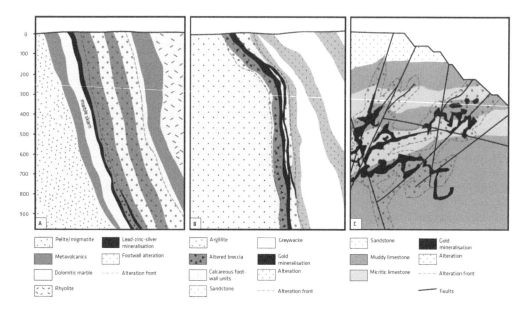

Figure 2. Simplified geological sections showing main lithological units as well as extent of alteration.

The alteration mineralogy of these systems is important with a view towards the effectiveness of the used admixture. The effectiveness can be fairly easy tested by performing yield stress measurements (using a viscometer or rheometer) or by measuring the flow table spread of different admixtures with a certain paste. A strong rheological impact on a paste can usually be linked to a strong plastification of the paste and hence a large increase of the flow table spread, slump values as well as much lower yield stresses for both rheometer and viscometer measurements. This selection process resulted in three different admixture types that were used for these deposits. The exact selection criteria will not be described in this paper, however, the presence of phyllosilicates in all three deposits as well as the dominating calcareous phases in deposit B and C, are strongly influencing the compatibility of certain polymers for these specific tailings.

An overview of the mineralogical composition and hence the overall chemistry of the deposits is shown in Table 1, where percentage composition for a certain mineral phase is given.

Particle size distribution of the three different mine tailings is shown in Figure 3. Mine A has the coarsest tailings and mine C the finest sieve fractions with 80% of tailings passing the 50 μm fraction.

4 INFLUENCE OF ADMIXTURES ON PASTE – RESULTS

4.1 Solid content

Once a suitable admixture has been identified for a paste backfill project, the reaction of the paste plant, once the admixture is dosed into the batched- or continuous filling process, is usually a strong decrease of torque in the twin shaft mixer. This decrease in torque is indirectly measured by the decrease in energy, the mixer draws to mix a certain volume of paste contained in the mixer. Modern paste plants have a fixed energy range the mixer is supposed to work in, in order to provide a paste consistency that is suitable to pump through the paste reticulation system. Once the energy draw falls below this pre-defined limit, water addition is automatically reduced. The same is the case once energy levels increase above the pre-defined limit. If this is the case, water is added to the mix in order to bring the viscosity of the paste into the desired window. As admixtures usually strongly reduce the yield stress of paste, mixing is usually easier once the admixture is added, the energy level drops, water is reduced and the solid content goes up.

Table 1. Mineralogical composition of the three deposits. Values indicate the volume percentage of contained minerals. Alteration minerals are indicated in dark grey. Minerals that might originate from the original host lithology as well as an alteration product are indicated by the light grey shading.

		A	B	C
		SEDEX Type	Orogenic Type	Carlin Type
Quartz	SiO_2	31.4	25	51.5
Sulphides	FeS_2, $CuFeS_2$, FeS, PbS, ZnS, FeAsS	1.9	1.3	
Muscovite	$KAl2AlSi_3O_{10}(OH)_2$	8		11.9
Biotite	$K(Mg,Fe)_3AlSi_3O_{10}(OH)_2$	10	1.3	
Chlorite	$(Fe,Mg,Al,Zn)_6(Si,Al)_4O_{10}(OH)_8$	7.2	5.9	
Calcite	$CaCO_3$		1.4	15.8
Dolomite	$CaMg(CO_3)_2$		15.7	15.7
Orthoclase	$KAlSi_3O_8$	22.8		
Albite	$NaAlSi_3O_8$		47.8	
Diopside (Mg-rich Pyx)	$CaMgSi_2O_6$	14.2		
Cordierite	$Mg_2Al_4Si_5O_{18}$	5.5		
Hematite	Fe_2O_3		1.7	3
Illite	$(K0.65)Al_2(Si_3Al0.65)O_{10}(OH)_2$			1
Bassanite	$CaSO_4\ 1/2H_2O$			7.1
	Admixture type	A	B	C

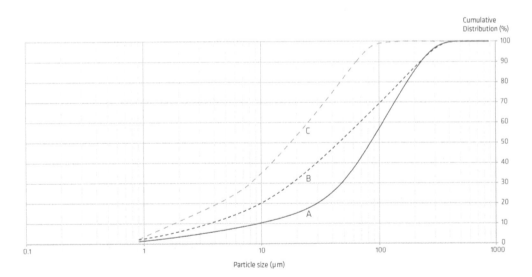

Figure 3. Particle distribution of the three mine tailings used for this study.

This fairly quick interaction is illustrated in Figures 4, 5 and 6 where parameters such as water addition and line pressure (both on surface and underground) as well as the admixture dosage and solid content are plotted along the time axis. It is easy to observe that admixture dosage and water addition to the mix strongly correlates. The more admixture is dosed into the system, the higher the reduction of added water, in some cases the water addition stops entirely and the solid content of the paste is approaching the solid content of the filter cake. Workability

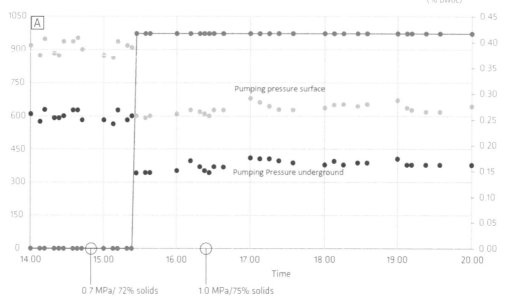

Figure 4. Development of pumping pressure for the SEDEX deposit (project A) for paste line pressure on surface and underground, before- and after the dosing of admixtures. Results for a paste mix containing 6% cement.

of the paste is often measured using a standard slump test at most mine sites. In all three sites, this slump test has been used to track workability of the paste throughout the observation span.

Remaining within these slump limits, which is usually between 7-10 inches on a standard concrete slump cone, was a requirement of all sites and was fulfilled at all times.

4.2 *Pumping pressure*

Once the admixture is added to a paste mix, the yield stress of the mix is usually decreasing strongly. As a response of the system, the water addition is subsequently reduced from the mix to reach the desired properties of the paste in terms of strength development and cement content, yield stress will increase again which is a well described correlation between the solid content of a slurry and its yield stress (Silva 2017, Sofra 2017).

Test results from large scale applications with admixture, in the three described paste plants vary, but most data recorded showed stable line pressure or decreasing line pressure after the plant reached a new equilibrium with a steady admixture dosage. The Admixture dosage usually range from 0.4 to 2% by weight of cement. This decrease in pressure can be nicely observed in Figure 4, 5 and 6. In all projects, the pressure actually decreases both on surface and underground, with increasing admixture dosage, despite increasing the solid content. Pressure spikes do occur once admixture dosing starts and the plant reacts on the lowered yield stress which is indicated by the lower torque value of the mixer with instant water reduction. These pressure spikes during the start-up phase are most likely related to the insufficient regulation of water addition/reduction into the mixer. This is discussed in more detail during the discussion of this paper.

4.3 *Strength*

Strength development of the paste with increasing solid content is very favourable, both in terms of the early strength development (after a few days) as well as the final strength after 28 and 56 days. Such strength gains allow for strong cement reduction in the fill to fulfil the designed strength requirements of the fill. Strength results for different cement contents with-

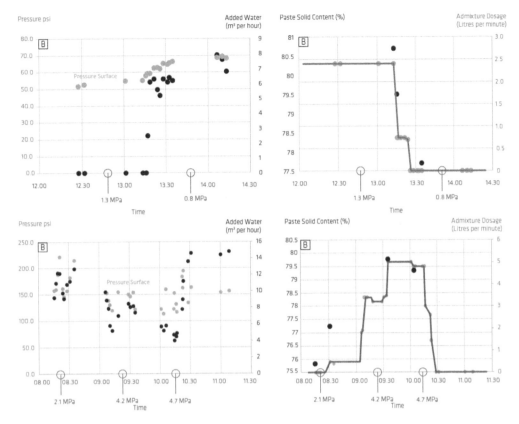

Figure 5. Development of pumping pressure on surface for the Orogenice Gold project (project B) with- and without dosing of admixtures. Results for a paste mix containing 5% cement (upper diagram) and 12% cement (lower diagram).

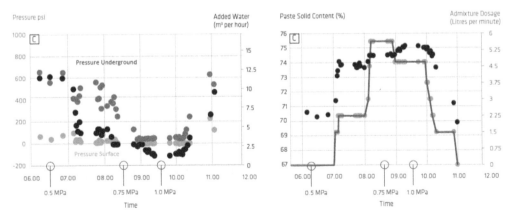

Figure 6. Development of pumping pressure on surface for the Carlin type Gold project (project C) with- and without dosing of admixtures. Results for a paste mix containing 8% cement.

and without admixtures are shown in Figure 7. The cement reduction potential observed in the described projects range from 20-50%.

Strength loss over time has been frequently observed in paste operations. This problem has not been observed for any of the three described projects and is not expected long term either,

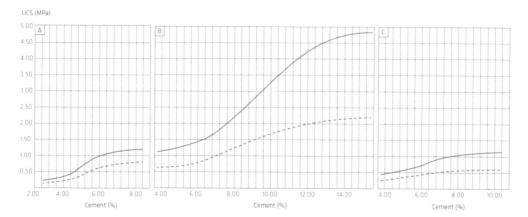

Figure 7. Strength results for project A, B and C for different cement contents in the paste fill. Dotted line displays results without admixtures, solid line the results with the use of admixtures.

as they have very low total sulphide content in the tailings and hence low SO_4^{2-} charges in the processing water. Dissolved SO_4^{2-} can have a strong effect on long term strength of the paste fill especially in fills containing a large proportion of sulphide minerals. Elevated SO_4^{2-} values in the pore water of the fill has a strong negative effect on long term strength of a cemented paste fill. Reduced water values in the fill will partly mitigate but no stop this issue.

5 DISCUSSION

The effect of admixture dosing into the paste stream showed to have a strong influence on a range of properties of the produced paste. High water to binder ratios in paste designs are common and water to binder ratios above 5 represents a challenge to achieve the needed strengths for a certain paste fill. This is mainly due to the oversaturated properties of such fills and the associated, limited strengths values even at high cement additions. Even in such water rich mixes, admixtures might still help as the dispersion force of cement and other particles in such a system is significantly increased when using admixtures (Lewis et al. 2000).

As the solid content of a paste mix is the driving factor for strength development at a given cement content, the ability of a suited admixture to strongly reduce yield stress of a paste is providing the precondition to reduce the water content of the fill. However, maintaining the workability of the paste and remaining within the pumping pressure limits of the reticulation system of the mine is critical when using admixtures. These workability limits were tracked at all three sites using a standard concrete slump cone to measure the slump of the paste mix continuously.

Despite the fact that the solid content was increased in all of the described cases, slump values tended to be elevated throughout the observation period, which indicates, that water could be withdrawn further. In two of the three plants, water addition to the mixer was reduced to almost zero with no further water reduction potential. Hence, the potential to increase the solid content and strength further reached the limit, unless improvements are made at the dewatering units (cone thickeners, vacuum disc filters) to produce a more dry filter cake. In two of the three cases described in this paper, the admixture overcompensates the effect, the solid content increase has on the line pressure of the reticulation system with pressures dropping well below the limits that have been observed prior to the addition of admixtures. This can be mainly related to the detrimental effect, excessive water has on a pumped paste mix. An oversaturated paste tends to segregate and free water is leading to elevated pressure levels compared with mixes that contain less water.

Pressure spikes have been recorded in all of the described operations. These pressure spikes were mainly related to the time period when the admixture dosage just started. This reflects the strong- and almost instantaneous impact the admixture has on the yield stress of the paste mix, which triggers immediate reduction of added water to the mixer. As this water addition is often controlled not sensitive enough (poorly controlled valves) and the system tends to over-mitigate changes of the mixer torque value, water might be withdrawn too quickly, causing short term pressure spikes. An example from case A: The mixer release time to the feed-hopper of the piston pump is linked to a certain paste viscosity which is, again, linked to the energy draw of the mixer. As admixtures lead to a strong fall-off of mixer energy, these release times were triggered much faster causing the hoper fill levels to rise and the pump to pump faster, which resulted in an increase of paste line pressure. This side effect was solved by fixing the hoper release time that resulted in a continuous pumping speed.

Strength gains across the described cases were significant and material cement reductions could be achieved with all projects. Cement reduction compensates well for the additional cost of the admixture and has a very positive effect on the overall cost performance of the paste fill system. Furthermore, strength development is faster due to better, more rapid and complete cement hydration which allows for faster stope cycle times and the fill efficiency rate increases as the cubic meter of fill contains a higher solid content and less water. Furthermore, as cement consumption is among the largest driver for CO_2 in underground mining operations, cement reduction in the paste fill will materially improve this important emission balance.

6 CONCLUSION

Paste backfill admixtures proved to have a powerful effect on paste mixes once a suitable admixture has been identified based on the physical and chemical properties of tailings. Admixtures are well suited to upgrade paste mix designs by increasing the solid content of the mix that usually goes hand in hand with higher achieved strength at the same binder content. Oversaturated fills with water to binder ratios of higher than 5 represents a challenge for a paste operation as elevated binder contents will not really improve the strength anymore.

The effectiveness of admixtures increases strongly with the ability of paste plants to pro-duces a dry filter cake of around 80% solids. Once the capacity of the dewatering units such as cone thickeners and filters are reached, admixtures can only reduce the amount of added water to the mix until water addition stops entirely.

All three cases presented in this paper showed strong cement reduction potential after admixtures had been implemented. Cement reduction potential can be as high as 20-50% of total cement used. Pumping pressures remained constant or were reduced despite the much increase solid content of the paste mixes. The cost performance of paste plants can be greatly improved by using latest admixture technology and by evaluating a well suited admixture. The overall economics strongly depend on available cement prices, admixture effectiveness and dosage and the ability of the paste plant to provide a good quality and dry filter cake.

BIBLIOGRAPHY

Aitcin, P.C. and Flatt, R. *Science and Technology of Concrete Admixtures*. 2016 Cambridge: Woodhead Publishing, pp. xxi
Aitcin, P.C. and Wilson, W. 2015. The Sky's the limit. *Concrete International 37*(1), pp. 53–58
Bakken, B.M. and Einaudi, M.T., 1986, *Spatial and temporal relations between wall rock alteration and gold mineralization, main pit, Carlin gold mine, Nevada, U.S.A.*: in Macdonald, A.J., ed., Proceedings of Gold '86, an International Symposium on the Geology of Gold: Toronto, Canada, pp. 388–403.
Erismann, F.J, Kurz, C. and Hansson, M. 2016. The Benefits of Incorporating Admixtures into Mine Paste Backfill. *19th International Seminar on Paste and Thickened Tailings: Proceedings of the Confer-ence on Paste and Thickened Tailings*. Santiago, Chile 5-8 July 2016. pp. 76–86
Erismann, F.J, Kurz, C. and Hansson, M. 2017. Translating Paste Backfill Admixture Results from the Laboratory into the Field. 20th International Seminar on Paste and Thickened Tailings: Proceedings of the Conference on Paste and Thickened Tailings. Beijing, China 15-18 June 2017. pp. 190–198

Hausen, D.M., and Kerr, P.F., 1968, *Fine gold occurrence at Carlin, Nevada, in Ridge*, J.D., *ed.*, Ore Deposits of the United States, 1933-1967, The Graton-Sales Volume: The American Institute of Mining, Metallurgical, and Petroleum Engineers, inc., New York, Ch. 46, pp. 908–940.

Lewis, J.A., Matsuyama, H., Kirby, G. and Morissette, S., 2000. Polyelectrolyte Effects on the Rheological Properties of Concentrated Cement Suspensions. J. Am. Ceram. Soc., 83 [8] pp. 1905–13

McKibben et.al. 1997 - McKibbe, M.A. and Hardie, L.A. 1997, *Ore-forming brines in active continental rifts:* In H.L Barnes (ed.), Geochemistry of hydrothermal ore deposits. Wiley Interscience, pp.877–930

Partington. C.A and Williams, P.J. 2000. Proterozoic lode gold and (iron)-copper-gold deposits: a comparison of Australian and global examples. *Reviews in Economic Geology, 13,* 69-101.

Radtke, A.S., Rye, R.O., and Dickson, F.W., 1980, *Geology and stable isotope studies of the Carlin gold deposit, Nevada*: Economic Geology, v. 75, pp. 641–672.

Ramachandran, V.S., Malhotra, V.M., Jolicoeur, C., Spiratos, N., 1998. *Superplasticizers. Properties and Applications in Concrete.* Materials Technology Laboratory, CANMET, Ottawa, Canada.

Silva, M., Contribution to Laboratorial Determination of Rheological Properties of Paste Backfill, Zinkgruvan and Neves-Corvo Case Studies. 2017. *Master Thesis*, IST Press, 82 pp. *Technico Lisboa, Lisbon, Portugal.*

Sofra, F. 2017. *Rheological Properties of Fresh Cemented Paste Tailings.* Yilmaz, E. and Fall, M. (eds) Paste Tailings Management, Springer, Cham, pp. 33–57.

Minefill 2020-2021 – Hassani et al (eds)
© 2021 Taylor & Francis Group, London, ISBN 978-1-032-07203-6

Paste fill stiffness investigation

James De Villiers Wickens
Paterson & Cooke Consulting Engineers (Pty) Ltd, South Africa

Stephen Wilson
Paterson & Cooke UK Ltd, England

SUMMARY: When conducting a backfill test campaign based on the selected mining method, there are various types of tests that are usually conducted to understand the in-situ behaviour of the material when placed underground. The unconfined compressive strength (UCS) and modulus of elasticity are the two most commonly measured parameters when dealing with a cemented paste fill material. However, when dealing with complex ore bodies or extraction techniques, additional testing, including triaxial and consolidation can be required to better understand the material's behaviour when exposed to a sustained ground load. This is of particular interest in yielding or closing ground conditions. The modulus of elasticity results, measured by conducting a UCS test, provides information on the strength of the material and the ability to resist deformation under load when the fill is exposed vertically and horizontally. However, when dealing with long term stability and support, it is important to understand the stiffness of the material and how it will resist deformation when exposed to different loading conditions when confined.

This paper will present selected triaxial and consolidation test data for a range of material blends, considerations that defined the testing program, and the nature of the tests performed. Conclusions will be presented as to the significance of the selected testing methodologies on the results obtained and their applications for industry.

Keywords: unconfined compressive strength, triaxial, backfill, modulus of elasticity and modulus of stiffness

1 INTRODUCTION

Before commencing with a backfill test campaign, it is important to understand how the backfill will be used and how it should perform. Key to the selection of a backfill type, other than the well established commercial drivers, is an understanding of how the material is required, or expected to behave within the proposed mining method or application. It is also important to develop a material testing program to support the engineering and project development process. Such a program must give consideration to the material available for backfill, the mining methodology and the anticipated ground conditions, including the development of vertical and/or horizontal ground stresses that may influence the backfill, and against which the backfill performance specifications must be developed.

The data presented within this paper are derived from a large, multi-phased test work program undertaken by Paterson & Cooke in its Cape Town laboratory. The project requirements for backfill included:

- Bulk mining extraction methods,
- High ore body extraction ratios,

DOI: 10.1201/9781003205906-2

Table 1. Test matrix.

Number	Constituted mixture	Concentration	Water/Cement ratio	Cement addition
1.1	75SN/25TA	75%m	4.76/1	7%
2.1	60SN/40TA	71%m	5.83/1	7%
3.1	60SN/20TA/20TB	76%m	4.51/1	7%
3.2	50SN/30TA/20TB	74%m	5.02/1	7%
4.1	50SN/40TA/10TB	72%m	5.56/1	7%
4.2	40SN/40TA/20TB	71%m	5.83/1	7%
5.1	60SN/20TAC/20TB	81%m	3.35/1	7%
5.2	60SN/5TA/15TAC/20TB	80%m	3.57/1	7%
5.3	60SN/10TA/10TAC/20TB	79%m	3.81/1	7%

- Modest mining horizons extending up to 800 m below surface, and
- A rockmass with a tendency for convergence or closure.

As a consequence of the above project conditions, it was necessary to investigate a backfill not only with a target strength, but also a target stiffness, and indeed the stiffness target predicates the backfill strength specification, it being the more onerous condition. The anticipated sustained vertical and horizontal loading forecasts for the project indicated a target stiffness in the order of 100 MPa.

The project was able to consider a number of component backfill products, including two tailings types and an imported sand product. Further more, the option to cyclone one of the tailings products was also available for consideration. In order for the test work program to adequately investigate and develop a backfill product to meet these criteria, the following tests were considered for this test program:

1. Material properties tests,
2. Rheology tests,
3. Unconfined compressive strength and modulus of elasticity tests,
4. One-dimensional consolidation testing and
5. Unconsolidated-Undrained triaxial tests.

1.1 *Test matrix*

Table 1 presents the test matrix and mix reference. The following abbreviations are used:
- SN: Sand
- TA: Tailings A
- TAC: Tailings A Cycloned
- TB: Tailings B

Development of this test matrix followed a number of previous test campaigns which enabled iteration and refinement of the mix designs towards attaining the target specification.

2 MATERIAL PROPERTIES TESTS

2.1 *Solids density*

The solids density of the samples was determined using a helium pycnometer, which measures the skeletal solids density. Table 2 presents the measured solids densities.

Table 2. Solids densities.

Sample reference	Solids density (kg/m^3)
Tailings A	2654
Tailings A cycloned	2641
Sand	2753
Tailings B	2495
Cement	3015

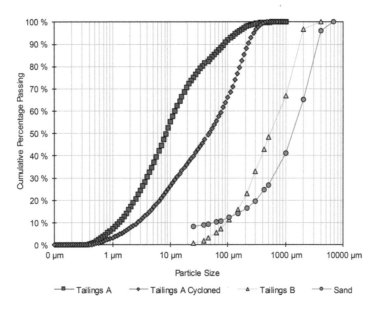

Figure 1.   Particle size distributions.

2.2   *Particle size distributions*

The particle size distributions (PSD) of the as-received materials were determined by wet sieving and laser diffraction. Figure 1 presents the measured PSD data.

2.3   *Cement quality*

A CEM III/B 32.5 N cement was supplied for the test work. Before commencing with the test work, cement mortar tests, according to EN 196-1:2005 Edition 2, were carried out to determine the quality of the cement. The 28-day results showed a compression strength of 34.8 MPa which exceeds the minimum required compressive strength of 32.5 MPa indicating that the cement complied with the standard requirements.

3   RHEOLOGY TESTS

An Anton Paar Rheolab QC rotational viscometer was used for the test work. The slurries were tested and analysed according to the infinite bob and cup method described by Chhabra & Richardson (1999). The data were analysed by applying the Bingham plastic model which is a two-parameter model describing the slurry rheology.

Figure 2 shows the rheogram for the various mixes. The data show that the yield stress varies from ~120 to ~260 Pa for the various constituted mixes.

4   UNCONFINED COMPRESSIVE STRENGTH AND MODULUS OF ELASTICITY TESTS

4.1   *Cement addition and water/cement ratio*

The percentage cement addition and water/cement ratio are calculated using the following formulas:

$$Percentage\ cement = \frac{mass\ of\ cement}{mass\ of\ solids + mass\ of\ cement} \tag{1}$$

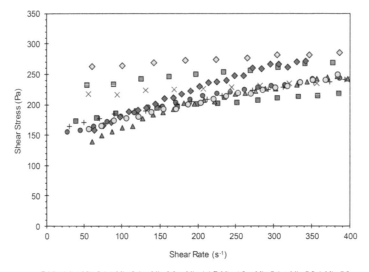

Figure 2.   Rheogram for Mixes 1.1 to 5.3.

$$Water : Cement\ Ratio = \frac{mass\ of\ water}{mass\ of\ cement} \tag{2}$$

4.2   *Test results*

Table 3 shows a summary of the w/c ratio at which the mixes were constituted, the cylinder (fill) density, the UCS and the modulus of elasticity. Figure 3 shows the UCS and modulus of elasticity test results versus the constituted mixtures. The data show that the mixes with a w/c ratio lower than 5/1, yield the highest results in terms of UCS and modulus of elasticity. Also evident from the data is that the proportionality between UCS and modulus of elasticity is not uniform, confirming the assertion that the different material blends, and importantly, the associated changes in overall particle size distribution of the final mixture, influence the geomechanical response of the backfill.

Table 3.   UCS and modulus of elasticity test results.

Mix Number	Water/Cement Ratio	Cylinder Density (kg/m³)	28 Day UCS (kPa)	28 Day Modulus of Elasticity (MPa)[1]
1.1	4.76/1	1927	2212	341
2.1	5.83/1	1845	2067	333
3.1	4.51/1	1923	2369	140
3.2	5.02/1	1859	2686	327
4.1	5.56/1	1712	1085	84
4.2	5.83/1	1786	1696	119
5.1	3.35/1	2024	3025	366
5.2	3.57/1	2009	2928	336
5.3	3.81/1	1955	2851	322

[1]Secant modules based on 25 and 75% of UCS.

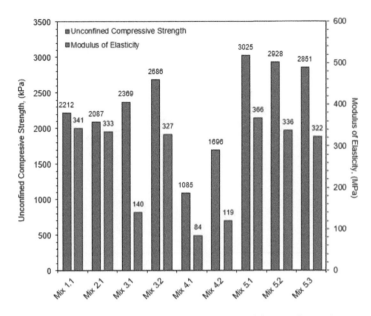

Figure 3.   Unconfined compressive strength and modulus of elasticity vs mix number.

5   ONE-DIMENSIONAL CONSOLIDATION TESTS

One-dimensional consolidation tests are conducted at axial loading pressures ranging from 10 to 25000 kPa to simulated ground load conditions which is based on the mining plan. For this particular test work program, the maximum axial load considered was 6400 kPa. Figure 4 and Figure 5 show the axial stress versus axial strain and void ratio versus axial stress repectively. The axial stress versus axial strain data indicates that the voids are compressed up to an axial stress of 800 kPa. This is evident as an exponential increase in the stress versus strain is seen, where a more linear relationship is seen at axial stress values above 800 kPa.

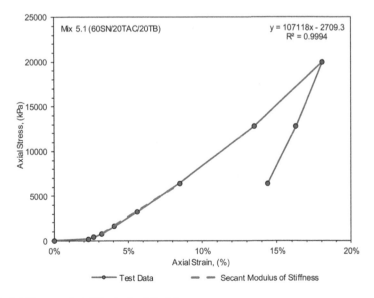

Figure 4.   Axial Stress vs Axial Strain for Mix 5.1.

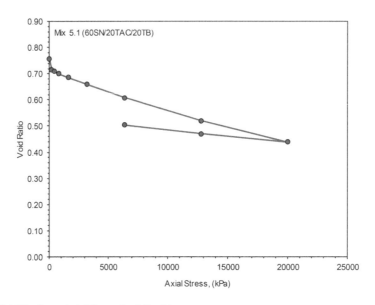

Figure 5.   Void Ratio vs Axial Stress for Mix 5.1.

Table 4.   Secant modules of stiffness results.

Mix Reference	Water/Cement Ratio	Axial Stress Range (kPa)	Secant Modulus of Stiffness (MPa)
1.1	4.76/1	800 – 6400	59
2.1	5.83/1		54
3.1	4.51/1		69
3.2	5.02/1		67
4.1	5.56/1		36
4.2	5.83/1		46
5.1	3.35/1		107
5.2	3.57/1		106
5.3	3.81/1		92

   Table 4 and Figure 6 show the average modulus of stiffness results for each mixture, with Figure 7 showing the change in void ratio at different axial loads for each mixture. As seen from the UCS and modulus of elasticity test results, the lower the water/cement ratio and subsequent denser fill, the higher the starting void ratio which results in a higher modulus of stiffness.

6   UNCONSOLIDATED-UNDRAINED TRIAXIAL TESTS

Table 5 and Figure 8 present the unconsolidated-undrained triaxial results. Mix series 3 and 5 yield the highest results in terms of effective cohesion, thus aligning with the UCS results presented earlier. The effective angle of friction varied from 18° at the lowest to 39° at the highest.

7   CONCLUSIONS

This paper presents the results of a series of tests conducted using various blends of materials in order to determine the mixture with the highest modulus of stiffness. The main conclusions are as follows:

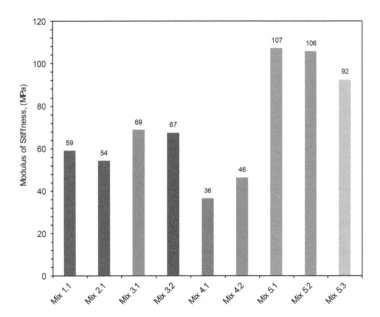

Figure 6.   Modulus of stiffness vs mix number.

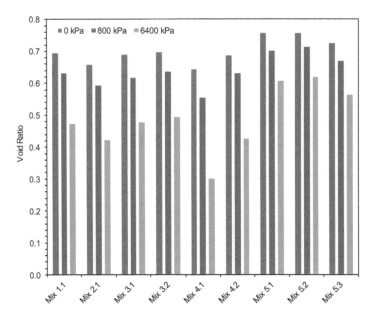

Figure 7.   Change in void ratio at different axial loads vs mix number.

1. Before commencing with any test work, it is important to understand what type of mining plan is to be used and what the purpose of the backfill should be, as well as how the host rock will behave under loading conditions. Once this has been established, the test work program should be developed to measure the required parameters in order to ensure that the backfill will meet the requirements. These parameters can vary from the UCS and

19

Table 5. Unconsolidated-undrained triaxial test results.

Mix Reference	Water/Cement Ratio	Effective Cohesion c'	Effective Angle of Friction φ'
1.1	4.76/1	680 kPa	26.5°
2.1	5.83/1	710 kPa	18.0°
3.1	4.51/1	850 kPa	24.5°
3.2	5.02/1	835 kPa	23.0°
4.1	5.56/1	223 kPa	29.0°
4.2	5.83/1	487 kPa	29.0°
5.1	3.35/1	658 kPa	37.0°
5.2	3.57/1	804 kPa	33.0°
5.3	3.81/1	820 kPa	31.0°

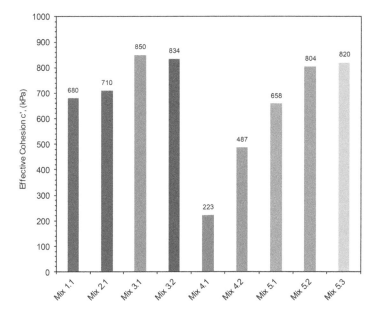

Figure 8. Effective cohesion versus mix number.

modulus of elasticity (understand how the backfill will behave when vertically exposed) to the modulus of stiffness (long term stability and support).

2. The tests data indicated that the mix series five yielded the highest results in terms of UCS, modulus of elasticity and modulus of stiffness. The starting void ratios for these mixes were above 0.7 where all the other mixes were below 0.7. At an axial load of 6400 kPa, the void ratios for these mixes were only compressed to above 0.55, where all the others were compressed to below 0.50.

3. The unconsolidated-undrained triaxial test results indicated that very similar effective cohesion results were measured for the majority of the mixes, except for mix series four. Mix series five had the highest effective angle of friction of above 31° compared to the other mixes where effective angles of friction of 29° and lower were recorded.

Overall the test work program sought to obtain a mix design that could deliver a stiffness in excess of 100MPa. Through the test work program, the data were able to substantiate that mix series 5 provided the best opportunity to attain this, and it is notable that this mixture contained the lowest fines content and returned the highest in-situ density of all the mixes prepared.

ACKNOWLEDGEMENT

The authors would like to acknowledge the contribution of Paterson & Cooke Cape Town laboratory staff for their contribution in assisting with the test work.

REFERENCES

ASTM International 2000b, *ASTM D2166 Standard Test Method for Unconfined Compressive Strength of Cohesive Soil*, ASTM International, West Conshohocken.

ASTM International 2002, *ASTM D4832 Standard Test Method for Preparation and Testing of Controlled Low Strength Material (CLSM) Test Cylinders*, ASTM International, West Conshohocken.

European Standard 2005, *EN 196–1 Method of Testing Cement – Part 1: Determining Strength*, Brussels.

British Standard 1975, BS 1377 Methods of Test for Soils for Civil Engineering Purposes – Part 5 and 8, Chiswick London.

Minefill 2020-2021 – Hassani et al (eds)
© 2021 Taylor & Francis Group, London, ISBN 978-1-032-07203-6

Pulsation dampening systems for piston pumps operating at backfilling installations

Tobias Lutz & Peter Peschken
Putzmeister Concrete Pumps GmbH, Aichtal, Germany

ABSTRACT: During metallurgical mining, most of the mined material is of no use for the industry. Only a very small portion can be used to receive the desired precious metals. The waste material is called tailings and must be transported to surface dams or mixed with additives to become a backfill paste which is transported back to the stopes. Hydraulic driven Piston pumps are one solution to deliver this high volume over long distances in a safe and economical way. However, the design principle of a hydraulically driven piston pump leads to pulsation in a delivery pipeline which is caused by the fact that the general design is based on a discontinuous working principle. The pumps are pumping one cylinder after another into the transportation pipeline. This results on one hand in flow fluctuations and consequently in pressure losses.

For optimal performance of the system it is recommended to use a pulsation dampening system with the Piston Pump. The reason is to create an approximately continuous flow of the material which is preventing pressure peaks resulting from water-based high-density solids (so-called Water Hammer Effect) with a low air content being pumped at great speed against a high pressure. Furthermore, the limited variances in flow and pressure are resulting in lower forces inside the pipeline system which leads to lower operating costs and a safer working environment for the employees on the mine site.

This paper identifies and describes the major three innovations and their different working principles of Pulsation Dampening of Paste- and Slurry Systems and discusses their application in Detail.

First System is the use of a hydraulic driven Seat valve pump with continuous Flow system. Second possibility is the use of piston pumps with hydraulic driven Pulsation Dampening System and finally, the use of Ventilated Pulsation Dampening Systems.

The job site reports show how the dampening systems are implemented at various job sites.

1 INTRODUCTION

1.1 *Working principle of hydraulic driven piston pumps*

Pulsation in a delivery pipeline of a reciprocating pump is caused by the fact that the general design is based on a discontinuous working principle as hydraulic-driven piston pumps are two-cylinder, single-acting, positive displacement pumps.

The pumping process can be described as follows: The Hydraulic cylinder H1 is connected to Material Piston P1 which is moving backwards and forwards in the material Cylinder M1. The same arrangement is applicable to the second hydraulic and material cylinder (Figure 1). During the pumping process, Hydraulic Cylinder H1 is moving backwards and sucking material into the material cylinder M1. At the same time hydraulic cylinder H2 is doing the counter movement and pushing the material in cylinder M2 into the delivery line. Both cylinders need

DOI: 10.1201/9781003205906-3

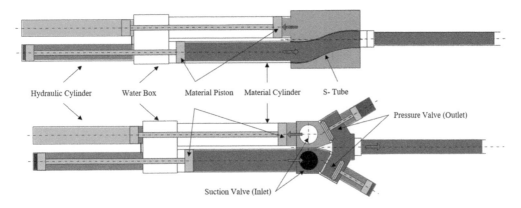

Figure 1.   Schematic Layout of an S- Tube and Seat Valve Piston Pump.

to be connected to the delivery pipeline which can be either a conical shaped pipeline; which swivels in front of both cylinders (so called S- Tube); or an arrangement of hydraulically operated seat valves (see Figure 1 for illustration).

Depending on machine size and set up of the piston pump the change over of the S- Tube or the seat valves is done within 0,5 – 1.0 sec. Consequently, the material in the pipeline sees an immediate drop in the material flow which can be indicated in the pressure reading (Figure 2). Consequently, the static water column (along with the dissolved air) expands slightly from its compressed state. As soon as this pressure wave hits an upstream interference (valve, elbow etc.), the wave will be reflected like an "echo" and can cause a Water Hammer. During this time, the velocity of the hydraulic drive cylinders is zero.

Low volumes and/or low operating pressures of piston pumps can be operated easily with pulsations if the pipeline is designed with proper wall thickness and fixation. However, it can be recognized that there is an increasing Trend in operating volumes and operating pressure in Tailings Handlings Applications (Peschken, 2018, p. 308). This leads to the fact that Pulsation dampening is becoming an important factor for the economical evaluation of a Tailings Pipeline. Pump manufacturers for Tailings Handling Systems have to provide solutions to prevent pressure peaks, especially those resulting from water-based high-density solids (so-called Water Hammer Effect) with a low air content being pumped at great speed against a high pressure. The savings in investment are substantial if the pipeline layout can be done with only a limited pulsation therein.

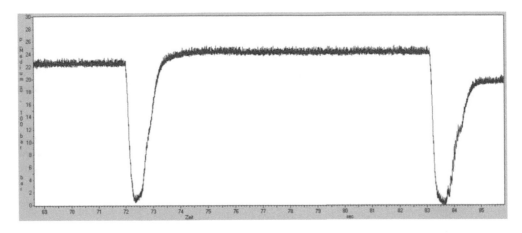

Figure 2.   Typical Pressure reading of a Piston Pump.

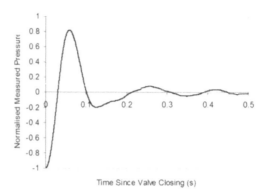

Figure 3.   Normalized Pressure reading of a Water Hammer (Donebythesecondlaw, 2009).

1.2   *The water hammer effect*

The Water hammer effect is an example of conservation of energy and results from the conversion of velocity energy into pressure energy (Bregman, 2017). It is created by stopping and/or starting a liquid flow (of a Newtonian Fluid) suddenly, which creates shock waves that travel back and forth within the pumping material. A practical example of a water hammer is turning off the shower at home immediately. "The hammer occurs because an entire train of water is being stopped so fast that the end of the train hits up against the front end and sends shock waves through the pipe. This is like a real train, instead of slowing to a stop, it hits into a mountain side. The back of the train continues forward even though the front cannot go anywhere. Since the water flow is restricted inside the pipe, a shock wave of incompressible water travels back down the pipe." (OMEGA Engineering, 2018).

This shock wave can be indicated as a variation in pressure (Dp). It is creating a pressure wave almost like sound waves, traveling upstream the pipeline has "all of its kinetic energy of motion and that due to compression is converted into pressure energy" (Evans, 2011).

"The first person to describe this effect was the Russian scientist Joukowski. He showed both theoretically and experimentally that there is a maximum pressure that can be produced, known now as the "Joukowski head" or "Joukowski pressure" depending on the units of its expression (Ord, S.C.; Stopford Projects Ltd, UK, 2006).

It is given by the following formula

$$h = \frac{v * c}{g}$$

where
c = speed of the sound wave in the pipe, known as the "wave celeric", [m/s[1]]
v = initial velocity of the liquid, [m/s]
g = 9.81 [m/s²]
h = Joukowski head, [m]

Thus, for water in a rigid pipe, where the speed of the sound wave in the pipe is (for example) 1500 m/s, for an initial water velocity of 2,5 m/s, then suddenly stopping the flow with a rapid valve action will result in a head of about 380 m, or 38 bar. The maximum closure time that will give this head is given by

---

1. in water it is 1.484 m/s. However, for Slurry or even paste the value is unknown and project specific

$$t = \frac{2 * L}{c}$$

where

L = length of pipe in [m]

t = time, which is the time it takes for a pressure wave to travel the pipe length and back again [sec].

Hence if the pipe is 2.500 m long, the maximum closure time is about 3.33 s, ie a closure in this time or less produces a Joukowski head. "For closure times up to 10 times this, [ie 33.3 s], the pressure is reduced in about inverse proportion which would result" (Ord, S.C.; Stopford Projects Ltd, UK, 2006, p. 4) in a pressure increase of 3,8 bar for the previous example.

Due to the pressure discontinuity, the passing of the wave front produces a force to the pipeline and its support at the next bending of the pipe.

Thus, for a 150 mm diameter pipe, the passage of the 38-bar wave described above produces a force of

$$F = \frac{\pi}{4} * D^2 * P$$

where

F = force, [N]

P = the pressure equivalent of the Joukowski head [Pa]

In our example, the Pipeline would see a Force of 67 kN, which is a significant force and could lead to damage of the pipeline, support or even the pumping system. "Sadly, that's not the full story. The forces and pressures calculated above are for specific, fast events with water as medium and therefore for ideal Newtonian Fluids. There are lots of shades of grey between a Joukowski event and a gradual change in fluid speed that gives no perceptible surge forces. So, in many cases, a more complex analysis or even tests have to be done to ascertain if a pipe is at risk" (Ord, S.C.; Stopford Projects Ltd, UK, 2006, pp. 2 - 5). For different material characteristics such as different rheological behaviors (thixotropic pastes; Slurries; cake pastes etc.) it is even more difficult to predict whether the risk of a Water Hammer exists or not. Nevertheless, it can be stated that the more and the bigger the solids (by volume) are within the pumping media the lower the risk of a water hammer as the speed of the sound wave in the pipe will be lower.

1.3 Consequences of pulsations in the pipeline

Pulsation in the pipeline is an unwanted circumstance of every designer and every operator of Slurry- and Paste Systems. The reasons are at hand. As described in the previous chapter, the pulsation may cause mechanical vibrations and fatigue failures which can lead to serious damage at the pipeline and its related components such as foundations and supports due to the high level of fluctuating forces. The lifetime of valves, flanges and other apertures is limited as they are affected by the pulsations as well. If the pulsations are considered during the design phase, the entire pipeline becomes an even higher capital investment of the operator. Consequently, it is a calculation between adding a pulsation dampening system to the pump or additional supports to the pipeline. In addition to that, the pulsations in the pipeline are leading to higher noise emissions to the environment of the pipeline which can cause issues for surface Pipelines (from Tailings Ponds and Ash ponds).

2 PULSATION DAMPENING SYSTEMS FOR HYDRAULICALLY DRIVEN PISTON PUMPS

2.1 Nitrogen charged Diaphragm Dampener

The first use of pulsation dampening systems was done with Nitrogen charged Diaphragm Dampeners as shown in Figure 4. Nitrogen charged Diaphragm Dampeners are common practise in hydraulic fluid systems or other fluid pumping systems like metering systems. The function of this pulsation dampener is based on the compression of a nitrogen gas Diaphragm which is usually designed in butyl- or fluoride rubber (Lutz-Jesco GmbH, 2018).

During the discharge of the piston pump the pumping material is squeezed into the Dampener and is compressing the Nitrogen Diaphragm. During the changeover of the S- Tube or seat valve, the material is pushed into the pipeline by the Nitrogen pressure in the Pulsation Dampening Diaphragm. As the volume of this bladder type Pulsation Dampeners is only limited, the quantity of installed dampeners must be increased for high flow rates. Consequently, big Tailing installations can be equipped with >10 of these Diaphragm Dampeners.

During the operation of various Slurry Systems with Piston pumps, there have been several issues reported (Hövemeyer & Zey, Industrial Pump Technology - Treatment, pipe transport and storage of high-density substances, 2019). The main concern is that the entire Diaphragm is sucked into the pipeline due to the high velocity and rapid change over of a piston pump. Furthermore, the system is very sensitive towards foreign bodies as stones or other spiky particles can harm the Diaphragm which lead to excessive maintenance work at the dampeners. For this reason, this technology is not used for new designed installations and plants. The industry improved the dampening methods towards three different working principles. First, the modification of the Piston pump which leads to an almost Pulsation Free operation. Secondly to a pulsation dampening system which includes an additional delivery cylinder and finally a ventilated dampening system which is a modification of the Nitrogen charged Diaphragm Dampener.

2.2 Constant Flow

The Constant Flow System is the best in class pulsation dampening system for seat valve – or ball valve pumps as it is a "no pulsation" dampening system. It is a pulsation avoidance system because it changes the way of operation of a piston pump, so the pulsation is not even created. Consequently, no additional mechanical components must be installed within the delivery pipeline. The only required Equipment is a Seat Valve or Ball Valve Piston Pump equipped with a Constant Flow Hydraulic Power Pack. With this system the material flow is kept at an almost even level which consequently reduces the pressure peaks significantly. This continuous material flow is achieved as the connection between the hydraulic cylinders is eliminated. Every cylinder of the piston pump is connected to an independent Hydraulic Oil Pump and acting as a single, independent piston pump. This enables the machine to perform the suction stroke faster than the pressure stroke to pre-compress the material in the material cylinder

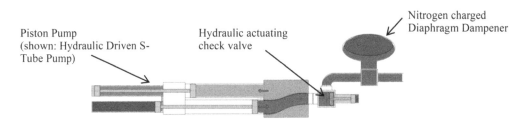

Nitrogen charged
Diaphragm Dampener

Piston Pump
(shown: Hydraulic Driven S-
Tube Pump)

Hydraulic actuating
check valve

Figure 4.   Schematic Layout of a Piston Pump with Nitrogen charged Diaphragm Dampener.

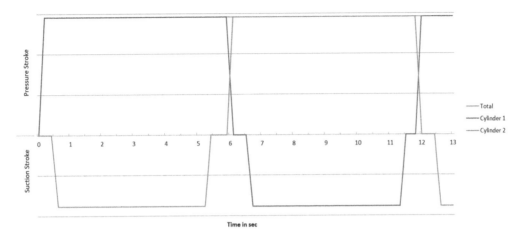

Figure 5. Constant Flow system - Working Concept.

before the pressure valve is opened. The Constant Flow System working principle is shown in Figure 5.

Cylinder 1 performs its pumping stroke, cylinder 2 starts compressing the material in the material cylinder. As piston 1 nearly reaches its end position, the speed will be reduced and piston 2 softly takes over the load. At this moment the speed of both cylinders is equal. After reaching the end position, piston 1 moves quickly back and sucks paste into the delivery cylinder while piston 2 pushes paste into the pipeline and performing its pumping stroke. The related pressure valve will close and now cylinder 1 does its pre-compression while cylinder 2 is finishing its cycle (Hövemeyer & Zey, 2019).

This Control Philosophy leads to an almost even Flow and Pressure Curve of the pump, directly after the pressure outlet as shown in Figure 6.

Besides the Pulsation Elimination, the Constant Flow system has additional benefits for the operator. The system doesn't need any additional mechanical or electrical components in the pipeline or in the pumping material. Consequently, no additional components must be maintained and operated. In addition to that, the system can be used for various, different materials with different flow rates up to a pipeline pressure of 150 bar without adjustment thereof. It doesn't matter whether the paste or slurry contains cement and tends to harden as the

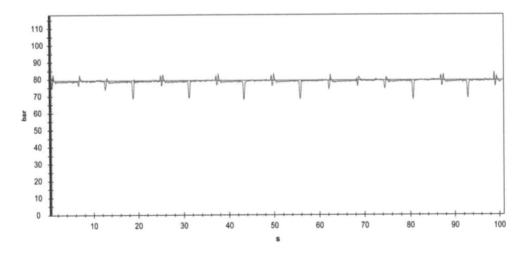

Figure 6. Constant Flow Pressure reading.

27

Constant Flow system can be cleaned out completely. This has been proven by operators in multiple plants worldwide since the invention of the Constant Flow system. Due to the continuous flow of the material, the paste or slurry doesn't need to be accelerated with every stroke from 0 to full speed. This results in a lower pressure level in the pipeline compared to the operation with Piston Pumps without Constant Flow system. Furthermore, the energy consumption of the electric motor(s) is homogenized as the load is not dropping during the changeover as it is with Pump Systems without Constant Flow. This relieves the variations in the power grid and improves its reliability. Finally, the piston velocity during the pumping stroke is slower as compared to machines without Constant Flow control (due to the changeover). This results in a lower wear rate and higher lifetime of the entire system.

2.3 *Hydraulic Pulsation Dampening System (HPD)*

The Hydraulic Pulsation Dampening System is the first real pulsation dampening system for any kind of Piston Pump application and is used when slurry or paste contains sand or coarse material. The Hydraulic Pulsation Dampening System literally acts as a third delivery cylinder, which is connected to the delivery pipeline by a T-flange immediately behind the pump outlet flange. It is dampening the pulsation of the system as it delivers material into the pipeline during the changeover of the S- Tube or the valves. It is either connected to the Hydraulic Power pack of the Piston Pump or is equipped with an independent one.

Tailings with a maximum grain size of < 25mm can be pumped with Seat Valve pumps or S- Tube Pumps. As soon as the grain size exceeds 25mm, the material should be pumped only with S- Tube piston pumps. If the Hydraulic Pulsation Dampening System is used with a Hydraulic Driven S- Tube Piston Pump, the pump has to be equipped with a hydraulic actuating check valve directly before the Hydraulic Pulsation Dampening System to prevent any backflow of the material into the wrong direction of the pipeline (cp. Figure 7). The Hydraulic Pulsation Dampening System is charged during each delivery stroke of the pump. The content of this third cylinder is actively pushed into the delivery pipeline during the changeover of the Piston pump. This closes the delivery gaps created by the change-over process and creates an almost even flow of material and consequently a much smoother pressure reading within the pipeline.

The Hydraulic Pulsation Dampening System is designed as a stand-alone unit and driven by the common power pack of the piston pump or driven by a dedicated, independent power pack. Hence it is very interesting for customers with existing paste plants as it can be used for new installations or as a retrofit for existing pump lines. The working principle of a Piston pump with the Hydraulic Pulsation Dampening System Pulsation dampening system can be described as follows. One material cylinder of the piston pump performs its pumping stroke.

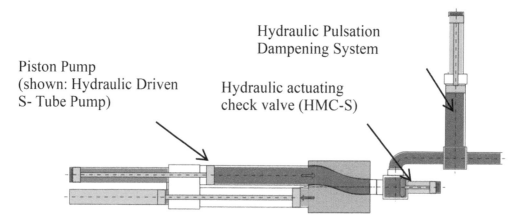

Figure 7.   Schematic Layout of a Piston Pump with Hydraulic Pulsation Dampening System.

Figure 8.   Hydraulic Driven S- Tube Pump with Hydraulic Pulsation Dampening System.

After reaching the end position, the Hydraulic actuated check valve is closing, and the Hydraulic Pulsation Dampening cylinder is extracting (and consequently pushing material into the delivery pipeline). As soon as the shift over of the seat valve or the shift over of the S-Tube is finished, the Hydraulic actuated check valve is opened, and the material cylinder of the piston pump is delivering material in the pipeline. At the same time the Hydraulic Pulsation Dampening System stops extracting material into the pipeline and reversing its direction, so it is recharged for the next change over. The pressure result can be seen in Figure 9:

Hydraulic Pulsation Dampening System Pulsation dampening systems can achieve a dampening rate of approx. 60 – 70%. The dampening rate is calculated by the ratio between the nominal pressure and the pressure drop during the shift over as indicated and described in Figure 9. This shows that the pulsation is not completely eliminated but water hammers are not possible anymore. Consequently, the design forces for the pipeline and its supports are tremendously lower. The advantage of this System is that it is the best and most reliable pressure dampening solution for hardening material (like cemented paste or slurry) with coarse material

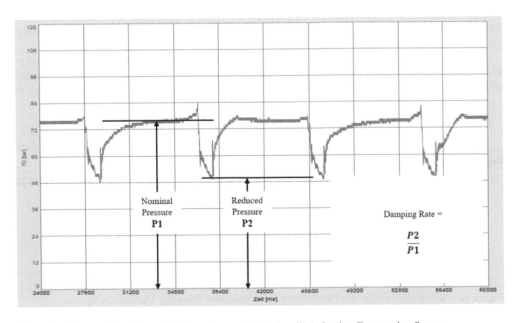

Figure 9.   Pressure Reading of a Piston Pump with Hydraulic Pulsation Dampening System.

(grain size: above 25mm). As it is designed for concrete material, the combination of S- Tube piston pumps with Hydraulic Pulsation Dampening System is insensitive towards foreign bodies like stones. Consequently, it is the proven choice of operation for Backfilling plants with coarse concrete and Ash handling systems where Bottom- and Fly- Ash are pumped as a mixture to the stope or Ash pond. As it can be operated with an independent power pack the Hydraulic Pulsation Dampening System can be easily retrofitted to any Piston Pump System.

2.4 *Ventilated pulsation dampening system*

The Ventilated Pulsation Dampener has been developed to reduce pressure drops in the pipeline with a maximum efficiency and a minimum required additional energy (Freitag, 2015). Like the Hydraulic Pulsation Dampening System, the Ventilated Pulsation Dampener literally acts as third (or multiple) delivery cylinder(s), which is connected to the delivery pipeline by a T-flange immediately behind the pump outlet flange. It is dampening the pulsation of the system using compressed air which is acting as a spring.

The Ventilated Pulsation Dampener consists of the actual dampening unit (which consists of multiple dampeners), an air accumulator and distribution unit, a high-pressure compressor and the patented intelligent control Unit (see Figure 10).

Important note: Due to the design of the Ventilated Pulsation Dampener it can be only used for non-hardening slurries and paste. Cemented paste must not be pumped through this system. Like the previously described Hydraulic Pulsation Dampening System, the Ventilated Pulsation Dampener is designed as a stand-alone unit. Therefore, it can be used for new installations or as a retrofit for existing pump lines as shown in Figure 11.

During the pump stroke of the pump, the pre-compressed air in the dampeners gets further compressed by the medium through which the medium rises in the dampeners. During the changeover of the seat valves or the S-tube, the compressed air presses the medium downwards into the conveying pipe, whereby the pressure collapse is reduced. The amount of air needed is detected by a pressure sensor in the dampening unit, calculated by the patented control system, generated from the compressor and provided from the accumulator and distribution unit (Freitag, 2015). As soon as the system is operating at a constant level, the compressor is automatically turned into stand-by mode and no additional energy is consumed anymore. Small changes in variations are compensated by the accumulator and distribution unit. As soon as the flow/pressure increases significantly, the compressor is turned on again and recharging the accumulator and distribution unit.

In regards of the Dampening rate we can rely on the physical law where more dampening volume will achieve higher dampening rates and consequently lower pressure drops during the changeover phase of the piston pump. From an economical point of

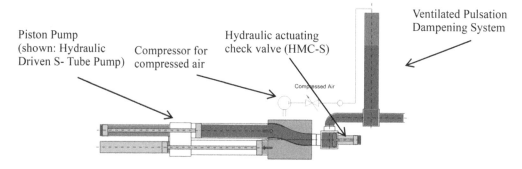

Figure 10.   Schematic Layout of a Piston Pump with Ventilated Pulsation Dampening System.

Figure 11. Installed Ventilated Pulsation Dampening System for Tailings Handling.

view, it can be outlined that the layout of the system should be done with a dampening rate of approx. 70% (see Figure 12).

The main advantage of the Ventilated Pipe Dampener is the low consumption of energy combined with no additional wear parts in the paste or tailings flow as it is using compressed air as a spring that stores energy during continuous operation (the compressor is only needed if the output or the pressure are changing). With an intelligent Control System, the Ventilated Pulsation Dampener can be automatically operated in different pressure levels without any manual adjustment. Due to the proven technology, the system is reliably eliminating water hammer in the pipeline while it is also easy to clean and maintain as no wear parts (like membranes) are necessary. The Ventilated Pulsation Dampener is a perfect Pulsation Dampening System for non-hardening pumping materials and applicable for new installations as well as an easy retrofit for all kinds of existing pumps.

3   PROJECTS WITH PULSATION DAMPENING SYSTEMS

Bulyanhulu Gold Mine: At Bulyanhulu up to 80 m³/h 10 inch slump paste were pumped over 2500 m with a maximum delivery pressure of 80 bar. The seat valve pump pushes the paste through a DN 200 pipeline. Pressure dampening was achieved by a HPD driven by the power pack of the main pump.

Eti Bakir Copper Mine: At Eti Bakir up to 65 m³/h cemented paste is pumped 830 m horizontal, then 290 m down and then over different distances underground to various chambers. The maximum delivery pressure is 150 bar. The seat valve pump of

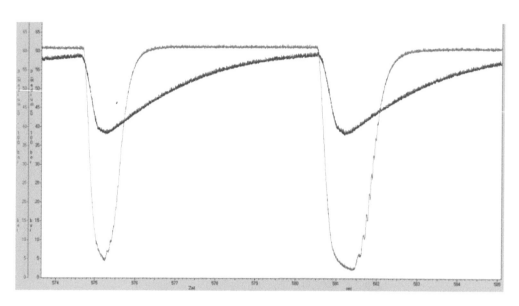

Figure 12.   Pressure reading of a Piston Pump with and without Ventilated Pulsation Dampener.

Figure 13.   Underground Backfill Plant at Kali & Salz.

type HSP 2180 HPS, driven by a Hydraulic Power Pack with 315 kW motor, is equipped with the PCF pressure dampening system. The paste process is typical for a gold or copper mine. The fine grained tailings with a maximum grain size of 110 µm is put into a thickener; the thickener underflow is pumped into a surge tank and the slurry from the surge tank is again thickened by disc filters. The cake from the disc filters is mixed with cement and slurry and then pumped underground with the piston pump into the chambers for stabilizing the mine.

Figure 14.   Fly- and bottomash pumping station at Kogan Creek.

Figure 15.   Tailings pumping plant at Coemin with VPD.

Kali & Salz: Seat valve pump and power pack with PCF system are placed underground. Up to 80 m³/h fine grained potash with brine and binder is pumped underground for stabilisation. The parts in contact with this aggressive slurry are made of stainless steel.

Kogan Creek: At Kogan Creek flyash with up to 50 mm bottom ash particles are pumped over 5.500 meters back into the open pit mine. The KOS 25.100 HP pump, driven by a Power pack of type HA 250+250 E conveys up to 152 m³/h at a max. delivery pressure of 69 bar. Pressure dampening is done by a HPD 280/750 driven by the main power pack. The pipeline diameter is 150 mm to achieve the necessary speed to avoid sedimentation. One pump system is working, one standby.

Coemin: At the Coemin Copper Mine up to 162 m³/h fine grained tailings are pumped at a maximum delivery pressure of 94 bar. Three HSP 25.100 HPS, each one driven by a HA

630 E, are installed. Pressure peak dampening of each pump is done by a Ventilated Pipe Dampener with 2 pipes each.

4 CONCLUSION

Pulsation Dampening is an essential element for Tailings handling and Paste Backfilling applications and will become even more important for future projects. Due to increasing delivery volumes, -distances and pressures it is going to be crucial to install the right dampening solution to secure a reliable and safe working condition of the piston pump system. Furthermore, pulsation dampening systems are helping to reduce the overall installation costs as the wall thickness of the delivery pipeline and its safety factors could be optimized accordingly. This paper outlined three different pulsation dampening solutions to avoid or to reduce the shocks in the pipeline significantly.

The Constant Flow System, which is obviously no dampening system as it avoids the pressure pulses in the pipeline due to its independent working cylinders. This system is the preferred solution for all new designed cemented Paste- and Slurry Systems and can be operated until 150bar continuous delivery pressure. A ventilated pulsation dampening system Ventilated Pulsation Dampener is the economically friendly solution for tailings handling if the material is not hardening. The additional benefit is that this system can be retrofitted to any existing Piston pump station. Due to the patented intelligent control, the ventilated pulsation dampening system is only consuming a minimum of energy during continuous operation. Finally, the paper presented the hydraulic pulsation dampening System Hydraulic Pulsation Dampening System which is the only solution for Backfilling systems with coarse material that are operated with S- Tube Pumps. It can be supplied for new project but as well be retrofitted to existing pumping stations.

REFERENCES

Bregman, A. (2017, May 9). Water Hammer. Retrieved July 15, 2019, from valvemagazine.com: https://www.valvemagazine.com/magazine/sections/back-to-basics/8418-water-hammer.html

Donebythesecondlaw. (2009). Wikipedia. Retrieved February 27, 2018, from Water hammer pressure pulse: https://en.wikipedia.org/wiki/Water_hammer#/media/File:Water_hammer_pressure.jpg

Evans, J. (2011). Pumps & Systems. Retrieved February 27, 2018, from The Causes of Water Hammer (Part One): https://www.pumpsandsystems.com/topics/pumps/pumps/causes-water-hammer-part-one

Freitag, U. (2015, November 25). TI 151124 - Ventilated Pulsation Damper (VPD).

Hövemeyer, D., & Dr.-Ing. Richter, A. (2014, April 11). Challenges of pump and process engineering for the transport of solids with hydraulic piston pumps. (B. Ernst & SohnVerlag für Architektur und technische Wissenschaften GmbH & Co. KG, Ed.) Mining Report (150 No 1/2). doi:https://doi.org/10.1002/mire.201400003

Hövemeyer, D., & Zey, W. (2019). Industrial Pump Technology - Treatment, pipe transport and storage of high-density substances. (Karl Schlecht Stiftung (KSG), Ed.) Aichtal.

Lutz-Jesco GmbH. (2018, November 5). Pulsation Damper PDM. Retrieved from http://www.lutz-jescoasia.com/LJ-Pulsation_Dampener_PDM-EN-MB.pdf#targetText=A%20basic%20distinction%20is%20made,is%20compressed%20to%20damping%20volume.

OMEGA Engineering. (2018). OMEGA Engineering. Retrieved February 27, 2018, from https://www.omega.co.uk/techref/waterhammer.html

Ord, S.C.; Stopford Projects Ltd, UK. (2006). Water Hammer – Do we need to protect against it? How to predict it and prevent it damaging Pipelines and Equipment. IChemE(SYMPOSIUM SERIES NO. 151), 1–20.

Peschken, P. (2018). Hydraulic Driven Piston Pumps for the Transport of Pastes and Slurries in the Mining Industry. Mining Report Glückauf(No. 4), pp. 306–317. Retrieved July 22, 2019, from https://mining-report.de/wp-content/uploads/2018/08/MRG_1804_KOSK_Peschken_180811.pdf

Putzmeister. (2014). Putzmeister Industrial Technology in Mining. Retrieved April 25, 2019, from Putzmeister: https://www.putzmeister.com

The Process Piping. (2017, September 4). Retrieved from Water Hammer in Process Plant: https://www.theprocesspiping.com/water-hammer/

Minefill 2020-2021 – Hassani et al (eds)
© 2021 Taylor & Francis Group, London, ISBN 978-1-032-07203-6

Paste quality control benchmarks

David Stone
President, MineFill Services, Inc., Bothell, USA

SUMMARY: Routine sampling and testing of paste mixes with uniaxial compression test cylinders is an accepted industry practice for routine monitoring of paste quality and strengths. However, unlike the concrete industry, there is no industry accepted quality control standard for assessing the results, or for determining what actions, if any, need to be taken. This paper sets out the groundwork for an industry standard for interpretation of routine quality control testing of paste mixes. The paper will examine the variability of several key benchmarks based on mine site quality control databases, and provides a draft assessment criteria.

Keywords: paste, quality, sampling, strength

1 INTRODUCTION

In the absence of an industry standard, most mines rely on concrete standards to design and interpret their backfill quality control (QC) programs. In the USA these standards are developed and published by organizations such as ASTM (American Society of Testing and Materials) and ACI (American Concrete Institute). However, the application of these standards to cemented backfills can be very difficult owing to the high degree of variability in backfill mixes, and the resulting variability in QC test results.

To illustrate, concrete industry standards denote an acceptable (passing) test result occurs when the average of the duplicate test cylinders exceeds the target strength, and no single test result deviates from the target strength by more than 10 percent (ACI 214R). Remedial measures must be taken if three consecutive tests fail. With mine backfills, it is safe to say that these criteria are rarely achievable.

A further, more dramatic, example is the coefficient of variation for concrete QC cylinders. According to ACI 214R, the coefficient of variation (CV) can range from 3 percent (excellent rating) to 6 percent (poor rating). Applying this analysis method to several paste QC databases from operating mines, the author notes that a coefficient of variation in the order of 20 to 30 percent is more typical.

Clearly there is a need to develop a better industry standard for assessing paste backfill quality to assist mine operators with interpreting test results and developing action plans in the event of undesired outcomes.

2 PASTE FILLS VERSUS CEMENTED ROCKFILLS

Routine quality control sampling is commonplace for both cemented rockfills and cemented paste fills but there are some differences. First, cemented rockfills are highly variable owing to the aggregate gradings which can vary dramatically over even short periods of time (see Stone

DOI: 10.1201/9781003205906-4

(2007, 2019)). Cemented paste fills, on the other hand, tend to be more consistent because the mix ingredients are better controlled. As such there tends to be far less variability in paste mix quality control test data, compared to cemented rockfills.

Another consideration is the number of mix recipes in use, and hence the strength benchmarks for interpreting the QC test results. In cemented rockfills a typical mix plant might be programmed with 4 or 5 recipes. Entire stopes are filled with a single recipe. With paste fills the mix recipes can vary on a daily basis, and a single stope may utilize several different recipes (e.g. plug pours versus bulk pours). This complicates the strength assessment of the QC test data as the target strengths are constantly changing.

In concrete all of the QC strengths are benchmarked to a 28 day cure. The same is largely true for cemented rockfills. Even though the stope turnaround may be as few as 7 days, ultimately the QC data, and the backfill specifications, are assessed on the basis of 28 day strengths.

This is not necessarily true for paste fills where 3 day, 7 day and 14 day strengths may be the only strength benchmarks that are set. Paste generally does not follow the traditional 7, 14, and 28 day curing curve seen in concrete hence paste strengths may not benchmarked to a 28 day cure. An example is the paste plug pour which is typically benchmarked to a 24, 48 or 72 hour cure. In fact, the rapid strength gains in paste can further complicate QC testing as the samples must be broken on a very regular schedule to avoid de-rating the results.

3 FACTORS THAT IMPACT PASTE UCS

Not all pastes behave the same and hence the rheology and strength of paste mixes can be highly variable. Table 1 shows a range of average paste strengths from QC test cylinders at 6 operating mines. The strength ranges across these six projects can be ten-fold or more. This table is based on several thousand QC strength tests.

The main culprit for the variability in the strength of paste mixes is typically a combination of the paste rheology (yield stresses in the paste mix) and the tailings mineralogy. Some of the factors that can impact paste strengths in the laboratory are described in more detail in the following sections.

Table 1. Average paste strengths from selected mine databases.

	Binder Type[1]	2.5% Binder	3.0% Binder	5.0% Binder	6.0% Binder
Mine A – 7d	OPC	0.72	0.54	1.21	
Mine A – 28d	OPC	1.50	1.39	2.26	
Mine B – 7d	OPC/FA	0.06	0.06	0.10	0.16
Mine B – 28d	OPC/FA	0.18			0.33
Mine C – 7d	OPC		0.21	0.25	0.26
Mine C – 28d	OPC		0.29	0.45	0.51
Mine C – 7d	MINCEM	0.22	0.24	0.52	
Mine C – 28d	MINCEM	0.36	0.41	0.88	
Mine D – 7d	MINCEM	0.45	0.40	0.60	0.74
Mine D – 28d	MINCEM	0.48	0.65	0.80	1.14
Mine E – 7d	OPC			0.21	
Mine E – 28d	OPC			0.27	
Mine E – 7d	BFS		0.41	0.35	
Mine E – 28d	BFS		0.47	0.55	
Mine F – 7d	OPC/FA		0.06		
Mine F – 28d	OPC/FA		0.10		

Note 1: OPC = Ordinary Portland, FA = Fly Ash, BFS = blast furnace slag. MinCem = OPC/BFS blend.
Source: MineFill Services client databases

3.1 *Factors that impact paste UCS – mix water quality*

The chemistry of the trim water or mix water used to prepare paste can have a direct impact on the paste quality. The prime culprit typically is dissolved sulphates (SO_4) from the oxidation of sulphide minerals. During metallurgical processing, ores with high sulphide contents, such as pyrite, can result in process water with very high sulphate values.

In the authors experience, the threshold for paste degradation is around 2,500 ppm SO_4 which is similar to the threshold for concrete. Sulphate degradation of cement is a well-documented phenomenon, and in paste backfills as it can lead to a rapid and significant loss of paste strengths in the laboratory after 28 days (see Pierce et al (1998), Benzaazoua et al. (2002)). There is however, some debate as to whether this is a laboratory induced issue due to oxidation during laboratory curing, or whether this extends to curing of paste confined in a stope. However, projects with very high sulphides will typically extend the QC testing to 56 or even 90 days to check for sulphate reactions.

Hyper-salinity in groundwater used for paste mixing is an issue for some mines. Hyper-salinity also occurs in soda ash and trona paste mixes due to salts dissolved in the process. Hyper-salinity has two significant impacts on paste: one is on the rheology, and the other is on the cement hydration and ultimately the strength of the paste.

Typically, the higher the salt content, the lower the yield stress. Tests by Jiang and Fall (2017) show a 40 percent reduction in yield stress for paste mixes with 100,000 mg/L of salt compared to mixes with zero salinity. It is important to understand that dissolved salts also affect the density of the carrier water that suspends the solids in the paste. At the Frogs Leg mine in Australia, for example, the paste carrier water was found to have a density of 1.15 t/m^3 due to 210,000 mg/L of dissolved salts (Mgumbwa and Nester (2014)). Hence interpreting rheology results on the basis of solids content alone can be misleading. Yield stresses should be correlated with the total density of the slurry and not just the solids content.

The strength of paste mixes that use hypersaline mix waters decreases as the salt content increases, however these mixes do show increasing strengths with increased curing. Tests by Jiang and Fall (2017) show an 85 percent reduction in 28 day paste strengths for paste mixes with 100,000 mg/L of salt compared to mixes with zero salinity. The loss in strength is attributed to the replacement of Ca^{2+} with Na^+ in the C-S-H cement hydration products.

3.2 *Factors that impact paste UCS – tailings mineralogy*

The single biggest influence on the strength of paste is the mineralogy of the tailings. This is particularly true for tailings with a very high specific gravity of solids such as sulphide rich tailings. Pyrite has a specific gravity of about 5.0 and pyrrhotite has a specific gravity of 4.5. Paste mixes with significant sulphide contents can have densities of 2.5 tonnes per cubic meter or more.

The significance of the specific gravity of the solids of the tailings is related to the binder additions which are always measured as a percentage of the weight of the dry solids. The result is that paste mixes with high density solids will contain a higher binder content on a volumetric basis. This then leads to higher strengths since the binder hydration reactions are based on volumes not solids content (see Fall et al. (2007)).

Paste batched with high specific gravity tailings are also susceptible to a rapid settling of solids in QC test cylinders. This creates a significant water bleed which lowers the water to cement ratio in the cured cylinders, which increases the laboratory measured density of the paste.

The presence of micas in tailings can be problematic because these minerals are hydrophilic and can absorb water into their crystal matrix. This can result in a gradual reduction of the free water in the paste, resulting in significant changes in the yield stress of the paste mix. These paste mixes are generally thixotropic, and care must be taken to avoid time dependant changes in the yield stress.

The processing of rare earths can result in oxidation products in the tailings that can be especially troublesome. The calcining of rare earths can produce calcium oxide CaO or quicklime which is unstable in the presence of water. In a recent laboratory program the quicklime was seen to undergo a rapid and violent exothermic reaction which not only boiled the water in the paste, but was also accompanied by a 200 percent volumetric expansion.

3.3 *Factors that impact paste UCS – sampling location*

Routine sampling of paste for QC testing is commonly done at mixer discharge at the paste plant. Since the majority of paste plants are located at the surface, the samples are generally transferred to a temperature-controlled surface laboratory. Some paste plants are located underground, in which case the samples are cured underground, or else transferred to the surface laboratory.

Less commonly, the paste sampling can be done at the discharge into the stope. Typically these samples would be cured and tested at an underground paste laboratory.

Testing has shown that the location for collection of the paste QC samples has little to no impact on the strengths recorded in QC testing. Hughes et al (2013) reports on an investigation at the Stillwater mine in Montana to address concerns about variances in paste strengths due to the long reticulation distances. Stillwater collected 86 comparative samples at both the paste plant and underground at the stope. The testing showed no statistically significant difference in the strength of QC cylinders collected at the surface paste plant compared to cylinders collected underground at the stope discharge.

Donovan et al. (2007) carried out a similar exercise at the David Bell mine in Canada, and likewise their results did not show any particular bias to one method or the other.

In cemented rockfills the human factor can play a significant role in the final strength of QC test cylinders (Stone (2007, 2019)). The variability introduced by different operators can range from samples that are over-tamped and on the verge of being concrete, to cylinders with excessive voids, poor end preparation, or in some extreme cases, cylinders that are not even full. Generally, the majority of reject cylinders will be consistently produced by one operator. Fortunately, this is not an issue in paste sampling because there is no sample preparation other than ensuring the cylinders are full and capped properly to prevent moisture loss.

3.4 *Factors that impact paste UCS – sample size effects*

De-rating of sample strengths due to the sample size is common with both concrete and CRF, but does not appear to be needed for paste fills. Hughes et al (2013) reports that no conclusive trend could be found between sample size and UCS. The authors own project databases likewise show that the UCS of QC test cylinders is not sensitive to the cylinder size. NIOSH did find that samples less than 38mm in diameter can over-estimate the paste strength. This was attributed to a possible accelerated curing effect in small diameter samples which have a higher surface area to volume ratio.

Scaling down the sample size has a number of positive benefits. The most notable is the size and volume of the curing cabinet needed to store samples. A typical paste operation would collect 4 to 6 samples per shift (for 3, 7, and 28 day breaks) resulting in the need to store upwards of 75 samples each weighing approximately 11 kg each. Hence most paste operations have scaled their QC cylinders down to a much more manageable 50 x 100mm or 100 x 200mm samples. The smaller sample volume takes up less space, and is far easier and safer to handle than full size concrete samples.

3.5 *Factors that impact paste UCS – curing conditions*

The curing conditions can have a significant impact on the QC test results. The ASTM C31 standard calls for an initial 48-hour cure at 16 to 27 °C in an environment preventing moisture loss. After the initial cure, the samples are to be stripped from the molds and cured in a 100 percent humidity environment at 23 °C (or stored in a water tank).

Most paste sampling is carried out remote from the QC laboratory hence the samples have to be moved. Ideally these samples would follow an initial cure period in conformance with the

ASTM guidelines. Typically, the curing of paste samples is done in a humidity cabinet, a humid controlled room, or even sealed plastic bags with a wet cloth.

There are two issues unique to the curing of paste samples. The first is the result of stripping the samples from the molds after the initial curing period. In pastes that contain appreciable sulphides, stripping the samples exposes the sulphides to air, which in turn triggers the production of sulphates and low pH. Because these reactions are exothermic, the heat of oxidation results in accelerated hydration of the cement. This can lead to some very interesting high early strengths. In one example the mine QC testing produced 7 day strengths of over 1 MPa, and 28 day strengths of over 2 MPa, with just 3 percent binder. This phenomenon is particularly troublesome in samples with high pyrrhotite contents. Unfortunately, these high early strengths are typically not sustained in the stope environment hence in-stope strengths can be found to be much lower than the laboratory strengths. An interesting comparison between samples that were sealed for 112 days versus samples stripped after 28 days, but cured for 112 days, can be found in Pierce et al. (1998).

The second issue is the result of allowing the samples to lose moisture due to improper storage. This can be witnessed in hot jurisdictions where the paste laboratory is housed in a standard shipping container without a proper temperature or humidity controls. The result, again, is abnormally high strengths as a result of a loss of the water needed to hydrate the cement, or accelerated curing from the excessive heat.

4 INTERPRETATION OF QC TEST RESULTS

The majority of mines enter the QC test data into an EXCEL spreadsheet and then produce a basic quality control chart such as an x-y plot of UCS versus time (Figure 1). When the data falls outside of the desired limits it should serve as a trigger that some form of action is needed. The reality is that most mines do not understand the trigger mechanisms coming from this type of plot because so many QC cylinders fail.

The preferred method for periodic review of backfill QC test data is the cumulative sums (CUSUMS) chart in accordance with ACI 214R-02. The cumulative sums chart is a cumulative summation of the deviation from an average for a series of tests over time. The CUSUMS chart

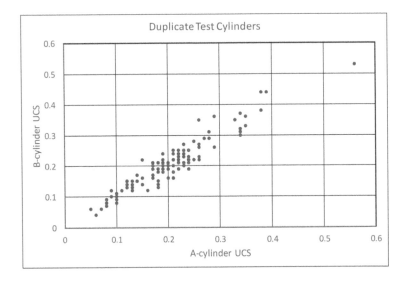

Figure 1. Variability in duplicate test cylinders.

Source: MineFill Services client databases

provides a method for detecting relatively small changes in the average strength of QC cylinders. It can help to identify when the change occurred and the approximate magnitude of the shift in strengths. As noted in ACI-214R, the charts provide greater sensitivity in detecting small, systemic changes in the average strength as compared to the basic quality control chart. A good example of the power of a CUSUMS chart is given in Stone (2007).

5   VARIABILITY IN SAMPLE DUPLICATES

A well run backfill QC program includes the collection of duplicate QC test cylinders, or in some cases, triplicates. It has been noted that concrete standards require the average of both cylinders to exceed the target strength, and no cylinder can deviate from the target strength by more than 10 percent.

Very few mines setup their databases to allow a direct comparison of the results from duplicate test cylinders hence this is a very difficult benchmark to track. However, a review of client databases showed deviations between duplicate cylinders (A versus B) that ranged from 6 percent to 46 percent. A plot of typical A versus B cylinder results is shown in Figure 1. Typically, the correlation coefficient comparing datasets for duplicate samples is very high (>95 percent) indicating a very strong correlation.

6   COEFFICIENT OF VARIATION

Under normal conditions the QC test results are expected to fit into a normal frequency distribution curve clustered around the mean. The dispersion of the test data is typically measured by calculating the standard deviation. However, in the case of backfill, a better measure of the variability of the test data is the coefficient of variation (CV) or relative standard deviation. The CV tends to be less affected by the variation in mean strength values and hence provides a more useful measure of the variability over a wide range of strengths.

As noted in the introduction, CV values for concrete can range upwards of 6 percent which is notably different than mine backfills. Reviewing client QC databases for a number of paste projects, CV values were found to range from a low of 15 percent, to a high of 60 percent. Most of the data (see Table 2) is clustered around an average CV value of about 35 percent.

Table 2 does give some insight into the factors that affect the CV value. The CV does vary according to the curing period (e.g. one set of samples does not exhibit the same CV at various curing periods), however no consistent trends are evident. There does appear to be a trend related to the type of binder. Much of the data in the table suggests that non-Portland Cement binders result in higher CV values: in other words Portland Cement is far more predictable and consistent as a binder.

7   LABORATORY STRENGTHS VERSUS IN-STOPE STRENGTHS

One of the critical benchmarks for paste backfills is the relationship between the strengths measured in laboratory QC samples compared to in-stope paste strengths. In cemented rockfills it has been established that in-stope strengths are typically lower than the laboratory strengths, in-part because of a strength de-rating due to sample size (see Stone et al (2019)). It has been shown herein de-rating of strengths does not apply to paste fills, hence it should not be a surprise that laboratory QC strength values are typically indicative of short term (6 month) in-stope paste strengths.

A number of investigations have shown that the strength of cured paste is a function of the confining pressure, hence paste strengths generally increase with depth in a stope. There is also a lot of evidence that paste continues to gain considerable strength after the standard 28 day cure. Bloss and Rankine (2005), for example, present the results of laboratory testing of Cannington paste samples subjected to long term curing. Their results showed the paste

Table 2. Coefficient of variation from selected mine databases.

DataBASE	Binder Type[1]	2.5% Binder	3.0% Binder	5.0% Binder	6.0% Binder
Mine A - 7d	OPC	36.5%	35.8%	30.6%	
Mine A - 28d	OPC	24.8%	21.0%	27.6%	
Mine B - 7d	OPC/FA		33.3%		31.9%
Mine B - 28d	OPC/FA				40.9%
Mine C - 7d	OPC		24.2%	42.0%	41.5%
Mine C - 28d	OPC		31.9%	46.5%	30.2%
Mine C - 7d	MINCEM	51.9%	53.3%	48.6%	
Mine C - 28d	MINCEM	58.9%	45.0%	34.4%	
Mine D - 7d	MINCEM	35.4%	31.5%	31.0%	32.4%
Mine D - 28d	MINCEM	27.5%	22.8%	58.3%	22.5%
Mine E - 7d	OPC			45.8%	
Mine E - 28d	OPC			43.8%	
Mine E - 7d	BFS		35.9%	17.1%	
Mine E - 28d	BFS		51.7%	13.4%	
Mine F - 7d	OPC/FA		21.2%		
Mine F - 28d	OPC/FA		17.4%		

Note 1: OPC = Ordinary Portland, FA = Fly Ash, BFS = blast furnace slag. MinCem = OPC/BFS blend.
Source: MineFill Services client databases

undergoes two curing cycles: an initial 28 day cure, followed by a second cycle after about 6 months for low binder content samples, or 4 months for high binder content samples. The second curing cycle was observed to double or even triple the 28 day strengths over a period of 12 months.

The Cannington observations are supported by a number of studies with in-stope block sampling and coring of paste. Typically these samples yield paste strengths far in excess of the 28 day strengths measured in the QC samples. Recent coring of paste at an operation in the Red Lake area of Canada, for example, showed in-stope paste strengths averaging 3.2 MPa after 78 days of curing, compared to laboratory QC cylinders which averaged 1.05 MPa after 56 days. This is strength gain of over 3-fold.

The question of paste degradation due to the oxidation of sulphides was recently addressed at a mine in Peru. In this case, Standard Penetration Testing (SPT) with a cone was conducted in paste filled stopes up to 30 m high, for paste with ages ranging from 60 days to 300 days. As expected, the SPT results showed increasing strengths with curing time, and also with depth in the stope. The laboratory QC test results at various curing times were then compared with the results of intact block samples collected in test pits and from subsequent fill exposures. In most instances the block sampling results were significantly lower than the laboratory QC test values, but in some stopes the results were very similar. This is not considered definitive proof of degradation of the paste as intact block sampling of paste has been shown to be very difficult and the results are typically disappointing.

8 EVALUATING QC DATA ON A DAILY BASIS

The case has already been made that backfill quality control test results typically exhibit considerable scatter, and also exhibit a significant number of "failures" (cylinders with UCS values under the target strength). This is normal in a mine environment. Given that test failures are accepted, guidelines are needed to determine if the fill being placed is acceptable. The ultimate goal is to prevent the placement of large quantities of poor quality fill.

QC test data can be separated into two groups. The first is daily samples collected during operation of the paste plant. These would consist of duplicate sets of some combination of 3, 7, 14 and 28 day test cylinders. During continuous operation of the paste plant the QC laboratory would be receiving and breaking these cylinders on a daily basis in accordance with the curing schedule.

The second grouping of these same samples is assigning the results to an individual stope. This topic is discussed later in the paper.

So what are the consequences of failure of both of the duplicate sample cylinders tested on the same day? In concrete this would be considered a failure. However, after reviewing several mine QC databases it would appear that this occurs in about 10 percent of the cylinders.

What about the failure of both samples over 2 consecutive days? However, again this seems to be a fairly common event occurring in about 5 percent of the cylinders. The same is true for 3 consecutive days of QC testing.

However, it does appear that it is rare for QC testing to result in four consecutive days of failures in duplicate cylinders. Hence it would appear that any issues in the backfill plant that result in low paste strengths typically persist for about 3 days. This is likely a function of the volume of individual stope pours – in other words the issues are not realized or corrected until the end of the pour. It should be clarified that the sample failures we refer to herein relate to consecutive samples for the same cure period (e.g. consecutive 3 day or 7 samples) and not to the testing of 3 day samples, then 7 day samples and so on from the same paste batch.

9 EVALUATION QC DATA ON A PER STOPE BASIS

ACI and ASTM standards recommend a minimum of 15 duplicate samples in order to produce an adequate population of samples for statistical accuracy. For a typical 1,000 m^3 pour at 60 m^3/h this equates to one sample set every hour. For pours over 10,000 m^3 this equates to about one sample set per shift.

The key question in evaluating the QC strength data on a per stope basis is how many samples are allowed to fail before the stability of the backfill mass becomes a concern. The best answer to this question can be found in Helinski et al (2017) which presents the results of a reliability analysis conducted with FLAC3D on a hypothetical 50 m tall x 25 m wide vertical exposure of backfill. The model randomly assigned strengths to 2 m cubed blocks based on a Gaussian distribution over a range of standard deviations. The results showed that the fill mass would remain stable provided the CV value was less than 60 percent, and that 66 percent of the fill mass exceeded the target strength required for stability. As the CV is increased the volume of paste that exceeds the target strength also needs to increase.

Based on the Helinski models, the author recommends the following guidelines:

- the acceptable failure rate in laboratory samples for a given stope should be about 1 in 4% or 25%,
- the mean laboratory value for all QC testing in a given stope should be above the target strength. This implies the in-stope "mass" strength will be better than the target strength.

A back analysis of actual stope failures would help to refine this analysis. However, the long-term curing trend of paste makes these guidelines conservative for paste exposures that exceed a 28 day cure.

10 CONCLUSIONS

The variability of uniaxial compressive strengths in paste QC test cylinders is not unexpected given the broad range of factors that can have an influence. These factors include the tailings chemistry, the curing conditions, and the type of binder, just to name a few.

Applying concrete standards to the evaluation of backfill quality control samples has been shown to be problematic. Hence the author has developed the following assessment criteria based on a review of QC data from operating mines (Table 3).

Table 3. Recommended assessment criteria for paste QC cylinders.

Benchmark	Acceptance Criteria
UCS results	CV < 30%
A versus B Cylinders	Coeffecient Correllation > 95%
	<10% difference in UCS
Daily UCS	Duplicate cylinders on consecutive days fail for < 4 days in a row.
Per Stope UCS	Minimum 15 cylinders
	75% of cylinders exceed target UCS
	Mean UCS of all cylinders > target UCS

ACKNOWLEDGMENTS

The author acknowledges the many contributions of client databases for analysis and review in the preparation of this paper. This is a work in progress and the author hopes to update the acceptance criteria as more data becomes available.

BIBLIOGRAPHY

Benzaazoua, M., Belem, T, Bussière, B. 2002. Chemical factors that influence the performance of mine sulphidic paste backfill. [In] *Cement and Concrete Research, Vol 32*, Pergamon, pp 1133–114.

Bloss, M.L., Rankine, R. 2005. Paste fill operations and research at Cannington Mine. [In] Ninth Underground Operators Conference. Perth, 7-9 March 2005. AusIMM Melbourne. pp 141–150.

Donovan, J., Dawson, J., Bawden, W.F. 2007. David Bell mine underhand cut and fill sill matt test. [In] *Proceedings of the Ninth International Symposium on Mining with Backfill: Minefill 2007*, Montreal, Quebec, Canada, 29 April – 3 May 2007, eds. F. Hassani et al.

Fall, M., Benzaazoua, M., Ouellet, S. 2004. Effect of tailings properties on paste backfill performance. [In] *Proceedings 8th International Symposia on Mining with Backfill: Minefill 2004*. Beijing, China, 2004. pp. 193–202.

Jiang, H., Fall, M. 2017. Yield stress and strength of saline cemented tailings in sub-zero environments: Portland cement paste backfill. [in] *International Journal of Mineral Processing. Vol. 60*, Elsevier, pp 68–75.

Helinski, M, Merrikin, D. 2017. Reliability analysis of mine backfill exposures. [In] *Proceedings of the Twelfth International Symposia on Mine Backfill: Minefill 2017*. Denver, 19-22 February 2017. Society for Mining, Metallurgy and Exploration, pp 281–294.

Hughes, P.B., Pakalnis, R., Deen, J., Ferster, M. 2013. Cemented paste backfill at the Stillwater Mine. [In] *Proceedings of the 47th US Rock Mechanics Symposia*. San Francisco, 23-26 June 2013. ARMA pp 1230–1236.

Mgumbwa, J., Nester, T. 2014. Paste improvements at La Mancha's Frog Leg underground mine. [In] *Proceedings of the Eleventh International Symposia on Mine Backfill: Minefill 2014*. Perth, Australian Centre for Geomechanics, pp 281–294.

Pierce, M.E., Bawden, W.F., Paynter, J.T. 1998. Laboratory testing and stability analysis of paste backfill at the Golden Giant Mine. [In] *Proceedings of the Sixth International Symposia on Mine Backfill: Minefill 1998*. Brisbane, 14-16 April 1998, Australian Centre for Geomechanics, pp 159–165.

Stone, D.M.R. 1993. The optimization of mix design for cemented rockfill. [In] *Proceedings of the Fifth International Symposia on Mining with Backfill: Minefill 93*. The South African Institute of Mining and Metallurgy Symposium Series S13, Johannesburg, ed. H.W. Glen, 249–253.

Stone, D.M.R. 2007. Factors that affect cemented rockfill quality in Nevada mines. [In] *Proceedings of the Ninth International Symposium on Mining with Backfill: Minefill 2007*, Montreal, Quebec, Canada, 29 April – 3 May 2007, eds. F. Hassani et al.

Stone, D.M.R., Pakalnis, R., Seymour, B. 2019. Interpreting backfill QA/QC test data – Do we need an industry standard?. [In] Proceedings SME Annual Meeting. Denver 24-27 February 2019. SME Pre-Print 19-043.

Minefill 2020-2021 – Hassani et al (eds)
© *2021 Taylor & Francis Group, London, ISBN 978-1-032-07203-6*

Using paste backfill for stabilizing underground stopes at the Giant Mine remediation project

Bernie Ting & David Stone
MineFill Services Inc

Kenny Ruptash & Brandon Fabien
Nahanni Construction Ltd

SUMMARY: In 2018, a remediation program was executed to stabilize the underground voids near stopes that store arsenic trioxide waste at the Giant Mine in the Northwest Territories of Canada. The shapes of the voids were reconstructed through integrating multiple cavity monitoring system surveys. A number of mix recipes of paste backfill and self-consolidating concrete (SCC) were designed with various bleeds, set times, uniaxial compressive strengths, and flow properties. Respectively, a total of 16,686 m^3 and 53,389 m^3 of SCC and paste were backfilled by gravity.

Keywords: reclaimed tailings, remediation, self-consolidating

1 BACKGROUND

Historic mining activities at the Giant Mine in the Northwest Territories of Canada left behind underground voids that resulted in a weakened crown pillar and surface subsidence. The subsidence becomes an immediate concern when it occurs in the proximity of public highways and nearby Baker Creek. Furthermore, some of the underground stopes store a large quantity of hazardous arsenic trioxide waste. The possibility of releasing the arsenic trioxide into the environment added further implications to the risk of unfilled voids. The objective of this remediation project was to stabilize an area in the C5-09 stope complex using backfill.

The C5-09 stope complex is located beneath three arsenic trioxide filled stopes. The void space in this area is complicated. The stopes were partially backfilled with rockfill and other unidentified materials. Some of the loose material was washed out to the lower levels of the mine during a large fluctuation in the mine water level in 2007. The ground conditions beneath the three arsenic stopes became unstable. The voids were undercut by a number of vertical raises, sub-levels, and draw points which present challenges as possible seepage or leakage during the backfill program. Sloughing from the hanging walls and footwalls, as well as pillar loading and ground noises had been observed over the past several years. Therefore, access to the C5-09 stope complex was limited.

Other challenges included the logistics of moving large quantities of fill materials to the relatively remote site, and the extreme cold temperatures during the winter season. In order to stabilize the area, an estimated volume of 54,084 m^3 needed to be backfilled. The operation required a number of specialized paste recipes with varying set times, flowability, strength, and bleed water properties.

DOI: 10.1201/9781003205906-5

2 BACKFILL ENGINEERING

One of the major challenges in this project was the lack of an accurate map for the underground. The shape and volume of the underground voids were reconstructed in 3D using a cavity monitoring system (CMS). In order to stabilize the area, and provide structural support to the nearby arsenic chamber, the plan was to use a combination of paste backfill and self-consolidating concrete (SCC).

Prior to backfill, remote cameras were placed at strategic locations to observe the progress and monitor possible leakage. Timber barricades were constructed to form a containment for the fill. The next step was to place a foam plug at the bottom of the void. This is shown in dark blue in Figure 1.

The bottom of the void was made of legacy backfill material, which had a wide variation of particle sizes. Therefore, a high slump (8 to 9 inches), relatively quick curing paste backfill was placed at the bottom of the void to build an impervious platform. Subsequently, the paste was increased to 10- to 10.5-inch slump to ensure good flowability. The bulk of the paste averaged 3% cement by weight, with a design strength of 100 kPa after 28 days. The paste flowed by gravity creating a flat profile. When encountering a barricade, the cement content was increased to 7%. The paste was placed in one-meter lifts every 24 hours until the paste level exceeded the height of the barricade to form a plug.

The main structural component is made of SCC, which had to be continuously poured to construct a 16,700 m³ monolithic structure. The SCC with a minimum of 12 MPa UCS was delivered through five adjacent boreholes as shown in Figure 2. The placement was through boreholes in a '1, 3, 5, 2, 4' sequence to maintain a fresh layer of material on the surface in order to avoid cold joints between pours. To minimize segregation and to place the SCC evenly, the backfill was delivered in tremie pipes with the bottom end held at the level of the freshly placed material.

After the SCC structure was completed, the operation switched back to paste backfill with a target UCS of 100 kPa to top up the void.

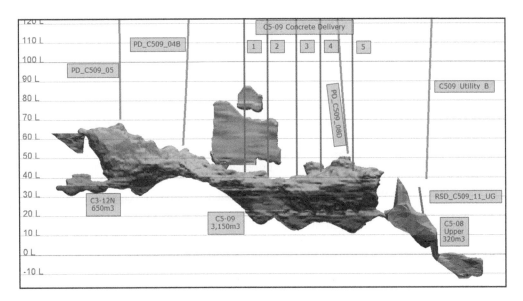

Figure 1. Underground void model reconstructed in 3D using CMS.
Source: C5-09 Stope Complex Work Plan (Nahanni Construction Limited).

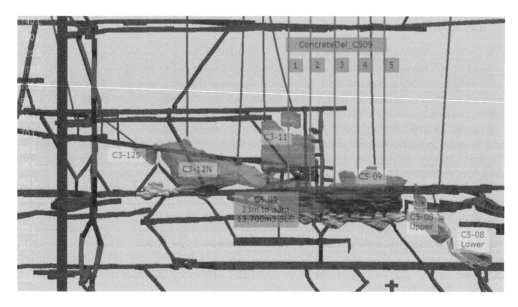

Figure 2.   Underground void model reconstructed in 3D using CMS.
Source: C5-09 Stope Complex Plan (Nahanni Construction Limited).

3   PASTE BACKFILL

3.1   *Reclaimed tailings*

Prior to the project, gold tailings were sampled from the tailings pond. The particle size distribution (PSD) varied significantly both vertically and laterally in the tailings pond. In the samples, the fraction passing 20 micron varied between 5% to 53%. In order to maintain the quality of the backfill, the tailings were reclaimed using a dozer, which scraped the tailings pond over a long distance. The tailings were screened to remove the foreign objects such as rocks and other organic materials. Then the piles were blended to reduce the fluctuations in tailings PSD.

3.2   *Paste mix design*

There were several criteria for paste design in this project. The uniaxial compressive strength (UCS) requirement was 100 kPa. Due to the expected fluctuations in tailings PSD, the target strength was set at 300 kPa to ensure all the paste meets specification.

In order to minimize the stress on the cement transportation to Yellowknife, the main type of binder used in this project was a 90/10 slag cement. Compared to backfill materials made with Ordinary Portland Cement (OPC), slag cement has a lower early UCS but a higher 28-day UCS. Nominally, the paste backfill operation used 120 tonnes of cement binder per day. When a quicker set time was required (such as when backfilling against a barricade), the cement content would be increased to improve early strength.

In addition, the paste needed to be a non-segregating material with a good flowability and minimal bleed water. An admixture was used to increase the solids content to 76% by weight while maintaining 10 to 10.5 inch slump. Figure 3 shows a summary of the UCS data during the first phase of the paste program. The fluctuation is a result of variation in PSD and cement content.

3.3   *Process description*

The paste backfill was pre-mixed using an auger-type mixer. The cement is added to the tailings based on the ratios in mix designs. The operators visually inspected the product to adjust

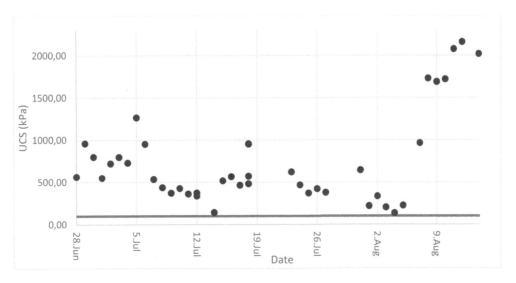

Figure 3.   UCS variation in the first phase of the paste program.
Source: Nahanni Construction Limited QC database.

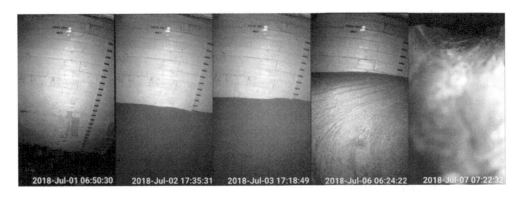

Figure 4.   Monitoring paste backfill through remote camera.
Source: Nahanni Construction Limited underground camera record.

the mix water flowrate. A quality control (QC) technician conducted slump tests every hour to provide feedback to the operators and ensure the paste met the specification.

The paste was poured into a 40 m³ mixing tank through a passive screen, which screened out rocks or clay lumps. The mixing tank had a single shaft paddle mixer that further mixed the paste. From the tank, a peristaltic pump was used to load the paste into concrete trucks, which transported the paste to the boreholes for backfill placement by gravity.

Throughout the project, the monitoring crew tracked the progress through remote cameras, periodic CMS surveys, and paste level measurements using a plumb bob. Figure 4 shows the progress of the paste backfill by looking at the paste level against the marking on the post in the background.

In total, 53,389 m³ of paste backfill was placed, with a maximum daily backfill of 1,750 m³ per day.

4 SELF-CONSOLIDATING CONCRETE AS BACKFILL

4.1 *Sourcing the aggregates*

The SCC consisted of two sources of locally available aggregates (ACE and Det'on Cho), reclaimed tailings, and cement binder. The rationale behind mixing multiple types of aggregate with tailings was to produce a desirable particle size distribution. More importantly, no local source of aggregate was sufficient in quantity for what was required in this project.

Comparing to the paste operation, the SCC program used more cement per day. At 600 m^3 per day of backfill, the operation required 225 tonnes of cement. At times, only 160 tonnes were delivered in a day, so the production rate had to be reduced to maintain a continuous pour.

4.2 *SCC mix design*

The SCC has a target strength of 12 MPa. The material was first designed at the on-site laboratory. The material was subjected to slump flow, segregation, bleed water and UCS tests in accordance to procedures outlined in the ASTM standards.

To examine the curing condition of the SCC, the maximum temperature and thermal gradient in the stope were calculated using computer simulation. Subsequently, full scale trial batching was carried out, and the SCC was placed on the surface to test the flowability as shown in Figure 5.

To direct the placement through the five boreholes, tremie pipes were lowered to the bottom. The pipes were clamped on the steel platform on the surface. Using underground cameras, the pipes were elevated with hydraulic jacks as shown in Figure 6. As the level rises, the pipes were trimmed to keep the chute low enough for the truck to unload into.

Next to the tremie pipes, two extensometers with temperature sensors were lowered and became embedded in the SCC when the backfill was completed. Figure 7 below shows an image of the arrangement taken from an underground camera.

The target daily backfill volume was 600 m^3/day, which was limited by the capacity of transporting cement to the site. When necessary, the production rate would be reduced to maintain a fresh layer of SCC on the surface. This was to minimize the possibility of forming a cold joint. Due to logistics, the operation was later forced to switch binder from slag cement to a fly ash cement. In anticipation of the switch, the team quickly designed another mix recipe that met the design criteria.

Figure 5.   Full scale SCC trial batching on the surface.
Source: MineFill Services project database.

48

Figure 6. Hydraulic jacks for raising the tremie pipes.
Source: MineFill Services project database.

Figure 7. Image of the tremie pipe and extensometer taken with underground camera.
Source: Nahanni Construction Limited underground camera record.

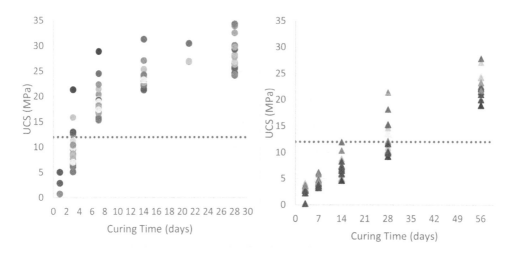

Figure 8.   UCS of SCC made with slag cement (left) and fly ash cement (right).
Source: Nahanni Construction Limited QC database.

During the continuous backfill, a sample of the SCC was collected at the borehole every 12 hours. The material was cast in 4 inch diameter by 8 inch high cylinders for UCS testing. As shown Figure 8, all the materials exceeded the 12 MPa UCS. As expected, the SCC made with slag cement developed a noticeably higher strength than that of with fly ash cement.

4.3   *Process description*

The SCC was produced using a conventional concrete batch plant. The aggregate bins ratio-ed the materials onto a conveyer belt using a loss-in-weight mechanism. The plant loaded the aggregates, cement, water, additives into concrete trucks. When the trucks arrived at the bore-hole area, they were parked in order to extend the mixing time for additional five minutes. After mixing was complete, the operator unloaded the SCC into the borehole.

In the SCC phase of this project, the daily consumption of cement varied between 110 to 310 tonnes per day. The process used between 383 to 1200 tonnes per day of locally sourced aggregate. The variation was largely dependent on the availability of the cement. Overall, a total of 16,686 m^3 of SCC was backfilled with a maximum backfill placement rate of SCC of 814 m^3 per day.

5   BACKFILLING IN COLD WEATHER CONDITIONS

Due to the challenge of receiving a sufficient quantity of cement, the project continued into December of 2018. The ambient temperature in Yellowknife lowered to an average of -17 degree Celsius and reached as low as -30 degree. Extreme cold temperatures can cause freezing on the equipment and particularly water lines. Furthermore, the freezing temperatures had a dramatic impact on the rate of cement hydration. There are several strategies for operating with paste backfill in these extremely low temperature conditions. Considering the logistical challenges and the locally available resources, the most reliable strategy was to heat the mix water. A simplified thermal calculation based on the heat transfer equation determined that the mix water needed to be above 25 degree Celsius for the paste to be above the freezing point.

In October, a diesel-powered water heater was brought to site and commissioned. Eventu-ally, the paste approached the freezing temperature without the heater. As shown in Figure 9, the paste started to freeze and became less flowable.

Figure 9.   Frozen paste caught on the screen.
Source: MineFill Services project database.

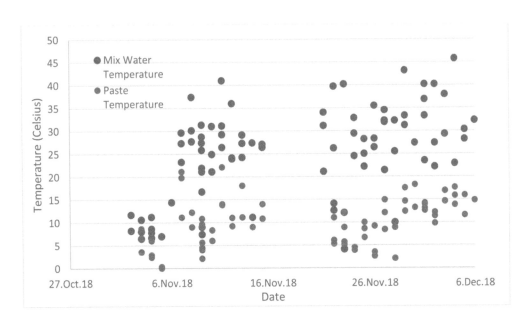

Figure 10.   Paste and water temperature.
Source: Nahanni Construction Limited QC database.

The temperatures of the mix water and paste were measured multiple times every day as shown in Figure 10. The very low paste temperatures triggered the use of the water heater. Even though the paste only contained 25 weight percent of water, the high heat capacity helped bring the overall paste temperature to above freezing. Note the fluctuation in the temperatures was a function of the time in the day and the rate of backfill production.

As with other cold weather operations, the cold weather required additional maintenance on the equipment to periodically clear frozen material. Winterizing the plant allowed the project to continue. The voids were filled to the collar of the boreholes and the project was completed on Dec 11th.

6 CONCLUSION

A combination of paste backfill and SCC was used to stabilize underground voids at the Giant Mine. CMS surveys were used to reconstruct the shape of the areas that needed to be backfilled. A number of paste and SCC mix recipes were designed with various flowability, set times, UCS, and bleed properties.

A continuous paste plant and a concrete batch plant were installed to produce the backfill materials, which were transported by truck to the boreholes. By gravity, the paste and SCC were placed underground, and totalled 16,685 m^3 and 53,388 m^3 respectively.

The process was monitored using remote camera, CMS surveys, and plumb bob measurements. Despite the challenges such as the lack of an accurate map of the voids, unstable underground conditions, extreme cold temperatures, and logistics of transporting materials to the site, the project was successfully completed on Dec 11th 2018.

ACKNOWLEDGEMENTS

The authors would like to thank Garrett Whipp, Steve Clayton, Public Services and Procurement Canada, and Golder Associates for their support in this project. The authors also want to express their gratitude to the entire crew who devoted their best effort to complete this challenging project.

BIBLIOGRAPHY

ASTM 2018, *C31 Making and Curing Concrete Test Specimens in the Field*. ASTM International. West Conshohocken.

ASTM 2018, *C39 Compressive Strength of Cylindrical Concrete Specimens*. ASTM International. West Conshohocken.

ASTM 2015, *C143 Slump of Hydraulic-Cement Concrete*. ASTM International. West Conshohocken.

ASTM 2016, *C940 Expansion and Bleeding of Freshly Mixed Grouts for Preplaced-Aggregate Concrete in the Laboratory*. ASTM International. West Conshohocken.

ASTM 2014, *C940 Slump Flow of Self-Consolidating Concrete*. ASTM International. West Conshohocken.

ASTM 2010, *D2216 Laboratory Determination of Water (Moisture) Content of Soil and Rock by Mass*. ASTM International. West Conshohocken.

NCL 2018. *Giant Mine C5-09 Remediation Project Quality Control Database*. Nahanni Construction Limited. Yellowknife, Northwest Territories, Canada.

NCL 2018. *Giant Mine C5-09 Stope Complex Plan*: Revision 3. Nahanni Construction Limited. Yellowknife, Northwest Territories, Canada. pp. 28.

Minefill 2020-2021 – Hassani et al (eds)
© 2021 Taylor & Francis Group, London, ISBN 978-1-032-07203-6

Investigation into the high transients experienced in Eleonore Mine's pastefill distribution system

Maureen Mcguinness & Kelvin Creber
Paterson & Cooke Canada Inc., Sudbury, Canada

Justin Jacobs
Paterson & Cooke USA, Denver, USA

Bernard Haley
Newmont Goldcorp, Rouyn-Noranda, Canada

SUMMARY: Eleonore Mine experiences high transient forces in their pastefill underground distribution system which results in significant damage to the pipeline supports. This investigation examines the causes of the transients in this pumped system and looks at their effect.

A transient flow model was developed and validated using operating data. The investigation showed that favourable transient conditions came from operation in full flow conditions. It was also shown that a misaligned hydraulic pulsation dampener can produce higher transients in the line than would be experienced without one. This investigation spurred design and operating strategy changes to combat transients in the underground distribution system.

Keywords: transients, positive displacement pump, backfill, piping

1 INTRODUCTION

An ongoing issue at Eleonore Mine has been high transient forces in the pastefill underground distribution system (UDS). These transients cause loud hammering underground with forces strong enough to cause damage to the pipeline supports. Pastefill production delays are common due to unexpected maintenance that is required to fix the UDS supports.

The goal of this investigation was to reduce the transients in the Eleonore UDS. In parallel to this work, improvements to alignment between the hydraulic pulsation dampener (HPD) and the positive displacement (PD) pump were on-going. It was acknowledged that there will always be times that the HPD is not operating at its optimal alignment. It was decided that the UDS should be upgraded to meet the maximum transient loads since they have resulted in damage to the system under the present conditions.

2 BACKGROUND

Transient forces in a pipeline are caused by fast moving pressure waves which in turn are a direct result of changes in flow rate. The mechanics of operation of a positive displacement piston pump inherently produces variation in flow rate. At every stroke change over, there is a short period of time in which there is no paste being pushed through the line. This sudden loss of flow causes a transient pressure wave to travel downstream in the pipeline. A HPD is

DOI: 10.1201/9781003205906-6

added to the system to counter the loss in flow by slowly accumulating paste during a pump stroke and then pushing paste into the pipeline during the switch between the pump cylinders.

This investigation sought to understand the causes and magnitude of the transients in the system, to develop a strategy for their mitigation, and - since the transient forces cannot be fully removed – to review anchor standard design loads. The process taken was to capture in-situ measurements of the transients in the system, and then model them — validating the data as appropriate. The model was then used to evaluate the effect that changes to the operating process has on the generated transient forces.

2.1 Eleonore pastefill distribution system

The pastefill distribution system at Eleonore is characterised by a short surface borehole after which the pipeline follows the ramp to 140 Level. Internal boreholes below 140 Level distribute paste to both the north and south side of the orebody as shown in Figure 1. The transient forces are reported to be more frequent and stronger in the upper mine, particularly above Level 140.

2.2 Hydraulic modeling

A typical hydraulic grade line (HGL) plot for a pour to the middle of the mine is shown in Figure 2. The distance between the HGL and the pipeline profile is an indication of the pressure in the pipeline at that point. This HGL plot indicates that the system requires a pump to

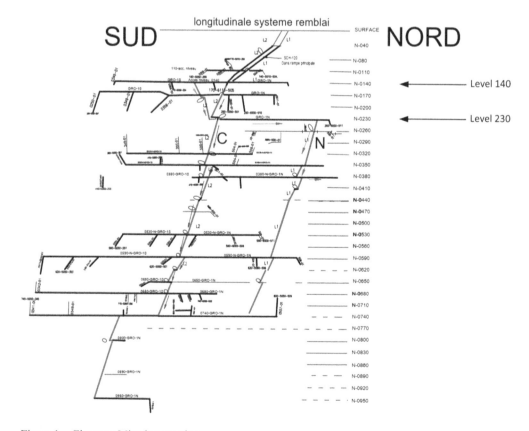

Figure 1. Eleonore Mine long section.
Source: Eleonore Mine 2019.

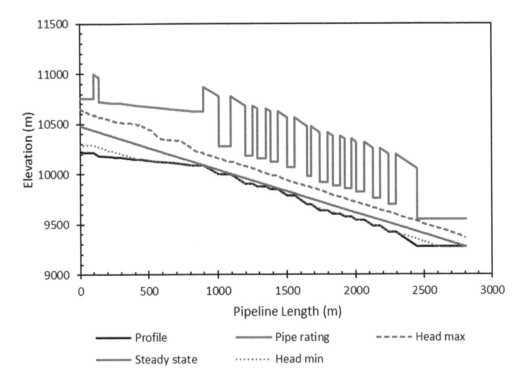

Figure 2.   Example of Eleonore hydraulic grade line plot.
Source: Study.

transport the paste the stope location[1]. The interception of the grade line and the pipeline profile indicates an area in which slack flow exists; in this case at the end of the 140 level.

The maximum transient pressures are shown on the HGL plot as a dotted line above the
HGL. The minimum pressure transient is shown below the HGL. These can add to the impact
loading, as the pipeline is temporarily placed in slack flow.

2.3   *Positive displacement pump and HPD*

Details of the Eleonore Mine paste pump and HPD are outlined in Table 1. The Putzmeister
KOS line of positive displacement pumps use an S-tube to switch between delivery cylinders.
The HPD accumulates paste during the cylinder stroke and then discharges it into the pipeline
during the short period of time that the S-tube switches to the other cylinder. This cycle is
depicted in Figure 3, where the blue dots indicate paste discharged from cylinder #1 and red
dots from cylinder #2.

2.4   *UDS supports*

The pipeline is supported with a series of hangers, guides and anchors. The anchors resist
forces acting in the axial direction of the pipeline, which includes the transient forces. An

---

1. For a system to flow by gravity only, the hydraulic grade line must intercept the y-axis. The distance above
   the pipeline profile that the hydraulic grade line intercepts the y-axis provides the pumping requirements to
   ensure flow.

Table 1. Eleonore positive displacement pump data.

Vendor	Putzmeister
Model	KOS 25100 HP
Delivery cylinder diameter	360 mm
Delivery cylinder length	2500 mm
Max delivery pressure	75 bar
Dampener Model	HPD (High Pressure Dampener)
Dampener cylinder diameter	279.4 mm
Dampener cylinder length	1500 mm

Source: Eleonore Mine 2019.

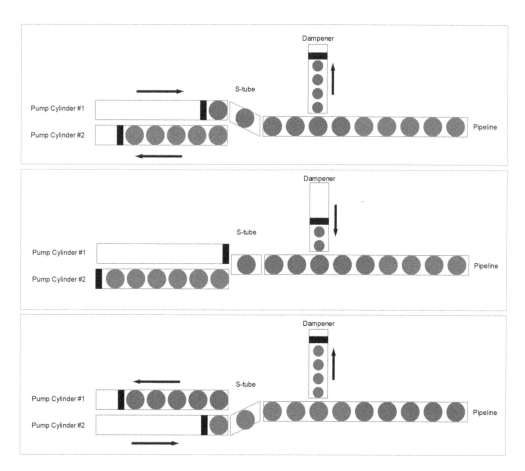

Figure 3. Cycle of paste flow during switch between PD pump cylinders and HPD.
Source: Based on schematic by Putzmeister.

example of an anchor used at Eleonore is provided in Figure 4. Two types of anchors are used: normal and heavy duty. The normal anchors were designed for 35 kN under the design assumption that an HPD would be used with the PD pump operation and that this would reduce the transient forces. The heavy duty anchor was designed at 150 kN. This was intended to be installed in the upper mine for the operation when the HPD was not yet installed.

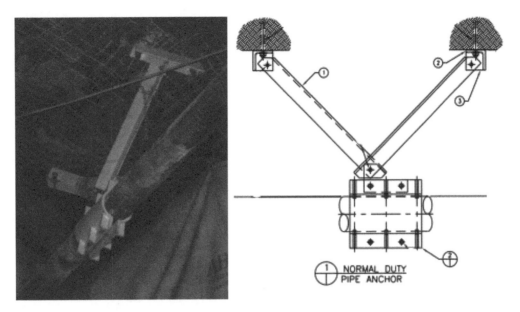

Figure 4.   Example of anchor support used at Eleonore.
Source: Study.

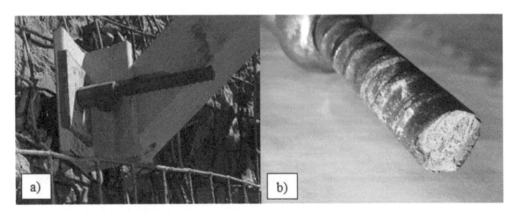

Figure 5.   A) Gap behind footing, b) Rock bolt sheared by transient forces.
Source: Study

Proper installation of the anchors and guides in the system is paramount to maintaining the integrity of the UDS system. In particular, the inclusion of grouting behind the anchor and guide foot plates was shown to have a ten-fold effect on the strength of the support. Without grout, the bolts which have a gap between the footplate and the wall (Figure 5a) are placed in a cantilever situation with the bolt not able to be put under tension. The result is that this bolt becomes the weak point in the system and can break prematurely due to impact loading and fatigue. An example of a broken bolt is shown in Figure 5b.

3   TRANSIENT MONITORING

To understand the behaviour and magnitude of the transients that were passing through the Eleonore UDS, they first needed to be recorded. Two high speed pressure gauges were

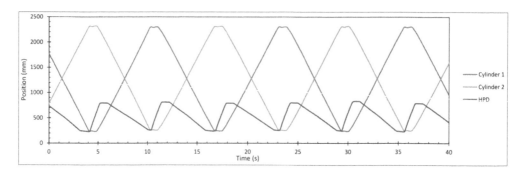

Figure 6.   Example cylinder position graph.
Source: Study.

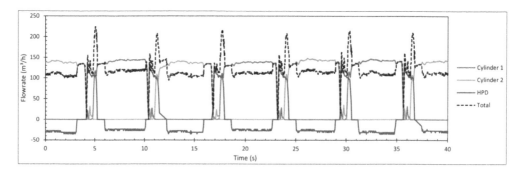

Figure 7.   Example flow profile of the paste.
Source: Study.

installed on level 140, which was one of the areas experiencing the most damage. This was at the bottom of the ramp from surface. Two gauges, with a known distance apart and elevation, were installed. There was also a regular, low speed pressure gauge in this area. Comparing the pressures recorded from the regular gauge to the high speed gauges, it was shown that the regular gauges were picking up the general trend of the reoccurring peaks in pressure but the magnitude of these peaks was reduced.

Since transients are caused by changes in flow rate, the behaviour of the flow in the pipe was evaluated using the recorded position of the PD pump cylinders. Given the known cylinder geometry, a flowrate over time could be determined. Two position curves are traced at any time; one for each cylinder of the pump. The cylinders are connected hydraulically and therefore the position graph of one cylinder is a mirror image of the other as shown in Figure 6. The HPD position was also recorded. It is shown to move forward only during the time when both pistons are at zero flow.

The flow profile in the pipeline can be developed by combining the flow contributions of the two cylinders and the HPD as shown in Figure 7. The negative flow in the HPD indicates the time in which it was drawing paste into the cylinder from the pipeline. The effect does not occur for the PD pump cylinders because those cylinders draw the paste from the hopper before it enters the pipeline.

4   TRANSIENT MODELLING

Each of the three cylinder flowrates are inputted into the transient model. The paste properties (yield stress, viscosity, % solids), pipe diameter and pipeline profile are also included in the model. The result of the modelling was a pressure profile over time at any point in the pipeline such as the one for level 140 pressure gauge location. This pressure profile was compared to the pressures measured in the field as shown in Figure 8.

The model was shown to have a similar result to the actual measurements. The amplitude of the pressure peaks aligned as did the frequency. This process was performed on three different occasions with similar results. From this validation, it was concluded that the model was representative of the Eleonore UDS system transients. This made the next steps of the study possible – to evaluate forces on anchors, and to evaluate the effect of operational changes on the transients experienced in the pipeline.

The transient pressures throughout the pipeline are then used to calculate axial force. A dynamic load factor (DLF) must be applied to all calculated loads to account for the amplification effect of the forces in dynamic conditions (ASME 2016). Based on past experience, and the natural frequencies of underground piping systems, a DLF of 1.75 was applied.

This procedure was followed for several pours spanning the range of the distribution system, including pours to the upper mine and lower mine both close and far from the main pipeline system. The maximum axial forces experienced for each pour are summarized in Figure 9.

Some key observations include that when the hydraulics are such that the 140 to 230 level borehole is in slack flow the transients are isolated in the upper portion of the pipeline, and that the transients in the upper mine sometimes exceed the 150kN load rating of the heavy duty anchors in the system. All cases are shown to exceed the 35 kN load rating of the normal anchor.

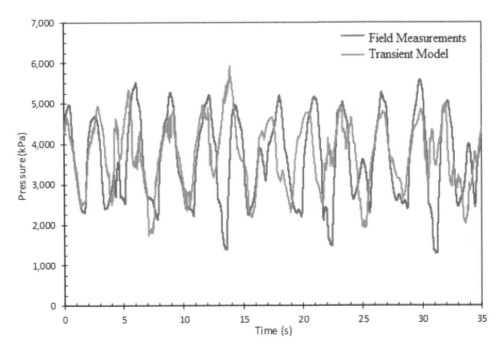

Figure 8.   Transient model pressure predictions compared to actual measurements.
Source: Study.

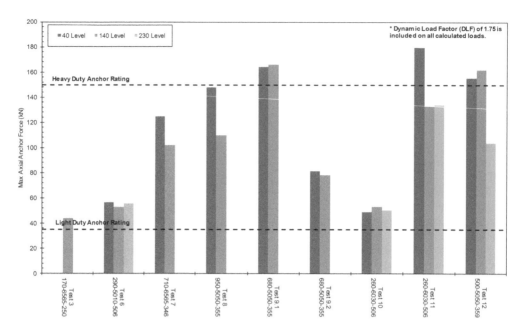

Figure 9.   Summary of modeling results of the axial forces at levels 40, 140 and 230.
Source: Study.

5   PROCESS PARAMETER EFFECT ON TRANSIENTS

Using the validated transient model, investigations were made to assess the impact of solids concentration, the cylinder/HPD flowrates and timing.

5.1   *Effect of solids concentration on transients*

The operating practice to date was to maintain a low mass concentration as it was the operational experience that the transient forces were larger when the solids concentration increased. This was in contradiction to the authors' experience where increasing the concentration of the paste usually results in lower transients in the system. To examine this effect, the cylinder position and operating parameters were measured for a series of paste concentrations, visualized in Figure 10. The model confirmed that the higher concentrations resulted in the lowest transient forces in the pipeline. However, the site experience was also confirmed by the modelling as it was shown that the transient forces initially increase with the solids concentration until a point at which they decrease well below the forces observed at low concentrations. This understanding was a key point in the development of an operational strategy for the system. In the graph, Test 9.1 and 9.1B have a section of pipeline in slack flow, not allowing transients to travel to the lower sections of the mine. As this slack flow section becomes smaller, the transient forces increase, until the pipeline no longer experience slack flow and the forces in the pipeline drop.

5.2   *Effect of flowrate on transients*

The effect of flowrate on the transients was also examined. For this test, the paste flowrate was gradually increased while keeping a constant paste concentration. The cylinder position and operating parameters were recorded and input into the transient model. The results were as expected; increased flowrate resulted in higher transients. It was not the magnitude of the

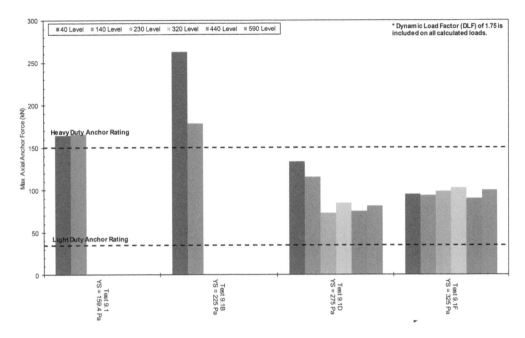

Figure 10.   Effect of solids concentration on transients.
Source: Study.

flow rate as such that caused the transients but the change when the flowrate was temporarily reduced. Slower speeds results in a smaller change in flow which corresponds to a smaller transient force in the system. In the case of this specific PD pump, the cylinder position data showed that the flow was coming to a stop temporarily, irrespective of the starting flowrate.

5.3   *Effect of HPD alignment on transients*

The interaction between the HPD and the PD pump cylinder was shown to have an important effect on the transients produced in the pipeline. The goal of the HPD is to push paste into the pipeline when the cylinders are switching over in an effort to maintain a steady flowrate in the pipeline. The HPD is activated based on the position of the cylinders and its timing can be modified to achieve the best overlap. The modelling showed that if the overlap is not well adjusted, the transient forces can be greater than if the HPD was not used. This is because the flowrates of the HPD and the cylinders are added together, potentially resulting in higher flowrates than the cylinder would provide. If that flowrate then drops to zero, an even greater transient is produced than normal.

Pipeline flowrates are compared in Figure 11 to show that the effect of HPD alignment. A well aligned HPD is successful in minimising transients as shown in (a). The flowrate in this example remained mostly within 25% of the target flowrate throughout the piston transitions. This flow profile can be compared to the case where no HPD is used (b) and where the HPD is activated too early (c) or too late (d). The benefit of the HPD is evident when compared to the flow profile when the HPD is absent, where the flowrate drops to zero at every cylinder switch, resulting in large changes of flow. However, the worst case scenarios occurred when the HPD timing was misaligned. When the HPD stroke started too soon (c), the flowrate in the pipeline momentarily doubles. The early start also caused the flow to end prematurely resulting in the high flowrate to drop to zero as the next cylinder had not yet started its forward movement. The resulting change in flow is greater than that seen when the HPD was not used at all. A similar problem is seen with a late start of the HPD stroke (d).

61

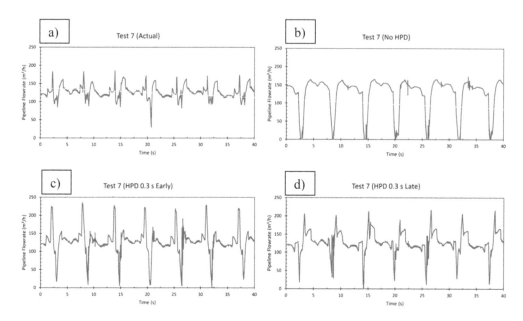

Figure 11. Flowrate profile for various HPD scenarios a) HPD aligned with PD pump strokes, b) no HPD, c) HPD starting early and d) HPD starting late.

Source: Study.

6 OPERATING STRATEGY TO MINIMISE TRANSIENT FORCES

6.1 *PD pump and HPD*

Since alignment between the pump and the HPD strokes were shown to have a large influence on the transients in the system, optimization of the timing is ongoing. Predictable, repeatable performance of the HPD is key to reducing the transient forces in the system and is currently under a parallel initiative on site.

6.2 *Anchor upgrades*

The anchor load capacity requirements were increased to the worst case scenario possible (misaligned HPD). Based on the transient analysis results, the heavy duty anchor (which is rated for 150 kN) will be used in the upper mine (level 230 and above). The normal duty anchor was upgraded to 100 kN and will be used in the lower mine (below level 230).

Upgrades to the normal duty anchor (shown in Figure 4) included adding a cross brace between the legs and strengthening the connection point between of support legs to the pipe bracelet. Emphasis was placed on a design which could be modified in-situ on already installed anchors. Grouting of all the footings that are not currently flush to the rock is also an important part of the system upgrade, as discussed in Section 2.3.

6.3 *Process modifications*

The investigation showed that the operation methodology has a large impact on the transient forces created in the system. While the UDS system is being upgraded to the new load capacity, there are some process modification that can be taken to reduce the transient forces.

Reduce paste flow rate: A lower flowrate will reduce the overall change in flow rate during times when the HPD timing is misaligned with the PD pump strokes. Similarly, reducing the

HPD flow rate will have a similar effect. While reducing the production rate is not a permanent solution, in the interim it will reduced the amount of downtime to repair UDS supports.

Increase solids concentration: It was clearly shown that higher solids concentration reduces the transients by putting the UDS system in a full flow condition. However, the transition to the higher solids concentration will cause higher transient forces. The strategy proposed is to start at lower concentration and lower tonnage then to increase the solids content to the desired amount. Once the system is stable and it has been confirmed that the pipeline is not in slack flow and causing large transient forces, the tonnage can then be increased. This will be a permanent operating strategy.

7 CONCLUSION

The investigation into the transient forces experienced in the Eleonore UDS system showed that they can be modelled based on the paste properties and the PD pump and HPD cylinder positions. A transient model was developed and subsequently validated with in-situ high speed pressure measurements. The model was used to evaluate the magnitude of the transient forces experienced in the UDS system. Under the current operating conditions, the transient forces were shown to approach and at times exceed the current load rating of the anchors.

Using the model in conjunction with field data, the effect of various process parameters on the transient forces was clarified. Key parameters included HPD timing, paste concentration and flow rate. It was found that misalignment of the HPD stroke with respect to the cylinder strokes can cause greater transients than when the PD pump is operated without a HPD at all. Temporary mitigation of transient forces can be achieved by reducing the flowrate of the pump. Increasing the solids concentration is key to maintaining a full flow pipeline, in turn lowering transient forces in the system.

Operational strategies were initiated on site to capitalize on the results of this investigation. These included upgrading the anchor support loads to match the expected transients in the system and ensuring proper installation of the support to benefit from the full load capacity of the anchors. A temporary solution is to decrease the flow rate in the system to keep the transient within the lower load capacity of the system until the proper upgrades can be completed. A permanent solution of operating with a higher paste solids concentration to ensure a full flow system has also been initiated. These changes are in addition to an ongoing effort to better align the dampener to the cylinder strokes to minimize the transient forces developed in the system.

ACKNOWLEDGEMENTS

The authors wish to express their gratitude to the Eleonore Mine for allowing the paper to be written and for providing support throughout the study.

BIBLIOGRAPHY

American Society of Mechanical Engineering 2016. *ASME B31.1: Dynamic Amplification of Reaction Forces*, Section II (3.5.1.3)

Study: *Eleonore Pastefill UDS 2019*. Completed confidentially for Newmont-Goldcorp by Paterson & Cooke Canada Inc.

Paste backfill measurements and testing

Minefill 2020-2021 – Hassani et al (eds)
© *2021 Taylor & Francis Group, London, ISBN 978-1-032-07203-6*

The effect of curing conditions on the strength development of cemented backfill samples

Brendan Dennis
Laboratory Supervisor, Outotec SEAP

Matthew Helinski
Process Technology Manager, Outotec SEAP

SUMMARY: Historically, strength development in underground backfill is estimated based on the results of historic unconfined compressive strength (UCS) testing. During both preliminary study phases and operational phases are batched to specific solids and binder contents and placed in curing chambers, with the intention that these chambers replicate the curing conditions experienced underground.

The rate of strength development in mining backfill can be greatly affected by variations in environmental factors including temperature & humidity. Therefore accurate representation of *in situ* curing conditions for cemented specimens is crucial.

This paper investigates the effect that differing curing conditions can have on the strength development of cemented mine backfill specimens. This paper explores how variations expected to be encountered in typical fill masses impacts the resulting strengths and also investigates how aspects that may appear subtle can actually have a significant impact on the resulting measured strengths. Given the findings of this investigation a series of recommendations are provided for improving the way mine backfill curing chambers are managed.

Keywords: cemented backfill, curing conditions, temperature, quality control testing

1 INTRODUCTION

Historic evidence from temperature monitoring within fill masses during curing show a consistent trend of increasing temperatures during the initial stages of curing followed by a plateau in temperature, with most cases plateauing at temperatures well in excess of 35°C. The increased temperature is expected to be a result of the exothermic cement reaction, while the sustained higher temperatures are expected to be due to the fills thermal mass. Examples of *in situ* temperature monitoring during filling, at various sites throughout the world are presented in Figure 1.

While it is widely recognised that curing temperature can significantly impact the strength of cementation materials, literature reviews show little record of controlled studies to quantify the impact of temperature on fill strengths in the context of how this may be relevant to onsite quality control testing.

This paper presents an investigation into the influence of curing temperature on the strength of different mine backfill types giving consideration to both; the magnitude of strength change that can occur as a result of temperature variations, and aspects that may appear subtle, but can actually have a significant impact on the quality control specimen curing temperatures, and therefore strengths measured in onsite quality control testing. The results from this testwork and analysis culminate in a series of recommendations for good practice quality control testing.

DOI: 10.1201/9781003205906-7

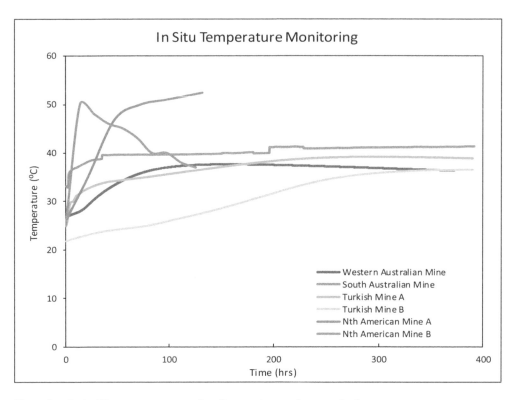

Figure 1. *In situ* fill mass temperature data from various underground mines.

2 CURING TEMPERATURE SENSITIVITY

To investigate this influence of curing temperature on fill strength, a series of cemented hydraulic fill mixes were batched with a range of different binder types and specimens for each mix. The samples were cured at temperatures of 23°C and 35°C. The hydraulic fill adopted in this work was a copper tailings material that was predominantly Quartz and Chlorite minerals. All mixes were batched to 69% solids with 10% binder addition. Specimens batched for each mix were tested for unconfined compressive strength after hydration periods of 2, 7 and 28 days. The fill mix matrix is presented in Table 1.

The results from this testwork are presented in Figure 2, which shows the measured unconfined compressive strength (UCS) after 2,7 and 28 days hydration. Mixes cured at 35°C are presented as solids bars, while the equivalent mix cured at 23°C are presented as open bars.

Table 1. Experiment mix matrix.

Mix No	Binder	Binder portion (%)	Curing Temp (°C)	Mix Solids Content (%)
1	General Purpose (GP) Cement	10	35	69
2	General Purpose (GP) Cement	10	23	69
3	GP 75% /Fly Ash 25%	10	35	69
4	GP 75% /Fly Ash 25%	10	23	69
5	GP 65% /Slag 35%	10	35	69
6	GP 65% /Slag 35%	10	23	69

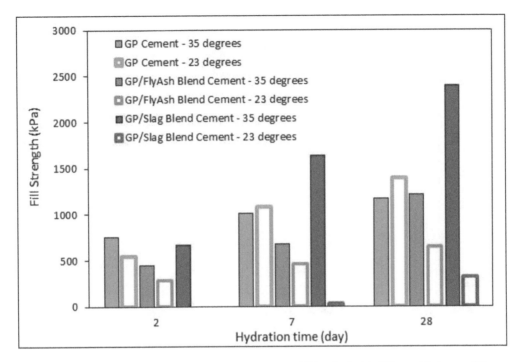

Figure 2.   Influence of curing temperature on strengths for fill batched with different binders at different curing temperatures.

Figure 2 shows a direct correlation between curing temperature and strength, with all mixes showing higher strengths when cured at higher temperatures. However, what is most interesting about these figures is that, when cured at 23°C the GP/Slag blend binder shows little strength gain. However, when the same mix is cured at the higher temperature (of 35°C) the resulting strengths are superior to all other mixes.

These results show that, compared with conventional concrete curing temperatures (of 23°C), not only does curing cemented backfill specimens at temperatures more representative of *in situ* yield higher strengths, but the elevated temperature also shows different binder types to be more efficient.

To investigate the sensitivity of more subtle temperature changes in curing temperature, over the range of temperatures measured *in situ*, a second testwork campaign was initiated. This campaign considered an Australian paste fill mix with high silica content tailings and a binder blend type that consists of 30% GP/70% Ground Granulated Blast Furnace Slag (Slag). The mix was cast at 70% solids with 4% binder addition. Immediately after casting, specimens from this mix were placed into curing chambers that were set to different temperatures. One chamber (Chamber 1) was set to 35°C, another (Chamber 2) set to 31°C and another (Chamber 3) set to a temperature of 43°C. Specimens were cast in 50 mm diameter moulds and subject to UCS testing after 3, 7 and 28 days hydration.

The measured strengths, after 3,7 and 28 days hydration, are plotted against curing temperature in Figure 3.

Whilst it is widely acknowledged that curing temperature can influence the strength of cured paste backfill, Figure 3 shows that, across the range of temperatures measured within fill masses, the influence of temperature on strength can be significant (at 30-200%). Given the variability of *in situ* temperatures presented in Figure 1 and the sensitivity of strength to temperature presented in Figure 3, a concerted effort should be dedicated to defining *in situ* curing temperatures for each individual site and ensuring that quality control specimens are cured under representative conditions.

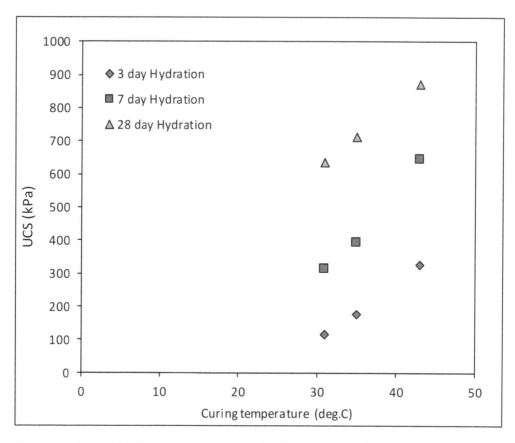

Figure 3. Influence of curing temperature on strengths of cemented paste fill samples.

3 QUALITY CONTROL CURING METHODS

Given the sensitivity of mine backfill strength to curing temperature, the authors set about to investigate how subtle aspects associated with onsite quality control curing methods may impact the measured strengths. A study was initiated to define the sensitivities around a typical curing chamber that may be used on site. The aim was to assess the influence of curing chamber size, and position within the curing chamber, on the curing conditions and resulting strength.

For this study, two curing chambers were utilised. Chamber #1 is a 1200 × 900 × 600 mm unit with one element and one thermostat. Chamber #3 is a larger 1600 × 1200 × 800 mm unit with three elements and one thermostat. Chambers are manufactured from 4 mm aluminium sheeting, but other models may be manufactured from stainless steel of similar thickness. Each curing chamber was set with the thermostat and heat coil submerged in 100 mm of water at the base of the chamber. Above the water line a horizontal mesh layer is inserted, and specimens are placed on this mesh. To meet capacity requirements specimens can be stacked in these chambers so some specimens may be cured up to 250 mm above the water line. This arrangement is typical of curing chambers adopted for onsite quality control testing in Australia. A photograph inside a curing chamber is presented in Figure 4.

For the first mix comparison, a total of 6 locations were selected within these chambers. These include L1, L2 and L4 in the larger Chamber #3 and L6, L7 and L8 in the smaller Chamber #1. L1, L6 and L7 were located on the chamber floor (immediately above the waterline), whilst L2, L4 and L8 were raised to a height of 250 mm above the waterline.

Figure 4.   Photograph of curing chamber used in experiments.

Images showing the location of these specimens on a plan view of the chambers is presented in Figure 5.

Paste for all specimens was batched in a single mix, which contained 4% binder batched to 70% solids. Samples were tested in duplicate for each location and hydration period. Samples were cast in 50 mm diameter moulds, and samples from each location were tested for UCS after 4, 7 and 28 days hydration. Specimens were also cast for long-term testing (at 240 days hydration), however these were yet to be tested at the time of compiling this document.

The measured strength is plotted against hydration time in Figure 6. Figure 6a presents the measured strength for specimens cured in the larger chamber (Chamber #3), while specimens cured in the smaller chamber (Chamber #1) are presented in Figure 6b. The results presented in Figures 6a and b are the average of the duplicate samples in each location, and it should be noted that the standard deviation of the duplicate samples was less than 10% for all cases.

The results presented in Figures 6a and b show that, for the same mix, in what may have been be considered identical curing environments, the measured 28-day strength varied by over 30%,

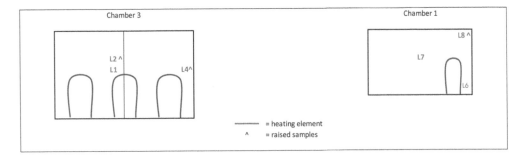

Figure 5.   Position of heating elements and specimen location within chambers.

across a given chamber and up to 60% across both chambers. Analysis of the test specimens showed that, across all specimens, dry densities were within 0.01 t/m³, while saturation levels were all between 98 and 100%, which suggests mix variability was not contributing to the strength discrepancy.

To investigate the source of this strength discrepancy environmental testing was undertaken at a range of different locations throughout each of the curing chambers. All probes showed consistently high relative humidity (in excess of 98%), but a notable variation in curing temperatures. The measured curing temperatures in Chamber #3 and #1 are plotted against time in Figure 7a and b, respectively. It should be noted that the zig-zag temperature profile is a result of the heating element engaging to raise the temperature.

Figure 7 shows a temperature variation or upto 14% across a given curing chamber and when comparing Figures 7a and b, this shows that even though both chambers had the same setpoint, temperatures varied by 16% between chambers.

To assess the correlation between curing temperature and strength, the 7 and 28 day strengths are plotted against measured curing temperature in Figure 8. Superimposed over Figure 8 is the strength verses curing temperature relationships previously presented in Figure 3.

Figure 8 shows a consistent trend of increasing fill strength with curing temperature, which is similar to trends presented in Figure 3. Given this consistent trend and other similar fundamental specimen properties (e.g. dry density and moisture contents) it is expected that the

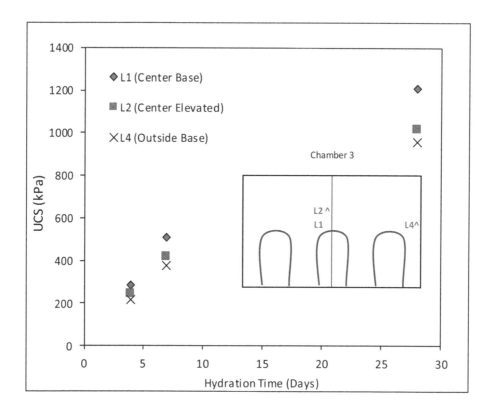

(a)

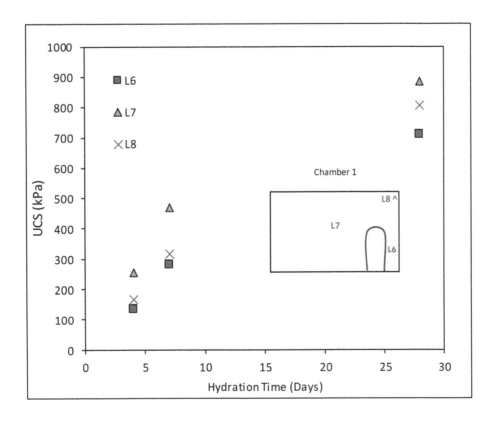

(b)

Figure 6.   Strength against hydration time for specimens cured in different locations of a (a) large and (b) small curing chambers.

observed variability in strength is attributable to the variation in curing temperature across the chamber/s.

Given that many Australian mines are located in arid environments where ambient temperatures can vary by over 20°C between day and night, and by over 40°C over the year, and that site curing chambers are not necessarily located in temperature controlled facilities, variation in curing temperatures both across the chamber and due to ambient temperature variations could be having a significant impact on measured quality control strengths.

4   CURING CONDITION MANAGEMENT

Given the sensitivity of strength to curing temperature and the observed variation in temperature across conventional site based curing chambers, insulating wrapping was applied to a typical chamber in an effort to minimise heat exchange. Insulation adopted was an Air-Cell shield consisting of foam cell core and reflective wrapping. Chamber 1 was fully insulated whilst an identical chamber (Chamber 2) was not. A photograph showing the fully insulated chamber is presented in Figure 9. This figure also shows the environmental condition monitoring probes at the centre and edge of the chamber.

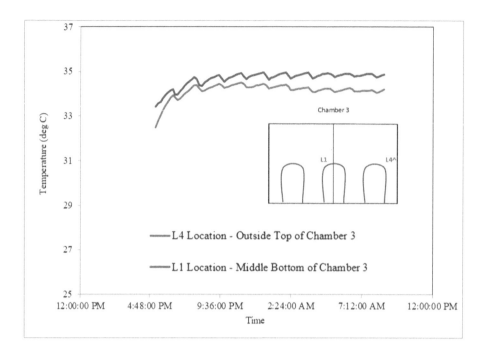

(a)

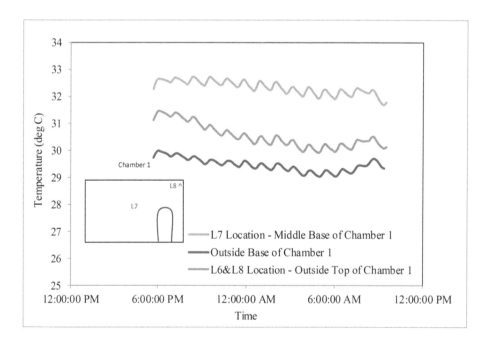

(b)

Figure 7. Hydration temperature against time for specimens cured in different locations of a (a) large and (b) small curing chambers.

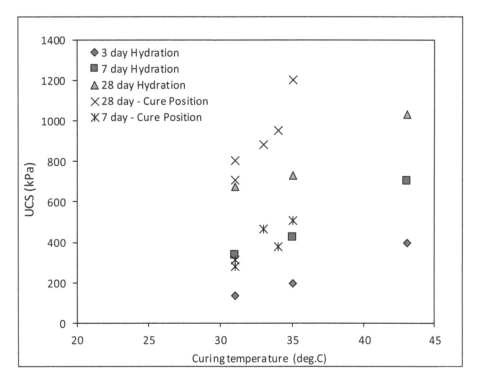

Figure 8.   Strength against curing temperature for samples within the same curing chamber superimposed over the results from Figure 3.

Figure 9.   Photograph of insulated curing chamber.

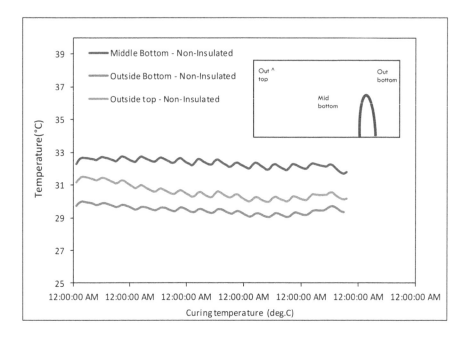

(a)

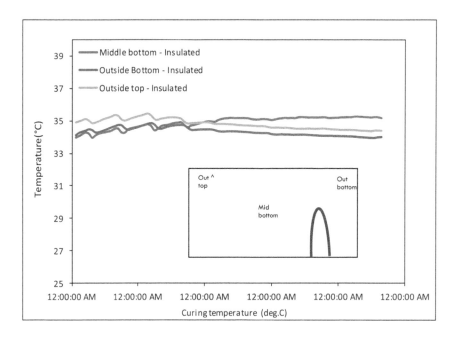

(b)

Figure 10.   3-location monitoring of chamber 1 (a) before and (b) after insulation.

Both curing chambers were set at the same thermostat set-point (of 35°C). Figure 10a and b present the temperature measurements at different locations within the uninsulated and insulated curing chambers, respectively.

Comparison between the temperature measurements in the insulated and non-insulated curing chambers show that, for the same thermostat setting, the insulated curing chamber provides a higher and more consistent temperature environment across the chamber.

To confirm the performance of specimens cured across the more consistent curing conditions a paste mix was cast and cured at locations L9-12 across the insulated curing chamber. The curing locations and measured strength (against hydration time) are presented in Figure 11.

The results presented in Figure 11 illustrate that the improved consistency of curing conditions across the chamber resulted in a significant improvement in the consistency of strength measured for specimens cured in different locations across the chamber.

The other notable feature about the increase in consistency of the curing conditions (from insulation) is the increase in temperature stability over time. Figure 12 presents a plot showing the variation in curing chamber temperature against time, for the insulated and uninsulated chambers. This figure shows the uninsulated chamber cycling in temperature in accordance with the ambient temperature during dan and night cycles. The insulated chamber shows a steady temperature regardless of the surrounding ambient temperature.

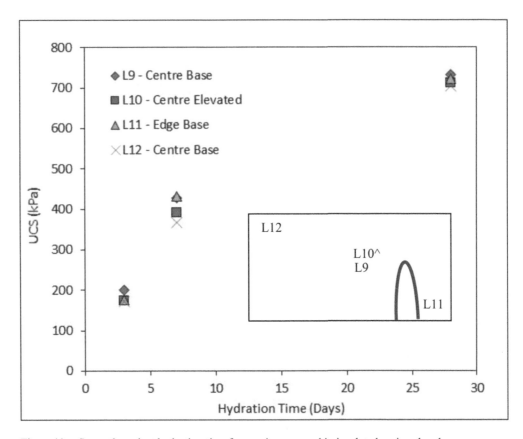

Figure 11.   Strength against hydration time for specimens cured in insulated curing chamber.

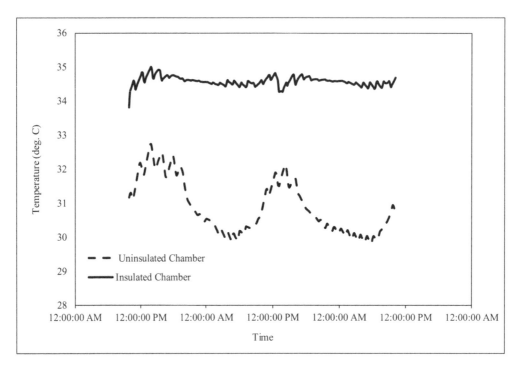

Figure 12. Curing chamber temperature cycles in insulated and uninsulated curing chambers.

5 CONCLUSION

This paper presents a study to investigate the influence of curing temperature on the strength of cemented paste backfill. While it is widely recognised that curing temperature has an influence on cemented backfill strengths and that insulation reduces heat transfer the interesting findings from this investigation showed that:

- Based on *in situ* data gathered from within cemented mine backfill masses, at various sites throughout the world, it can be seen that *in situ* curing temperatures are significantly higher (at 30-50°C) than recommended concrete industry quality control curing temperatures (23°C).
- Testing showed that, relative to specimens cured at recognised concrete industry curing temperatures, specimens cured at higher temperatures not only showed higher rates of strength gain, but also show different binder types (in this case GP/Slag blends) to be more efficient.
- Testing across the range of temperatures expected to be encountered in typical minefill masses (at 30-45°C) show that, across this range, the strength of cemented paste backfill can vary by 30-200%, with the most significant influence being at early hydration periods. On this basis it is considered critical that sites identify *in situ* curing conditions (temperature) and ensure that quality control specimens are cured at similar temperatures.
- Measurement of temperature variations across an uninsulated curing chamber showed that a thermostat feedback loop is insufficient for achieving steady curing conditions across the chamber. The magnitude of the temperature variation was shown to create a 30-60% difference in measured fill strengths depending on chamber size and specimen position. Given the observed variability, it is recommended that all laboratory and field based curing chambers are:
  - ○ fitted with thermostats and properly sized heating coils.

- ○ properly insulated to eliminate a variation in temperature both across the curing chamber and due to variations of external ambient temperatures.
- ○ fitted with multiple environmental sensors to confirm that the desired temperature and humidity are achieved throughout the chamber.

- *In situ* temperature measurements indicate a considerable change in fill curing temperatures over time. This aspect was not pursued as part of this investigation. However, given the sensitivity strength to curing temperature, it is recommended that future work be dedicated to developing rational strategies for curing quality control specimens under conditions that more accurately represent the evolution of curing temperature with hydration time.

Minefill 2020-2021 – Hassani et al (eds)
© 2021 Taylor & Francis Group, London, ISBN 978-1-032-07203-6

Experimental investigation on shear strength properties of interface between backfill and rock

Guangsheng Liu
BGRIMM Technology Group, Beijing, China

Weilv Wu
Beijing General Research Institute of Mining and Metallurgy, Beijing, China

Lijie Guo & Xiaocong Yang
BGRIMM Technology Group, Beijing, China

Zhihong Zhang
College of Architecture and Civil Engineering, Beijing University of Technology, Beijing, China

SUMMARY: Direct shear tests were carried out with planar interfaces between cemented backfill and rock to obtain cohesion and friction angle. Triaxial compressive tests with the same backfill were used to reveal proportions of shear parameters between interface and backfill. Nonplanar interfaces with different roughness were shear tested for further comparison. A constitutive model for interface was modified based on Mohr Coulomb criterion, which could used to investigate interaction between backfill and rock mass. Numerical simulated shear stress strain were used to validate the model.

Keywords: Backfill, Interface, Shear test, Roughness, Constitutive model

1   INTRODUCTION

With increasing requirements on mine safety and environment protection, mining with backfill has been widely used in underground mines worldwide. Stability of backfill in mine stopes is an important factor contributing to the safe mining, especially for the large scale and high efficiency stage open stoping with subsequent backfill mining method. After excavating secondary stopes, stability of exposed cemented backfill in primary stopes will affect not only mining safety, ore recovery and dilution rate, but also mining costs as numerous cement has to be added to the cemented backfill in primary stopes to achieve the stable self-standing.

Investigation on backfill stability relies on a reasonable evaluation of its stress distribution. As early as 1943, Tezaghi proposed that due to arching effect, actual vertical pressure at the bottom of soil was much smaller than the self-weight of overlying soil. Aubertin et al. (2003) introduced the arching effect from soil mechanics to mining with backfill to analyse the stress distribution of uncemented backfill in vertical narrow stopes. Li (2005) proposed a three dimensional analytical model for the stress distribution of backfill in vertical stopes and analysed influences from stope dimensions and backfill strength properties. But the model is mainly suitable for stress calculations of uncemented backfill. Liu (2019) further modified the

DOI: 10.1201/9781003205906-8

three dimensional analytical model which could be used for arching stress evaluation of the cemented backfill with practical values of cohesion.

Stress distribution is also the basis for required strength design of the backfill in mining. Mitchell (1989) proposed an analytical model to evaluate strength requirement of backfill in mine stopes, in which interface cohesion between side rock walls and backfill is emphasised. Li (2013) and Liu et al. (2018) extended the analytical models for the backfill strength requirements based on Mitchell method (1989), in which strength parameters of the interfaces between backfill and rock walls are key influential factors.

Liu et al. (2016, 2017) utilised numerical methods to investigate necessity of considering interface element between backfill and rock for better simulation of backfill stress distribution. It was found that when interface strength equals to the backfill, there is no difference on backfill stress in numerical models whether to consider interface element or not. However, when interface strength is lower than the backfill, the addition of interface element has significant influences on backfill stress, and the stress distribution is mainly controlled by the interface rather than the backfill.

To sum up, it can be seen that the interfaces between backfill and rock mass controls the backfill stress distribution, which will further affect backfill strength requirement design, thus determine the balance of mining safety and costs. Therefore, it is necessary to investigate the mechanical parameters of interface between backfill and rock.

Direct shear test on the fill-rock interface is an important method to obtain its mechanical characteristics. Up to now, Manaras (2009), Manars et al. (2011), Nasir and Fall (2008, 2010), and Koupouli et al. (2016, 2017) have carried out exploratory tests of interfaces. But there are some differences of influencing factors in these research, and some opposite conclusions.

Manaras (2009) referred to joint roughness coefficient (JRC) from rock mechanics to prepare backfill-rock interface samples for direct shear tests. Effects from curing age, cement content of backfill and interface roughness on shear strength of interfaces were studied.

Nasir and Fall (2008, 2010) used silica sand to replace mine tailings to prepare interface samples with different hard materials (concrete, brick, rock). The results show the friction angle and cohesion along interfaces increase slightly with the increase of curing age within 28 days. When curing temperature increases from 2 °C to 20 °C and 30 °C, the friction angle and cohesion of interfaces increase. The friction angle along interfaces is about 0.6 to 1.0 times of the friction angle of corresponding backfill, but no clear proportional relationship for cohesion.

Koupouli et al. (2016, 2017) used different lithologic materials (marble, shale) as hard materials, and prepared interface samples with unclassified tailings and cement (2.3% and 8.2% for cement content). After curing 3 to 7 days, direct shear tests were carried out with the interface samples. It concluded that the friction angle along interface is larger than the friction angle of the corresponding backfill, while interface cohesion is smaller than the backfill. This conclusion is different with the conclusions of Fall and Nasir (2010).

From above, it can be seen that there is not a complete consensus on the shear strength properties of interface between backfill and rock. However, in practical mining with backfill, interface strength parameters (friction angle and cohesion) and their relationships with corresponding backfill have significant influences on stress distribution and strength requirement design of the backfill. Therefore, it is necessary to investigate the shear properties along the interface between backfill and rock mass.

2   DIRECT SHEAR TEST ON PLANAR INTERFACE

Planar interface samples were prepared with backfill on the top and rock at the bottom. Granite was selected as rock material with size of length 150mm, width 150mm and height 102mm. Acrylic plates was used to build the mould surrounding bottom rock and forming the same volume of space on the top for pouring backfill slurry, showing in Figure 1(a).

(a) Typical moulds to prepare interface samples and cured planar interface samples

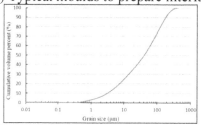 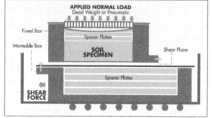

(b) Particle size distribution of tailings   (c) Schematic of interface shear box

Figure 1.   Planar interface preparation and direct shear test device.

Backfill was mixed with classified tailings from a nickel mine of China and ordinary Portland cement. The particle size distribution of tailings is shown in Figure 1(b). The density of backfill slurry was fixed to 72%, and cement contents (solid mass percentage) were set to 20%, 9.1% and 4.76% respectively. The curing age is 7 and 28 days in the curing room where temperature was fixed to 20±0.5 % and relative humidity 95±5 %.

A large scale interface shear device produced by Durham Geo Slope Indicator following ASTM D 5321, D 6243 was modified to carry out direct shear tests with interface samples. The vertical pressure applied perpendicular to the interface ranges from 50kPa to 500kPa.

After a series of direct shear tests on planar interface samples between backfill and rock, typical curves illustrating shear stress changes with horizontal displacement are given in Figure 2. It

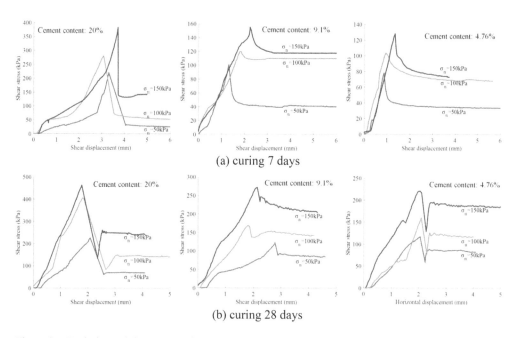

Figure 2.   Variations of shear stress along with horizontal displacement for planar interface samples.

also shows the variations along with cement contents and normal stresses perpendicular to the interface under different curing days.

It can be seen from the Figure 2 that:

(1) The curves typically contain four parts with the increase of horizontal shear displacement, including the first pore compaction stage, the stage of straight climb for shear strength to a peak point, the stage of quick fallen of shear strength after peak point, and the last stage that residual strength keeps constant with steady increase of displacement.
(2) Both peak shear strength and residual shear strength increase with the increase of normal stress perpendicular to the interface at the same cement content and curing age.
(3) Under the same curing age and vertical stress, with the increase of cement content, the peak shear strength increased significantly, but no significant differences for the residual shear strength with different cement content.
(4) With the increase of curing age, the peak shear strength of interfaces at the same cement content increased significantly, and the residual shear strength increased slightly.
(5) Under the condition of low normal stress, the shear stress curves show significant stress drop after peak point, leading to the residual shear stress is much lower than the peak shear stress.
(6) Before reaching the peak point, the shear strength is mainly composed of adhesive strength and internal friction strength. But after the peak point, there is almost no adhesive cohesion along interface, remaining frictional strength along the interface.

The tested shear strength of interfaces between backfill and rock shown in Figure 2 was further fitted with Mohr Coulomb criterion, i.e. $\tau_p = c_p + \sigma_{np} \times \tan \phi_p$ where τ_p (kPa) is the peak shear strength and σ_{np} (kPa) is the normal stress perpendicular to the planar interfaces. And then the planar interface cohesion c_p and internal friction angle ϕ_p could be obtained, as shown in Table 1.

It can be seen from the Table 1 that:

(1) With the increase of cement content, both the cohesion c_p and internal friction angle ϕ_p of the planar interface increased significantly at the same curing days.
(2) With the increase of curing age from 7 to 28 days, the cohesion c_p and internal friction angle ϕ_p of the planar interface increased slightly at the same cement content. And with the increase of the cement content, the magnitudes for the increasement of interface cohesion and friction angles will be enlarged.

3   DIRECT SHEAR TEST ON NONPLANAR INTERFACE

Nonplanar interface samples were prepared with the same backfill recipes, slurry casting method and direct shear test procedures. But before backfill slurry pouring, the upper surface of the rock material were cut to regular teeth to represent different roughness values of the interface, as shown in Figure 3. The roughness is quantified by angle $\alpha \in (0°, 90°)$ of each tooth. The bigger for angle α, the larger for quantified interface roughness between backfill and rock.

Table 1. Planar interface cohesion c and internal friction angle ϕ calculated with shear test results.

Binder content	Curing days	Linear fitting formula	Correlation coefficient	Cohesion c_p (kPa)	Internal friction angle ϕ_p (°)
20%		$\tau_p = 95.3 + 0.78 \times \sigma_{np}$	0.918	95.3	38
9.1%	7	$\tau_p = 73.8 + 0.54 \times \sigma_{np}$	0.998	73.8	28.5
4.76%		$\tau_p = 53 + 0.50 \times \sigma_{np}$	1	53	26.7
20%		$\tau_p = 162.1 + 1.04 \times \sigma_{np}$	0.908	162.1	46
9.1%	28	$\tau_p = 80.1 + 0.58 \times \sigma_{np}$	0.920	80.1	30.1
4.76%		$\tau_p = 62.5 + 0.52 \times \sigma_{np}$	0.989	62.5	27.4

The teeth angle was set to 10°, 30°, 45°, 60° respectively to prepare the rock base with different interface roughness, and then the backfill slurry with different cement contents was poured to the upper volume and cured for 28 days to carry out the direct shear tests. The direction of applied shear force is perpendicular to the direction of teeth alignment, not along with the groove. When the teeth angle α equal to 0°, it will come to the planar interface.

After a series of direct shear tests on the nonplanar interface samples between backfill and rock, the relationships between peak shear strength τ_r and the normal stress σ_{nr} acting on equivalent horizontal contact surface were obtained. Then the equivalent cohesion c_r and interface friction angle ϕ_r along the rough interface were calculated with linear fitting method based on Mohr Coulomb criterion varying with the teeth angle α.

Figure 4 shows relationships of peak shear strength τ_r, equivalent cohesion c_r and internal friction angle ϕ_r of rough nonplanar interface with teeth angle α. The density of backfill slurry is also fixed to 72% and the cement content showed in Figure 4 is 9.1%.

It can be seen from the Figure 4 that:

(1) The peak shear strength increase linearly with the increase of equivalent normal stress acting on the nonplanar interface from 50kPa to 300kPa.
(2) When the teeth angle is smaller than 30°, the peak shear strength increases with the increase of the teeth angle under the same normal stress.
(3) When the teeth angle is bigger than 30°, the peak shear strength increases very slightly under the same normal stress. This is corresponded with the simulation results of Liu et al. (2016) that when the nonplanar interface becomes rough enough, the potential shear sliding will move from along interface to inside the backfill, which will make the shear stress determined by the backfill strength.
(4) The cohesion of the nonplanar interface increases almost linearly with the increase of the teeth angle. But, the friction angle of the nonplanar interface firstly increases and then decreases with the increase of the teeth angle, turning at the teeth angle of 30°.

Figure 3.   Nonplanar interface with different roughness and typical shear tested samples.

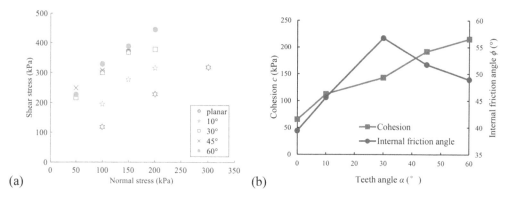

Figure 4.   Relationships of shear strength, cohesion and friction angle of rough interface with teeth angle.

84

The relationships between the interface and the corresponding backfill are important to backfill stress and strength requirement evaluation in mine stopes, because the ratios of cohesion and internal friction angle between interface and backfill are usually adopted in most previous studies (Mitchell 1989, Li et al. 2005, Liu et al. 2018).

Along with the planar and nonplanar interface samples between backfill and rock, cylindrical samples of the backfill with the same recipe were casted and cured for triaxial compressive tests, by which the cohesion c_0 and internal friction angle ϕ_0 of the corresponding backfill could be obtained. Figure 5 shows a part of the cured cylindrical backfill samples and the typical testing procedure with soil triaxial apparatus.

After the triaxial compressive tests of cylindrical backfill samples, the cohesion c_0 and friction angle ϕ_0 of the backfill with different cement contents and cured for 28 days are listed in Table 2. Meanwhile, the cohesion c_p and friction angle ϕ_p of the planar interfaces, and the cohesion c_r and friction angle ϕ_r of the rough nonplanar interfaces (teeth angle $\alpha = 60°$) are also given in Table 2 for comparison.

It can be seen from the Table 2 that:

(1) With the increase of cement content, all values of the cohesion and friction angle of the backfill, planar and nonplanar interfaces increase. This result partly responds to the controversy of whether the backfill friction angle is insensitive to the cement content but mostly determined by the particle size distribution of aggregate (unclassified tailings here).

(2) The backfill cohesion is always bigger than the interface cohesion, but the backfill friction angle is always smaller than the corresponding interface friction angle (planar and nonplanar). This result is similar to the conclusions shown in Koupouli et al. (2016, 2017).

(3) The cohesion c_p of planar interfaces is much smaller than the backfill cohesion c_0, and the increase of cement content will enlarge the differences. However the cohesion c_r of nonplanar interfaces (teeth angle $\alpha = 60°$) is slightly smaller but very close to the corresponding backfill cohesion c_0 at different cement contents. It indicates that the increase of roughness for the interface will increase the interface cohesion, but with the backfill cohesion as the upper limit.

(4) The friction angle ϕ_p of planar interfaces is slightly bigger than the backfill friction angle ϕ_0. And with the increase of interface roughness, the differences will be enlarged. The friction angle ϕ_r of nonplanar interfaces is much bigger than the backfill friction angle ϕ_0. This is because the frictional resistance along interface is composed of combined action of backfill and rock. The contributions from rock with much higher friction angle makes the equivalent fricton angle of interface increase obviously. Besides, the shear dilatancy effect along raised teeth of nonplanar interfaces helps to expand the frictional resistance and lead to a higher equivalent fricton angle.

Figure 5.   Casted cylindrical backfill samples for triaxial compressive tests.

Table 2. Comparison of cohesion and friction angle of backfill with planar and nonplanar interfaces.

Binder content	Triaxial compressive tests of cylindrical backfill		Direct shear tests of planar interface		Direct shear tests of rough nonplanar interface ($\alpha = 0°$)	
	cohesion c_0 (kPa)	friction angle ϕ_0 (°)	cohesion c_p (kPa)	friction angle ϕ_p (°)	cohesion c_r (kPa)	friction angle ϕ_r (°)
20%	718.6	31.4	162.1	46	661	52.9
9.1%	256.6	27.5	80.1	30.1	215	48.9
4.76%	155.8	26.3	62.5	27.4	125	45.3

(5) It is not suitable to separately use ratios of interface cohesion to backfill cohesion and interface friction angle to backfill friction angle at one model to evaluate backfill stress or strength requirement, because it is not simple proportional relationship for shear strength parameters (cohesion and friction angle) between the interface and the backfill.

5 MODIFIED SHEAR STRENGTH MODEL FOR INTERFACE

According to the shear test results for interfaces and triaxial test results for corresponding backfill under 28 curing days, the conversion relations of cohesion c_p and friction angle ϕ_p for planar interface compared with cohesion c_0 and friction angle ϕ_0 for backfill can be seen in equations (1) and (2) respectively.

$$c_p = 0.177c_0 + 34.798 \qquad (1)$$

$$\phi_p = 3.734\phi_0 - 71.595 \qquad (2)$$

The conversion relations of cohesion c_r and friction angle ϕ_r for rough nonplanar interface compared with cohesion c_0 and friction angle ϕ_0 for backfill can be seen in equations (3) and (4) respectively.

$$c_r = 0.956c_0 - 26.893 \qquad (3)$$

$$\phi_r = 1.383\phi_0 + 9.737 \qquad (4)$$

From the above equations, it can be seen that for planar and nonplanar interfaces, it is not simple proportional relationship for cohesion and friction angle between interface and backfill. And these relations could be utilised to update the analytical models for stress distribution and strength requirement evaluation, e.g. Mitchell (1989), Li et al. (2005), and Liu et al. (2018).

In addition, the classical Mohr Coulomb criterion (i.e. $\tau = c + \sigma_n \times \tan\phi$ where τ (kPa) is the peak shear strength and σ_n (kPa) is the normal stress perpendicular to the failure plane of the material) which is mainly suitable for uniform cohesive materials has been widely used for backfill stress calculation. But there is no consideration about the interface between backfill and rock, and no parameter of interface roughness in the criterion. The shear strength criterion proposed by Barton (1977) for rock joints consider the roughness of failure plane, but there is no consideration of large interface cohesion from cemented backfill in the criterion, and the JCR values characterizing roughness are difficult to measure.

Here, incorporating equations (1) and (2) to the classical Mohr Coulomb criterion, the equation (5) could be obtained for the shear strength of planar interface based on the tested values of cohesion c_0 and friction angle ϕ_0 of cemented backfill.

$$\tau = \sigma_n \tan(3.734\phi_0 - 71.595) + 0.177c_0 + 34.798 \tag{5}$$

Similarly, incorporating equations (3) and (4) to the classical Mohr Coulomb criterion, the equation (6) could be obtained for the shear strength of rough nonplanar interface (teeth angle $\alpha = 60°$) based on the tested values of cohesion c_0 and friction angle ϕ_0 of cemented backfill.

$$\tau = \sigma_n \tan(1.383\phi_0 + 9.737) + 0.956c_0 - 26.893 \tag{6}$$

Empirical equations (5) and (6) obtained here with data from direct shear tests on interfaces and triaxial compressive tests on backfills could be used for the mechanical analysation of the interfaces between backfill and rock during the evaluation of stress, required strength and stability of cemented backfill in mine stopes.

For the validation of these modified equations based on Mohr Coulomb criterion, a series of numerical modellings were carried out with FLAC3D to simulate the direct shear tests on planar and nonplanar interfaces. The dimensions of the numerical models are corresponding with the fill-rock interface samples, and the backfill shear strength properties are based on the triaxial test results (The density of backfill slurry is fixed to 72% and the cement content is 9.1%).

Figure 6 shows the comparisons of peak shear strength of planar and nonplanar (teeth angle $\alpha = 60°$) interfaces at different normal stress perpendicular to the interface.

It can be seen from Figure 6 that the simulated peak shear stress with FLAC3D generally fit well with the tested peak shear stress along with the planar interfaces and nonplanar interfaces (teeth angle $\alpha = 60°$). This demonstrates that modified equations (5) and (6) based on Mohr Coulomb criterion to calculate the peak shear strength for planar and nonplanar interfaces are usable to some extent.

However, it can also be seen in Figure 6 that there are still obvious differences between the simulated and the tested peak stress. This might be caused by the nonlinear deformation of backfill under normal pressure and the different failure modes along interfaces due to shear dilatancy of nonplanar interfaces.

Besides, Figure 6 illustrates that the simulated peak shear stress will gradually flatten under higher normal stress (after 200kPa) for nonplanar interfaces (teeth angle $\alpha = 60°$). But the simulated peak shear stress for planar interface show a linear growth with the increase of the normal stress. This linear growth trend also happens for both planar and nonplanar interfaces during the direct shear tests. More tests and analyse should be carried out with nonplanar interface samples in different roughness to understand its differences with corresponding simulated results.

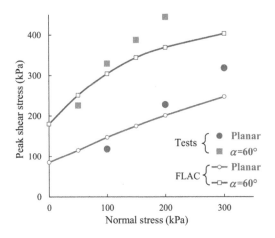

Figure 6.   Comparison of tested and simulated peak shear stress along planar and nonplanar interfaces.

6 CONCLUSIONS

In order to investigate shear strength properties of planar and nonplanar interface between backfill and rock, a series of interface samples with different roughness and backfill recipes were prepared and set to direct shear tests. Triaxial compressive tests were carried out to obtain the cohesion and internal friction angle of the corresponding backfill. The relationships about cohesion and internal friction angle between interface and backfill were compared and analysed. The key conclusions in the investigations are summarised bellow:

(1) The tested backfill cohesion is always bigger than the interface cohesion, but the backfill friction angle is always smaller than the corresponding interface friction angle (for conditions of planar and nonplanar interfaces).
(2) The increase of interface roughness will enlarge the interface cohesion and friction angle, but the corresponding backfill cohesion is the upper limit for the increasement of the interface cohesion.
(3) The relationships of cohesion and internal friction angle between interface and corresponding backfill are not simple proportional ratios which had been adopted in most previous studies, but conform to linear correlations.
(4) Empirical linear functions were given to calculate relations of cohesion and friction angle for planar and nonplanar interface compared with cohesion and friction angle for the corresponding backfill, respectively.
(5) New models were proposed based on Mohr Coulomb criterion for the calculations of peak shear strength of planar and nonplanar interface with the tested cohesion and internal friction angle of corresponding backfill. Numerical simulations validated the availability of the new models.

ACKNOWLEDGEMENTS

Financial supports from the National Natural Science Foundation of China (grant no. 51774040, 51804031) and the Youth Science and Technology Innovation Fund of BGRIMM Technology Group (grant no. QCJ201804) is gratefully acknowledged. Many thanks to the peer reviewers to give constructive suggestions on the paper.

BIBLIOGRAPHY

Aubertin, M., Li, L., Arnold, S., Belem, T., Bussière, B., Benzaazoua, M. and Simon, R. 2003. Interaction between backfill and rock mass in narrow stopes. [In:] *Proceedings of the 12th Panamerican Conference on Soil Mechanics and Geotechnical Engineering*, Essen, Germany, 22–26 June.

Barton, N. & Choubey, V. 1977. The shear strength of rock joints in theory and practice. *Journal of Rock Mechanics and Rock Engineers.* 10(1-2):1–54.

Fall, M., and Nasir, O. 2010. Mechanical behaviour of the interface between cemented. *Geotechnical and Geological Engineering*, 28(6):779–790.

Tezaghi, K. 1943. *Theoretical Soil Mechanics.* John Wiley and Sons: New York, NY.

Li, L., Aubertin, M., and Belem, T. 2005. Formulation of a three dimensional analytical solution to evaluate stresses in backfilled vertical narrow openings. *Canadian Geotechnical Journal*, 42(6), 1705–1717.

Li, L. 2013. Generalized solution for mining backfill design. *International Journal of Geomechanics*, 14(3), 04014006.

Liu, G.S., Li, L., Yang, X.C., and Guo, L.J. 2016. A numerical analysis of the stress distribution in backfilled stopes considering nonplanar interfaces between the backfill and rock walls. *International Journal of Geotechnical Engineering*, 10(3):271–282.

Liu, G.S., Li, L., Yang, X.C., and Guo, L.J. 2017. Numerical Analysis of Stress Distribution in Backfilled Stopes Considering Interfaces between the Backfill and Rock Walls. *International Journal of Geomechanics.* 17(2): 06016014.

Liu, G.S., Li, L., Yang, X.C., and Guo, L.J. 2018. Required strength estimation of a cemented backfill with the front wall exposed and back wall pressured. *International Journal of Mining and Mineral Engineering*, 9(1), 1–20.

Liu, G.S., Yang, X.C. and Guo L.J. 2019. Models of three-dimensional arching stress and strength requirement for the backfill in open stoping with subsequent backfill mining. *Journal of China Coal Society*, 44(5):1391–1403.(in Chinese)

Manaras, S. 2009. *Investigations of backfill-rock mass interface failure mechanisms*, M.A.Sc. thesis, Queen's University, Kingston, Ontario, Canada.

Manars, S., De Souza, E. and Archibald, J.F. 2011. Strength behaviour and failure mechanisms of backfill-rock mass interfaces. [In:] *Minefill 2011, 10th International Symposium on Mining with Backfill*, The Southern African Institute of Mining and Metallurgy, Cape Town, South Africa.

Nasir, O., and Fall, M. 2008. Shear behaviour of cemented pastefill-rock interfaces. *Engineering Geology*, 101(3/4):146–153.

Koupouli, N.J.F., Belem, T., Rivard, P. and Hervé, E. 2016. Direct shear tests on cemented paste backfill–rock wall and cemented paste backfill–backfill interfaces. *Journal of Rock Mechanics and Geotechnical Engineering*, 8(4):472–479.

Koupouli, N.J.F., Belem, T. and Rivard, P. 2017. Shear strength between cemented paste backfill and natural rock surface replicas. [In:] *Proceedings of the First International Conference on Underground Mining Technology*, Australian Centre for Geomechanics, Perth, Australia, 375–385 October.

Minefill 2020-2021 – Hassani et al (eds)
© *2021 Taylor & Francis Group, London, ISBN 978-1-032-07203-6*

The concept of obtaining backfilling material using the dredging method

Maciej Gruszczyński & Stanisław Czaban
Institute of Environmental Engineering, Wrocław University of Environmental and Life Sciences, Poland

Szymon Zieliński
KGHM Polska Miedź S.A., Hydrotechnical Unit, Poland

SUMMARY: One of the problems of the mining industry is the management of fine mineral tailings (with a solid phase diameter of less than 0.1 mm),which are the result of the enrichment process. Due to their physical properties, mainly thickening and compressibility, the fine-grained materials are not suitable for use in large quantities in the mining industry or other industrial branches. A number of experiments were conducted to develop a technology allowing the practical use of fine-grained post-flotation tailings as hydraulic or cemented hydraulic fill and also as a material for goaf tightening. The paper presents the results of the research conducted on obtaining condensed fine-grained tailings of copper ore flotation by the dredging method. During the experiments, the following parameters were determined: pump operation parameters, material density, linear head losses On top of this, the process of depositing material on the beach was assessed. During the research both traditional measurement methods and laser scanning techniques were used to assess the deposition of material on the beach. Experimental studies have confirmed the effectiveness of obtaining material by the dredging method that can be used in backfilling technologies or for goaf tightening.

Keywords: backfill material, paste, mature fine tailings, tailings storage facility

1   INTRODUCTION

In Poland, over 110 million tonnes of mining waste is generated annually. The main group of waste are by-products of non-ferrous metal ore enrichment processes from washing and cleaning minerals, and coal fly ash resulting from coal combustion (Figure 1). The management of generated waste is a major problem for the economy and the natural environment in Poland.

The subject of many studies at Polish universities and foreign scientific units was the use of generated waste (Gruszczyński, 2019, Kotarska, I. 2012, Plewa, 2006) as the main material or addition to mixtures used during: hydraulic grouting of caving areas, filling unused opening-out heading, underground cemented backfill, fire prevention and fire fighting Pierzyna, and Popczyk, 2014). In the coal mining technology, hydraulic grouting with fine-grained mixture of industrial waste such as fly ashes and flotation tailing is widely used (Pierzyna, and Popczyk, 2014). Scientific work is underway to apply coal fly ashes to solidify rock debris (Plewa et al. 2008). The use of fluidized bed combustion products in hydraulic backfill is being considered (Plewa et al. 2009).

The basic filling material is sand obtained from fossil deposits or mined from deposits hydraulically. The use of waste instead of sand will reduce the impact of opencast excavations and will allow for the management of significant amounts of waste that end up in tailings stor-

DOI: 10.1201/9781003205906-9

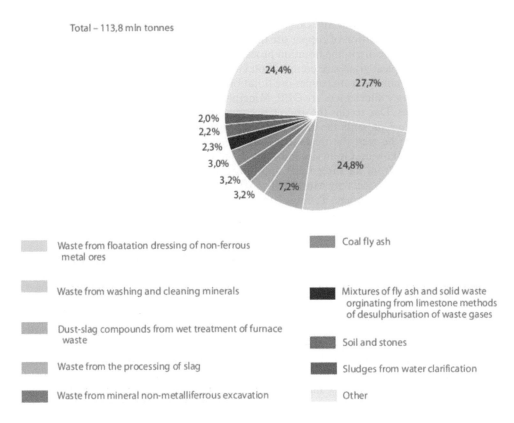

Total – 113,8 mln tonnes

24,4%
27,7%

2,0%
2,2%
2,3%
3,0%
3,2%
3,2%
7,2%
24,8%

Waste from floatation dressing of non-ferrous metal ores

Coal fly ash

Waste from washing and cleaning minerals

Mixtures of fly ash and solid waste orginating from limestone methods of desulphurisation of waste gases

Dust-slag compounds from wet treatment of furnace waste

Soil and stones

Waste from the processing of slag

Sludges from water clarification

Waste from mineral non-metalliferous excavation

Other

Figure 1. Structure of waste generated by waste type in 2017.

Source: Bochenek, Dzik, Górksa, Kiełczykowska, Kulasza, Nowakowska, Pawłowska, Rudnicka, Sulik, Szondelmejer, Wojciechowska, and Wrzosek, 2018. Environment 2018. Spatial and Environmental Surveys Department

age Facilities (TSF). The technology of hydraulic backfill has a number of advantages such as limiting the impact of mining operations on the land surface, the possibility of operating in protective pillars, the reduction of fire, water and gas hazards and the rockburst risk reduction (Popczyk, 2014). One third of the mining tailings generated in Poland is copper ore flotation tailings, generated in KGHM Polska Miedź S.A. Ore Enrichment Plants. Despite many research works, no technology has been implemented to date by the Company for the use of tailings generated during enrichment in mining technology. Copper ore flotation tailings have for many years been the subject of research for relocation or use in industries other than mining. The use of waste as a mineral binder, proppant, sealing material for goafs or as lithification material is considered (Łuszczkiewicz, 2000, Ratajczyk 2015, Speczik et al. 2003).

2 CHARACTERISTICS OF COPPER ORE FLOTATION TAILINGS

At KGHM Polska Miedź S.A. extraction is carried out in three mines, ie.: Lubin, Polkowice-Sieroszowice, Rudna. The Lubin Mine is the oldest of the mines operated in the enterprise and extracts red sandstone and carbonate-shale ore. The Polkowice-Sieroszowice mine extracts sandstone and shale carbonate ore, however, unlike the Lubin mine, most of the extraction is obtained from shale carbonate ore. The latest mine of the enterprise is the Rudna mine, which has the deposit with the highest mineralization from all KGHM mines. The

Rudna Mine extracts sandstone and carbonate-shale ore (Bartlett et al. 2013). Table 1 presents the mineralogical composition of copper ore flotation tailings from individual KGHM mines.

In order to enrich the ore in the flotation process, the ore requires significant fragmentation. Figure 2 shows a typical grain size composition curve for five different enrichment processes. The particle size of tailings from copper ore flotation processes oscillates within 0.1 mm.

At each of the three mines there are OEP. Mineral processing is adapted to the ore extracted in each of the mines. Due to this the grain size composition of the tailings differs for the individual OEP. Figure 3 presents the grain size composition curves for OEP Rudna, Polkowice and Lubin. As can be seen the thickest tailings come from OEP Lubin, while the smallest come from OEP Polkowice. Grain size indicator d_{50} of tailings is in the range 20-60 μm.

Flotation tailings from all three OEP flow through a pressure hydrotransport system to Tailings Storage Facility Żelazny Most (TSF ŻM) where the tailings are subjected to the process of disposal through storage. Tailings larger than 0.1 mm are used to build the dam body, while smaller tailings are used to seal the TSF.

3 CHARACTERISTICS OF THE TAILINGS STORAGE FACILITY ŻELAZNY MOST

The exploitation of TSF ŻM began in 1977 and it has been used and continuously enlarged to this day. The facility is the largest tailings pond of this type in Europe and one of the largest in the world. Figure 4 presents an aerial photograph of the object.

The facility is being built using the „upstream" method by depositing flotation tailings as a structural material for the dam body (Świdziński et al. 2015). About 10 years ago, KGHM decided to expand the existing TSF ŻM adding a Southern Quarter (SQ). The new quarter is to be adjoined to the southern part of the existing facility. The completion of the new pond is planned for 2021. The capacity of the new quarter is to be approximately 1/3 of the volume of the main pond. The new quarter is to be exploited using thickened tailings technology so as to make the quality of the process water, which is pumped back from TSF to the OEP, independent from deposition area and meteorological conditions.

4 RESEARCH ON THE POSSIBILITIES OF OBTAINING AND RELOCATION OF FINE TAILINGS.

In TSF ŻM in which non-thickened tailings technology is used, fine tailings accumulate in the central part of the pond, while coarse tailings are most often used as a constructing material for the dam body and the beach (N. Dhadli *et al.*2012)]. The distribution of particles is caused by the occurring natural process of segregation and sedimentation. Figure 5 shows a typical

Table 1. Mineralogical composition of copper ore flotation tailings.

Mineral	KGHM mine/Content [%]	
	Lubin, Rudna	Polkowice
Calcite	29.95	58.3
Quartz	44.46	6.85
Dolomite	7.70	7.82
Kaolinite	4.76	3.29
Gypsum	1.65	4.70
Biotite	3.30	1.12
Feldspars	1.35	0.82
Clays	8.35	4.78
Desirable minerals	1.10	1.32

Source: Kotarska, 2012. Odpady wydobywcze z górnictwa miedzi w Polsce - bilans, stan zagospodarowania i aspekty środowiskowe, Cuprum, vol. 65, no. 4, pp. 45–64

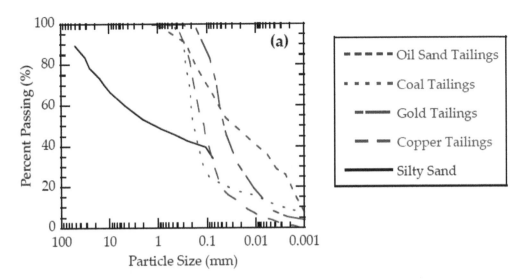

Figure 2. Grain size composition of flotation tailings after the enrichment processes: oil sand, coal, gold, silty sand.

Source: Gorakhki and Bareither. 2017. Sustainable Reuse of Mine Tailings and Waste Rock as Water-Balance Covers. Minerals, vol. 7, no. 7, p. 128

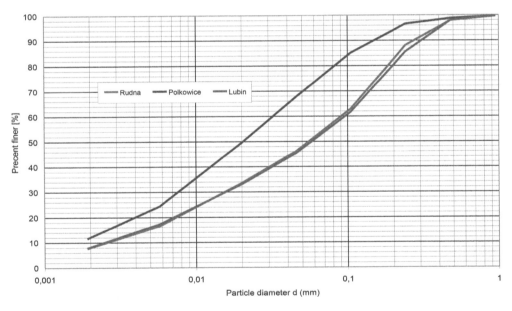

Figure 3. Average grain size compositions of tailings from OEP Rudna, Polkowice and Lubin.

Source: own study

cross-section of a TSF with marked zones of deposits of individual types of tailings. In the central part of TSF Mature Fine Tailings (MFT) deposit zone can be observed (N. Dhadli et al. 2012, Wang et al. 2014, Wells et al. 2011). MFT was the subject of the research presented in the article.

During the research conducted in 2010, the presence of MFT in TSF ŻM was detected. Considering the operating technology of the new facility, i.e. the southern quarter, it was

Figure 4. Tailings Storage Facility Żelazny Most seen from a bird's eye view. 1-2 regions of the dredger work. 3-4 deposition areas. Photo taken during drone flight.

Source: GeoPixel 2017. Wykonaliśmy nalot fotogrametryczny nad zbiornikiem unieszkodliwiania odpadów wydobywczych Żelazny Most – GeoPixel UAV dla geodezji. [Online] http://www. geopixel.pl/wykonalismy-nalot-foto grametryczny-nad-zbiornikiem-unieszkodliwiania-odpadow-wydobywczych-zelazny-most/. [Accessed: 20-Jul -2018].

Figure 5. MFT deposition zone in cross-section of TSF.
Source: Beier, and Sego, 2008. The Oil Sands Tailings Research Facility. Geotechnical News, no. June.

proposed that MFT be used as thickened tailings transferred from the main facility ŻM to the new quarter. This solution would reduce the costs of thickening of tailings in the thickening station. The TSF ŻM would be used as a large thickener with an existing deposit. MFT collection and transport had to be resolved. The Institute of Environmental Engineering of Wrocław University of Environmental and Life Sciences carried out the research on the dredging, transport and deposit of MFT tailings. During the research and analysis of the collected results, the concept appeared of using the acquired MFT as a material that could be used to produce a backfill paste or as a material for grouting of caving areas.

The aim of the study was to assess the behavior of the MFT deposit in the reservoir of the TSF ŻM under continuous material collection in a period of more than 24 hours. Semi-technical (field) measurements of the collecting, transporting and deposition of copper ore flotation tailings were carried out as well as measurements of geometry, volume and depositing efficiency (Figure 8) using high resolution laser scanning techniques. Both the stability of tailings inflow to the dredging pump and the dynamics of rheological parameters of pumped material were examined. The variability of concentration of dredged material as a function of time was assessed. The behavior of the slope of deposited tailings and its garin size composition were examined. During the research the Watermaster Classic IV dredger was used. Areas of the dredger's work during the tests and the location of the depositing areas is shown in Figure 4. The laser scanning technique was used to determine the behavior of the tailings deposited on the TSF ŻM beach. The volume of deposited tailings was determined, as well as the slope of the deposited material. Measurement methodologies were developed and measurements installation were made. The installation diagram is shown in Figure 6. During the experiments carried out on-line such parameters as flow rate, flow drag and density of dredged MFT were noted.

A sampling system from the pressure collector was designed and installed. For laboratory tests, the samples were taken from the material deposited on the plot and from the discharge pressure collector of the dredger. Samples were taken from various dredger work positions shown in Figure 7. Laboratory tests of the grain size composition of collected tailings were carried out (Figure 8). No difference in composition as a function of location was observed.

Geodetic data collected during measurements with a laser scanner was collected, and a spatial numerical model of individual stages of deposition was built. Fuel consumption during the dredger operation, position change and pressure in the discharge pipeline were also examined.

The paper presents the results of a test lasting 31 hours of continuous dredging of a MFT deposit. During the tests, the density of the transported mixture varied from 1069.9 to 1292 kg·m⁻³. The concentration by volume of the transported mixture varied from 0.0387 to 0.1618. Table 2 gives the range of changes in volumetric flow rate of the mixture. The average hourly flow rate of the mixture was in range from 147 to 394 m³·h⁻¹. Solid phase volume flow rates were calculated from (1):

$$V_s = c_v Q_m \tag{1}$$

The solids flow rate was in range from 10.0 to 53,5 m³·h⁻¹. The average flow rate was 28.2 m³·h⁻¹. Mass flow rates of water in the mixture were calculated from (2):

$$G_w = (1 - c_v) Q_m \rho_w \tag{2}$$

The water flow rate was from 124,5 to 340,5 Mg·h⁻¹. The average flow rate was 223.8 Mg·h⁻¹. Mass flow rates of solids in the mixture were calculated from (3):

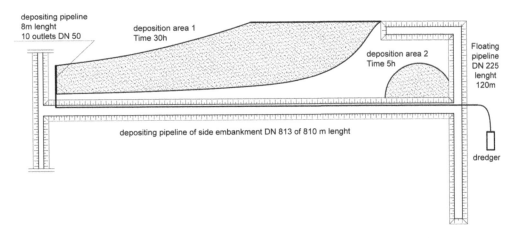

Figure 6.   Scheme of transport, measuring and deposition installation of tailings.
Source: own study

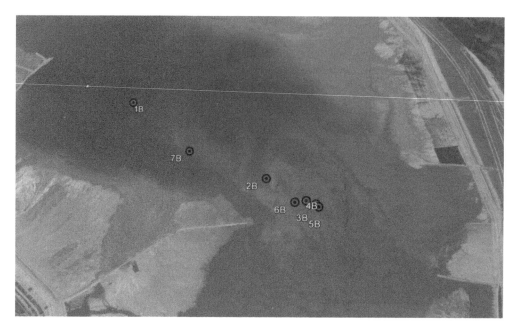

Figure 7. Location of dredger during the tailings excavating.
Source: own study (Google Earth)

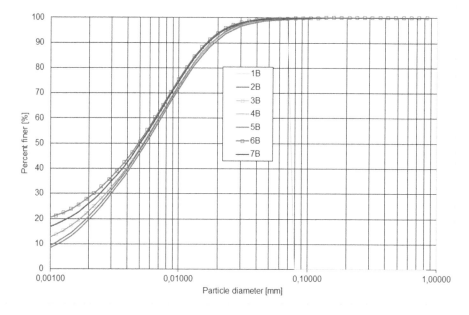

Figure 8. Grain size compositions of tailings collected in different positions of the dredger.
Source: own study

Table 2. Parameters of the dredging vessel between 11:00 the first day and 18:00 the second day of the experiment.

Time	Mixture ρ_m	Concentration by volume C_v	Flow rate Mixture Q_m	Flow rate Solids V_s	Mass of water G_w	Mass of solids G_s	Mass of mixture G_m
hour	$kg \cdot m^{-3}$	-	$m^3 \cdot h^{-1}$	$m^3 \cdot h^{-1}$	$Mg \cdot h^{-1}$	$Mg \cdot h^{-1}$	$Mg \cdot h^{-1}$
11	1245.1	0.1358	394.0	53.5	340.5	150.0	490.5
12	1271.3	0.1503	353.5	53.1	300.4	149.0	449.4
13	1267.7	0.1483	282.4	41.9	240.5	117.5	358.0
14	1253.1	0.1403	261.8	36.7	225.0	103.0	328.0
15	1292.0	0.1618	269.1	43.5	225.5	122.1	347.6
16	1257.4	0.1426	273.8	39.0	234.7	109.5	344.2
17	1269.4	0.1493	290.8	43.4	247.4	121.8	369.1
18	1250.0	0.1386	326.3	45.2	281.1	126.8	407.9
19	1220.3	0.1221	280.1	34.2	245.9	95.9	341.8
20	1161.5	0.0895	350.3	31.3	318.9	87.9	406.9
21	1069.9	0.0387	258.7	10.0	248.6	28.1	276.7
22	1099.2	0.0550	230.5	12.7	217.8	35.6	253.4
23	1111.1	0.0616	201.9	12.4	189.5	34.9	224.3
0	1111.4	0.0617	267.8	16.5	251.3	46.4	297.7
1	1165.8	0.0919	224.8	20.7	204.2	57.9	262.1
2	1133.4	0.0739	303.2	22.4	280.8	62.9	343.7
3	1196.2	0.1087	239.6	26.1	213.6	73.1	286.7
4	1196.9	0.1091	220.9	24.1	196.8	67.6	264.4
5	1163.7	0.0907	221.8	20.1	201.7	56.4	258.1
6	1078.0	0.0433	237.0	10.3	226.7	28.7	255.5
7	1140.2	0.0777	273.7	21.3	252.4	59.7	312.1
8	1147.7	0.0819	267.5	21.9	245.6	61.4	307.0
9	1160.5	0.0890	249.4	22.2	227.2	62.2	289.5
13	1276.2	0.1531	147.0	22.5	124.5	63.1	187.6
14	1266.8	0.1479	172.4	25.5	146.9	71.5	218.4
15	1281.2	0.1559	171.8	26.8	145.0	75.1	220.1
16	1278.6	0.1544	168.3	26.0	142.3	72.9	215.2
17	1265.7	0.1473	180.5	26.6	153.9	74.5	228.4
18	1264.1	0.1463	189.7	27.8	161.9	77.9	239.8
Minimum	1069.9	0.0387	147.0	10.0	124.5	28.1	187.6
Average	1203.3	0.1126	252.0	28.2	223.8	79.1	302.9
Maximum	1292.0	0.1618	394.0	53.5	340.5	150.0	490.5

Source: own study

$$G_s = c_v Q_m \rho_s \qquad (3)$$

The solid phase flow rate was from 28,1 to 150 $Mg \cdot h^{-1}$. The average solid phase flow rate during individual hours was 79.1 $Mg \cdot h^{-1}$.

The results of the measurement using laser scanning for the depositing region number 4 from Figure 4 Czaban et al. 2017). The Figure 9 presents the raw data of the measurements taken in the field for two cases – area before and after depositing.

Based on the collected point cloud, a numerical surface model was created before and after depositing (Figure 10).

By analyzing the changes occurring on the deposition area by comparing mathematical models, the average slope of the deposited MFT surface can be determined.

97

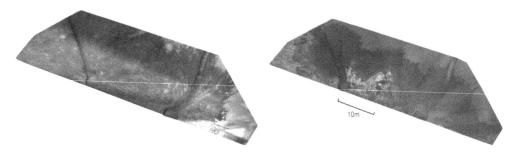

Figure 9. Point clouds of the analysis area of the tailings deposition area No. 4 for the state before depositing (state 0) and after deposition (state 1).

Source: own study

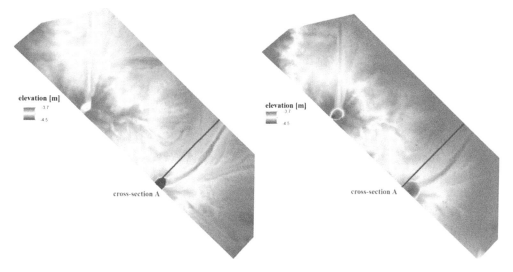

Figure 10. Triangulated Irregular Network (TIN) of research area. State before depositing (state 0) and after depositing (state 1).

Source: own study

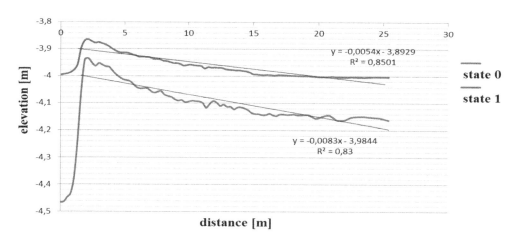

Figure 11. Cross-section A for the area of depositing tailings No. 4 – state 0 and 1.

Source: own study

5   RHEOLOGICAL RESEARCH OF MFT

Rheological tests were carried out on tailings collected while dredging. Two-parameter Bingham model (4) was chosen to describe the results of rheological measurements of the collected samples:

$$\tau = \tau_o + \eta_B \dot{\gamma} \tag{4}$$

12 tailings flow curves with different concentrations were analyzed. The analyses of the dependence of individual rheological parameters of the Bingham model in relationship to concentration by solids c_s were executed. The exponential equation [2][7][8], with was used to describe the relationship between concentration, yield stress (5) and viscosity (6) of Bingham model.

$$\eta_B = a \cdot c_s^{\,b} \tag{5}$$

$$\tau_o = c \cdot c_s^{\,d} \tag{6}$$

The approximation results (7) and (8) using the method of minimizing the residual sum of squared deviations (Gruszczyński, 2019):

$$\eta_B = 20,6 \cdot c_s^{\,7,83} \tag{7}$$

$$\tau_o = 11553 \cdot c_s^{\,7,05} \tag{8}$$

The fit of the model to the collected measurement data is shown in Figure 12.

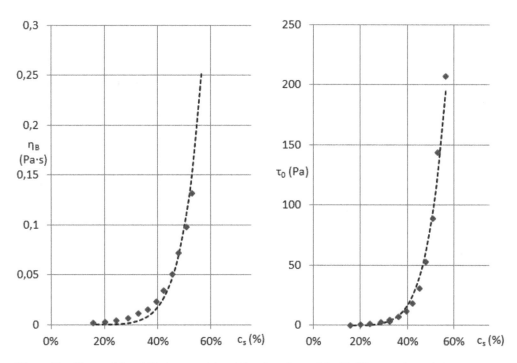

Figure 12.   Shape of the yield stress model and the viscosity model with 12 measuring points.
Source: own study

NOTATION

a, b, c, d Empirical parameter, result of approximation (-)

c_s Mass concentration of solid phase (%)

c_v Volume concentration of solid phase (%)

d_{50} Grain size indicator, particle diameter corresponding 50% cumulative undersize particle size distribution (μm)

G_m Mass flow rate of mixture ($Mg \cdot h^{-1}$)

G_s Mass flow rate of solids ($Mg \cdot h^{-1}$)

G_w Mass flow rate of water ($Mg \cdot h^{-1}$)

Q_m Flow rate of mixture ($m^3 \cdot s^{-1}$)

V_s Solids volume (m^3)

x Distance (m)

y Elevation (m)

$\dot{\gamma}$ Shear rate ($1 \cdot s^{-1}$)

η_B Viscosity of Bingham model ($Pa \cdot s$)

ρ_m Mixture density ($kg \cdot m^{-3}$)

ρ_s Solids density ($kg \cdot m^{-3}$)

ρ_w Water density ($kg \cdot m^{-3}$)

τ_0 Yield stress (Pa)

6 CONCLUSIONS

- The research confirms the existence of a Mature Fine Tailings deposit, completely deprived of sand fraction in the central part of Tailings Storage facility Żelazny Most
- The presented research results confirm the possibility of obtaining fine tailings thickened in the process of natural sedimentation from the central part of Tailings Storage facility Żelazny Most,
- The fine tailings can be used as a liquefaction additive, reducing the flow resistance to underground cemented backfill,
- The use of fine tailings obtained by the dredging method eliminates the need to use hydrocyclones to fractionate tailings stream from Ore Enrichment Plants,
- Rheological tests of samples taken during experiments have shown that the tested mixture is a non-Newtonian liquid,
- Bingham's rheological model satisfactorily describes the flow of the tested substance. It is possible to precisely describe the yield stress and viscosity of the Bingham model as a function of solid phase concentration for the tested density range,
- The obtained method of dredging fine-grained tailings naturally thickened can be considered to be the material for lithification,
- The experiments carried out on depositing of Mature Fine Tailings dredging method indicate that the slope of deposited material oscillates within 1%,
- Mature Fine Tailings obtained by dredging method can be considered the material for backfill process.

BIBLIOGRAPHY

[1] Bochenek, D., Dzik, M., Górksa, A., Kiełczykowska, A., Kulasza, A., Nowakowska, B., Pawłowska, T., Rudnicka, M., Sulik, J., Szondelmejer, K., Wojciechowska, M., and Wrzosek, A. 2018. *Environment 2018*. Spatial and Environmental Surveys Department.

[2] Barnes et al. 2014 - Barnes, H., Hutton, J. and Walters K. *An Introduction to Rheology*, vol. 43, no. 1. Elsevier Science Publishers B.V.

[3] Bartlett et al. 2013 - Bartlett, C. S., Burgess, H., Damjanović, B., Gowans, M. R. and Lattanzi, R. C. 2013. *Technical Report on the copper-silver production operations of KGHM Polska Miedź S.A. in the Legnica-Głogów Copper Belt area of southwestern Poland*, vol. 53, no. 9.

[4] Beier, N. and Sego, D. 2008. The Oil Sands Tailings Research Facility. *Geotechnical News*, no. June.

[5] Czaban et al. 2017 - Czaban, S., Gruszczyński, M., Pratkowiecki, R. and Skrzypczak Z. The measuring field system of fine tailings flow during dredging vessel," *18th International Freight Pipeline Society Symposium*. Czech Association of Scientific and Technical Societies

[6] GeoPixel 2017. *Wykonaliśmy nalot fotogrametryczny nad zbiornikiem unieszkodliwiania odpadów wydobywczych Żelazny Most – GeoPixel UAV dla geodezji*. [Online] http://www.geopixel.pl/wykona lismy-nalot-fotogrametryczny-nad-zbiornikiem-unieszkodliwiania-odpadow-wydobywczych-zelazny-most/. [Accessed: 20-Jul–2018].

[7] Gruszczyński et al. 2019- Gruszczyński, M.,. Błotnicki, J., Czaban, S. and Tymiński T. The effect of solid components on the rheological properties of copper ore tailings. *International Conference of Transport and Sedimentation of Solid Particles*. vol. TS 19, no. September, pp. 269–276.

[8] Gruszczyński, M. 2019. The assessment of the impact of the rheological parameters of fine-grained copper ore flotation tailings on friction head losses in pressure pipes. Wrocław University of Environmental and Life Sciences.

[9] Kotarska, I. 2012. Odpady wydobywcze z górnictwa miedzi w Polsce - bilans, stan zagospodarowania i aspekty środowiskowe, *Cuprum*, vol. 65, no. 4, pp. 45–64.

[10] Łuszczkiewicz, A. 2000. Koncepcje wykorzystania odpadów flotacyjnych z przeróbki rud miedzi w regionie Legnicko-Głogowskim," *Inżynieria Mineralna*. no. 1, pp. 25–36.

[11] M. H. Gorakhki and C. A. Bareither. 2017. Sustainable Reuse of Mine Tailings and Waste Rock as Water-Balance Covers. *Minerals*, vol. 7, no. 7, p. 128.

[12] N. Dhadli *et al.*2012. Technical Guide for Fluid Fine Tailings Management. *Oil Sands Tailings Consort. Canada's Oil Sands Innovation Alliance*, pp. 3–3 to 3–4,3–9,4–7.

[13] Pierzyna, P. and Popczyk, M. 2014. Odzysk odpadu energetycznego z metody mokrego odsiarczania spalin do likwidacji zbędnych wyrobisk górniczych. *POLITYKA Energetyczna – ENERGY POLICY J.*, pp. 341–354.

[14] Plewa et al. 2008 – Plewa, F., Strozik, G. and Mysłek Z. Zastosowanie odpadów energetycznych do zestalania rumowiska skalnego Wprowadzenie. *Polityka Energetyczna*, pp. 351–360.

[15] Plewa et al. 2009 – Plewa, F., Strozik, G. and Piontek P. Zastosowanie ubocznych produktów spalania z kotłów fluidalnych energetyki zawodowej w podsadzce hydraulicznej Wprowadzenie. *Polityka Energetyczna.*, vol. 12, no. 2/2, pp. 485–495.

[16] Plewa, F. 2006. – Plewa, F., Strozik, G. and Jendruś, R. Możliwość zagospodarowania odpadów drobnofrakcyjnych z energetyki w procesie doszczelniania gruzowiska zawałowego w warunkach kopalni. *Polityka Energetyczna*. pp. 457–466.

[17] Popczyk, M. 2014. Możliwości zastosowania do podsadzki hydraulicznej mieszaniny piasku z żużlem energetycznym. *Polityka Energetyczna - ENERGY POLICY J.*, pp. 405–416.

[18] Ratajczyk K. 2015. Spoiwo mineralne oparte na przetworzonych termicznie odpadach flotayjnych powstających w KGHM Polska Miedź S.A.," *Prace Instytutu Ceramiki i Materiałów Budowlanych*, vol. 21, no. 18, pp. 7–21.

[19] Speczik et al. 2003 - Speczik, S., Bachowski, A., Mizera, A. and Grotowski A. Stan aktualny i perspektywy gospodarki odpadami stałymi w KGHM Polska Miedź S.A. *WARSZTATY 2003 z cyklu „Zagrożenia Nat. w górnictwie"*, pp. 155–177.

[20] Świdziński et al. 2015 - Świdziński, W., Tschuschke, W., Świerczyński, W. and Wolski W. Obiekt Unieszkodliwiania Odpadów Wydobywczych „Żelazny Most" – olbrzymie wyzwanie geotechniczne. *Inżynieria Morska i Geotechniczna.*, vol. nr 3.

[21] Wang et al. 2014 – Wang, C., Harbottle, D., Liu, Q. and Xu, Z. Current State of Fine Mineral Tailings Treatment - A Critical Review on Theory and Practice. *Mineral Engineering*, vol. 58, no. c, pp. 113–131.

[22] Wells et al. 2011 - Wells, P., Revington, A. and Omotoso O. Mature fine tailings drying — technology update [In:] *14th International Seminar on Paste and Thickened Tailings*, Perth. 2011 5–7 April. Australian Centre for Geomechanics, pp. 155–166.

Minefill 2020-2021 – Hassani et al (eds)
© 2021 Taylor & Francis Group, London, ISBN 978-1-032-07203-6

The properties of the backfill mixtures based on own fine-grained waste

Radoslaw Pomykała & Waldemar Kępys
AGH University of Science and Technology

SUMMARY: Backfilling and filling voids in Polish underground mines has mainly two faces: the classic hydraulic backfilling and caulking gobs with use fine-grained mixtures. The first type is backfilling, in which the main material is sand has over 100 years of tradition in Poland, but in recent years it has virtually disappeared in hard coal mines. There are used for almost 30 years other methods - caulking and filling voids caving mainly with mixtures based on different types of fly ash. Fine-grained hard coal waste is eagerly added to such mixtures, which is the cheapest way to manage it. In recent years there have been a number of changes on the ash market. Their price increased significantly due to widespread use in the construction sector. Coal mines wanting to use known and proven technology at a low level of costs, are forced to look for alternative mixtures, based more on their own waste. The article describes the properties of caulking mixtures based on own (mining) waste with the addition of parameters improving the parameters. The possibility of producing binder mixtures with higher mechanical properties was also assessed. The scope of research concerned the properties of mixtures in a liquid state, including rheological properties, as well as the properties of mixtures after hardening.

Keywords: backfill, suspension, mining waste, sediment, tailings

1 INTRODUCTION

Several types of hydraulic backfills have been used in Polish hard coal mining for years. In the twentieth century it dominated classic sand backfilling used as a way of filling goaf, providing support for the floor, and a significant reduction of the surface subsidence. From the beginning of the 21st century it was gradually withdrawn due to high costs and the need to limit exploitation in areas requiring special surface protection. From the late 1980s to the 1990s, another type of backfill began to gain popularity - mixtures based on fine-grained materials. For over 20 years they have been a permanent element of mining technologies. The basic component of such mixtures - suspensions are fly ash and water (most often saline water from mine dewatering), For this purpose, fly ash of various types can be used: from dust and fluidized bed boilers, whether or not containing flue gas desulphurization products. Suspensions are also an effective method of managing various types of fine-grained mining waste (Pomykała et al. 2015, Popczyk, Jendruś 2019). As their components, waste from on-going coal processing and enrichment is usually used: flotation waste and sludge waste from jigs. Suspensions of particulate materials are used as solidified backfill, for filling voids and working liquidation, but usually as a mixture to the sealing and caulking of gobs caving in longwall mining system. Such use is primarily aimed at fire and methane prevention - by cooling gobs and their internal insulation.

DOI: 10.1201/9781003205906-10

In the last 10 years there have been a number of changes on the ash market. First of all, its use in the construction sector has increased significantly - as a component of cements, concrete and building materials, road foundations. This resulted in a decrease in the availability of cheap ashes for mines, especially in summer, and an increase in their prices to levels difficult to accept by mines. This increased the pressure to look for materials that would allow partial or full replacement of fly ashes, while ensuring the current scope of use of suspensions as solidified backfill and mixtures for gobs caulking.

2 ALTERNATIVE MATERIALS

Among the materials for which methods of development are constantly sought are various fine-grained mining waste deposited in surface settling tanks. Such wastes are sludge (sediment) from treatment of mine water accumulated in the surface mine water settling ponds, where sedimentation of fine gangue and coal particles occurs Gruchot et al. 2015).

The possibility of increasing the proportion of tailing waste in suspensions so that they constitute the main component of mixtures is also considered. This applies to both wastes from current production and deposited in tailings ponds.

Bottom ashes from coal combustion are another type of waste that can be used to prepare suspensions. While bottom ashes from dust boilers are widely used in construction, geotechnics and engineering works, management of bottom ashes from fluidized bed boilers is difficult (Pomykała 2013). The main reason is their chemical composition, in particular the content of CaO and SO3. The presence of these components is associated with the flue gas desulphurization process carried out in the fluidized bed boiler. As properly managed waste is based on its recovery, new possibilities and directions of bottom ash management from fluidized bed boilers are sought. Their use for making suspensions due to their grain size is not possible. On the other hand, the separation of small fractions may enable such a development direction.

The article presents the results of research on the use of own waste from hard coal mines for the preparation of waste-water suspensions. Flotation waste from coal enrichment processes and sediment from underground water treatment deposited in settling tanks were used for the research. As an addition to such mixtures, fly ash from a fluidized bed boiler, separated bottom ash as well as cement were used, which allowed to obtain the required binding properties.

3 MATERIALS AND RESEARCH METHODOLOGY

For the preparation of suspensions, energy waste from coal combustion in a fluidized bed boiler was used: fly ash - designated FA, and separated bottom ash - designated BA. The bottom ash was separated on a 0.315 mm sieve. The 0-0.315 mm fraction was used for the tests. The selected size of distribution resulted from the grain composition of the bottom ash, ensuring the right amount of separated ash (about 40% of the waste was the 0-0.315 mm fraction). The larger the grain size, as well as their share, promotes sedimentation in the pipeline, it can also result in uneven material distribution in the binding mixture. Flotation waste from coal enrichment was used to prepare the mixtures - designated as FT and sediments (sludge) from the underground water sedimentation tank, designated as M. Due to the significant share of water in the mine's own waste (humidity FT 45% and M 39%) and to create greater opportunities for of underground water from the mine, CEM 32.5R cement was also used.

The use of suspensions in mining technologies requires that the suspensions meet the relevant requirements specified in PN-G-11011: 1998 - Materials for solidification backfill and mixtures to caulking gob. Requirements and test methods (Gruchot et al. 2015). The basic requirements for the materials themselves are to meet the limits in leaching of chemical impurities and radionuclide content (Pomykała 2013, Popczyk, Jendruś 2019).

An important role in the preparation of suspension formulas is played by several parameters: consistency (possibility and safety of transport as well as ability to penetrate voids) ability to bind water and solidify, setting time and compressive strength (also in the wetness test - after seasoning samples in water).

For each of the suspensions, a series of tests was performed, which allowed determining the variability of properties depending on the composition and determining the possibility of their use in mining for solidifying backfill or caulking mixture. The following properties of suspensions were tested: apparent density, fluidity, volume of supernatant water, setting time - beginning and end of setting using a Vicat apparatus. The fluidity measurement is carried out by spreading a given volume of suspension on a flat surface by lifting the standard cone into which the suspension has been previously poured. The Rheotest 4.1 cylindrical rheometer was used for testing rheological properties. Tests of the strength of mixtures for uniaxial compression were carried out after 14 days.

According to the assumption, mining waste was the main components of the suspensions. Mixtures prepared with the participation of sediment from water treatment were designated as M1-M6, while with the participation of flotation waste: FT1-FT5 (Table 1). The contents of individual ingredients were measured by mass. The compositions of suspensions were selected so as to obtain the smallest volume of supernatant water and at the same time the consistency enabling safe transport through the pipeline and flow along the outlet from the pipeline. As a result, in mixtures with sediment from underground water treatment, the ratio of the mass of components S (M + FA + BA + CEM) to water W was S/W = 1.2 and 1.5. Due to the higher water content in the flotation waste, more water-binding components were used in the mixtures with their participation. As a result, the S/W ratio for suspensions with FT was S/W = 1.38; 1.75 and 2.0.

4 TEST RESULTS

The sediment (M) from underground water treatment was first mixed with water in a 1:1 weight ratio, and then fly ash was added in an amount of 0.5 sediment mass. A suspension M1 was obtained, the spreading rate of which was 140 mm. Because standard (Pomykała 2013) provides for the use of suspensions with a minimum fluidity of 180 mm, the next suspension was M3, in which half of the ash FA was replaced with cement. As a result, the fluidity of the suspension increased to 170 mm. By using separated BA bottom ash instead of FA fly ash,

Table 1. Composition of the suspensions.

| Mixture | Share of ingredients by weight % | | | | | |
| | mine waste | | ashes | | cement | |
	M	FT	FA	BA	CEM	S/W
M1	66.7	-	33.3	-	-	1.50
M2	80.0	-	20.0	-	-	1.25
M3	66.7	-	16.7	-	16.7	1.50
M4	80.0	-	10.0	-	10.0	1.25
M5	66.7	-	-	33.3	-	1.50
M6	66.7	-	-	16.7	16.7	1.50
M7	80.0	-	-	10.0	10.0	1.25
FT1	-	57.1	42.9	-	-	1.75
FT2	-	57.1	21.4	-	21.4	1.75
FT3	-	72.7	13.6	-	13.6	1.38
FT4	-	57.1	-	21.4	21.4	1.75
FT5	-	50.0	-	25.0	25.0	2.00

suspensions with higher fluidity properties of 150 mm (M5) and 190 mm (M6) were obtained. By reducing the proportion of S components in relation to water to the value of 1.25 and also by limiting the content of binding components (FA, BA and CEM), suspensions M2, M4 and M7, for which the spreading was 210-220 mm, were obtained (Table 2).

A small amount of supernatant water, up to 1.5%, was noted for mixtures with the sediment. According to the standard (Pomykala 2013), in solidified backfill only max. 7% of supernatant water content is allowed, and for caulking gobs mixtures - max. 15%. Such a small amount is very beneficial because there is no danger due to the formation of water reservoirs at the bottom in the mine, and pumping costs will be relatively low. It also allows reducing the costs associated with the discharge of saline water into surface watercourses.

The results of the early compressive strength test indicate that some suspensions (M1, M3 and M6) already after 14 days showed higher strength than required for solidified backfill (min. 0.5 MPa). Compressive strength depends on the proportion of binding components and their ratio to water. Similarly as in the case of building materials - slurries, mortars or concretes (Pomykała et al. 2013). In addition, suspensions with fly ash (FA) showed higher compressive strength than with separated bottom ash (BA).

Flotation waste (FT) was also mixed with water in a 1: 1 mass ratio before preparing the suspensions. Next, binding components were added, but in an amount greater than in the case of M1-M7 suspensions, because the humidity of FT was higher than M. As a result, FT1-FT5 suspensions were prepared, in which the mass ratio of all solids to water was S/W = 1.38 (FT3), 1.75 (FT1, FT2, FT4) and 2.0 (FT7). The fluidity of these suspensions ranged from 210 mm to 290 mm. The base suspension made of FA (FT1) had a fluidity rate of 210 mm (FT1). Replacing half of the FA ash with cement increased the fluidity to 230 mm (FT2). In turn, the use of BA instead of FA caused a further increase in fluidity to 270 mm (FT4).

Start of suspensions setting was varied depending on the type and content of individual binding components in relation to water and solids content. In general, fly ash suspensions started to set sooner than suspensions with separated bottom ash. The use of cement instead of some fly ash or separated bottom ash accelerated the setting of suspensions.

Early uniaxial compression strength (after 14 days) for all suspensions prepared with postflotation waste was higher than 0.5 MPa, ranging from 0.61 MPa (FT3) to 2.16 MPa (FT1). The highest strength was found for suspensions with the proportion of cement, the smallest - with separated bottom ash (Table 2).

The results of testing the rheological properties of the tested suspensions are presented in Figures 1 and 2 in the form of flow curves.

Table 2. Suspension properties.

Mixture	Properties of liquid suspensions			Start setting time [day]	Finish setting time [day]	Compression strength after 14days [MPa]
	density [Mg/m^3]	fluidity [mm]	Supernatant water [%]			
M1	1.31	140	1.0	3	5	0.57
M2	1.23	220	1.5	9	11	0.34
M3	1.32	170	1.0	2	4	1.09
M4	1.27	220	1.0	4	7	0.45
M5	1.30	150	0.4	9	12	0.37
M6	1.29	190	0.0	2	6	0.56
M7	1.24	210	0.0	5	10	0.24
FT1	1.34	210	0.0	3	4	1.66
FT2	1.33	230	0.3	3	4	2.16
FT3	1.23	290	1.8	4	8	0.61
FT4	1.34	270	6.9	4	6	0.65
FT5	1.34	250	7.3	3	5	2.71

Source: own study

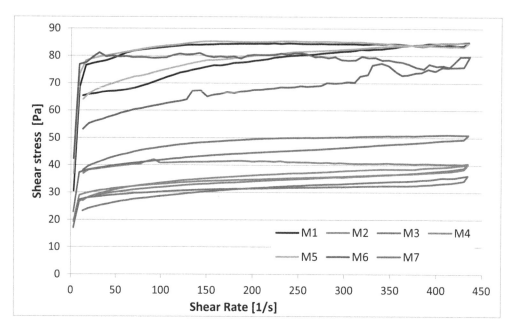

Figure 1.   Flow curves of suspensions with sediment (M).

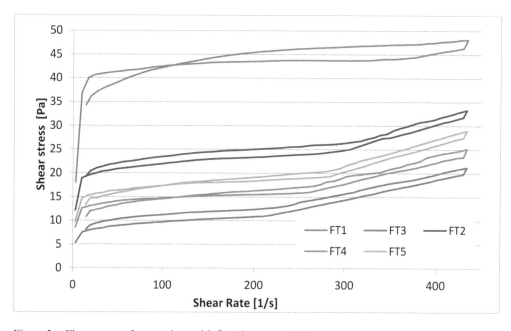

Figure 2.   Flow curves of suspensions with flotation waste (FT).

All tested wastes have a clear flow limit, and at the same time, as the shear rate increases, the increase in viscosity is relatively small. This shape of the flow curves indicates a large role of plastic properties, and is typical for flotation wastes from hard coal processing. The highest shear stress values were obtained for suspensions with fly ash. In the case of suspensions with the participation of sediment (M) a constant flow curve is observed in a wide range of share rates. In the case of suspended solids with flotation waste (FT), an increase in share stress is observed after exceeding the shear rate of 250 [1/s].

5 SUMMARY

The article presents the results of tests on the properties of suspensions, in which the main component was fine-grained mining waste in the form of sediment from underground water treatment and flotation waste from hard coal processing. In order to bind water and solidify the mixtures, the following ingredients were added to them: fly ash and separated bottom ash with a fraction of 0-0.315 mm, from coal combustion in a fluidized bed boiler, as well as cement.

One of the main objectives of the research was to indicate whether and to what extent separated bottom ash can replace fly ash in such suspensions, as well as whether and to what extent the addition of cement allows to increase the share of fine-grained mining waste. As the results of the research have shown, it is possible to select the appropriate composition of suspensions, in which the main component will be waste in the form of sludge from underground water settlers or flotation waste. This can increase the possibility of using this type of difficult material, which is often landfilled. In addition, the use of separated bottom ash from coal combustion can complement the shortage of fly ash, especially in summer.

Replacement of fly ash with other materials cannot take place without prior analysis. The use of bottom ashes, even their separated fractions, involves the risk of sedimentation in the pipeline and its failure. This involves the need for more complex planning of transport parameters in the pipeline. In turn, the addition of cement raises the costs of the filling itself. Nevertheless, increasing the amount of fine-grained mining waste used in mining technologies is now a necessity due to the limited possibilities of depositing it on the surface.

ACKNOWLEDGEMENTS

The tests were carried out based on statutory works No. 11.11. 100 482.

BIBLIOGRAPHY

Gruchot A., Zając E., Zarzycki J. 2015. Analysis of Possibilities for Management of Hard Coal Mine Water Sediments. *Annual Set The Environment Protection, Volume 17*, p. 998–1016 *[in Polish]*.

Pomykała R., 2013. Properties of waste from coal gasification in entrained flow reactors in the aspect of their use in mining technology. *Archives of Mining Sciences, vol. 58 no. 2*, s. 375–393.

Pomykała R., Kępys W., Piotrowski Z., Łyko P. Grzywa A. 2015. The temperature influence on the properties of the fine-grained suspension used in underground workings, *Archives of Mining Sciences, vol. 60 no. 4*, s. 1053–1070.

Popczyk M. Jendruś R. Impact of ash and water mixture density on the process of gob grouting in view of laboratory tests.2019. *Archives of Mining Sciences, 64 no 3*, 625–634.

Minefill 2020-2021 – Hassani et al (eds)
© *2021 Taylor & Francis Group, London, ISBN 978-1-032-07203-6*

Cemented paste backfill failure envelope at low confining stress

Murray Grabinsky & Andrew Pan
University of Toronto, Toronto, Canada

SUMMARY: A popular method to determine backfill strength for sidewall exposure assumes that the cohesive bond strength is equal to one-half the Unconfined Compressive Strength (c_b = ½ UCS). A new laboratory study uses direct shear, UCS, and novel tensile strength test results to show that this strength assumption is not valid for the tested CPB (and quite possibly many other CPBs). The implications of this finding in the context of strength for exposed backfill sidewalls is investigated using generalized design examples.

Keywords: Exposed backfill sidewalls, strength envelope, backfill design

1 INTRODUCTION

Stope sequencing of an underground ore body often results in previously placed backfills having their sidewalls exposed when an adjacent unmined stope is subsequently extracted. Mitchell et al. (1982) described some historical approaches to assessing the backfill strength required for sidewall exposure and developed a new solution. The solution's validity was verified through laboratory scale tests in which model backfills had their sidewalls exposed and the nature of the failure (if any) was assessed. The material strength at the time of model testing was also determined by direct shear testing on control samples. The simplest expression of the design formula, and the one recommended for use in practice, takes the form.

$$UCS = \frac{\gamma H}{1 + \frac{H}{L}} = \frac{\gamma L}{1 + \frac{L}{H}} \tag{1}$$

where UCS is the unconfined compressive strength, γ is the backfill unit weight, and H and L are the height and strike length (respectively) of the exposed backfill face. The first term in Equation 1 is the one presented in Mitchell et al. (1982) whereas the second term (which is equivalent) will be a more convenient form later in this work. The second form also obviates the singularity that exists for H ≫ L. This design formula is elegant in its simplicity and has been widely adopted. It is less clear how mining operations apply Factors of Safety (if any) to this limiting strength formula, and how they might implement QA/QC procedures to ascertain if the design strengths are being reliably achieved in their underground operations. Regardless, the formula's widespread and successful use can be taken as a *de facto* validation of its design suitability.

Despite Equation 1's popularity, the theoretical background to its development seems to be relatively poorly understood. A particularly unusual assumption is that the strength mobilized on the backfill's sidewalls, the so-called "bond strength" c_b, is equivalent to one-half the UCS: i.e., c_b = ½ UCS. This assumption appears to have gone unchallenged in the almost four

DOI: 10.1201/9781003205906-11

decades since publication, but for backfill it is not the typically expected form of failure criterion for a generally accepted frictional material. Therefore, the purpose of this paper is to present results of careful laboratory tests at low confining stress that show the tested material does not have a failure envelope conforming to the $c_b = \frac{1}{2}$ UCS assumption, and to explore the design implications of this finding. However, first a brief review of the background to Equation 1 is given to provide context to the laboratory test work.

2   BACKGROUND TO MITCHELL'S METHOD (EQUATION 1)

Mitchell et al. (1982) explain that prior to their studies it was commonly assumed that an exposed backfill height should be capable of supporting its own self weight (i.e., UCS = γH), or else it would be assumed that the backfill would fail according to the model of a vertical face in clay (in which case UCS = $\frac{1}{2} \gamma H$). The discrepancy between these approaches was deemed unsatisfactory and hence the motivation for their original work. They recognized that neither of these previous methods accounted for the stabilizing effect of bond strength developed between the backfill and the end walls (each with width w and height H, the distance w being perpendicular to the exposed face). They decided to incorporate this stabilizing effect through a form of "arching" analysis (Terzaghi, 1967).

The term "arching" is controversial, and it is generally recognized that the internal stress distribution in such "arching" phenomenon does not correspond to the mechanical behaviour of a structural arch (Handy, 1985). For the purposes of backfill analysis it is perhaps better to describe the phenomenon as "net weight" analysis. That is, the total weight of the backfill (γHwL) will be reduced by the sidewall stresses and result in a net weight which, when averaged over the horizontal area (wL) will give a vertical stress. If it is further assumed that the average horizontal stress at the backfill's base is zero (because the sidewall has been exposed) then this net vertical stress must be less than the material UCS. In the case of the Mitchell analysis, using the assumption $c_b = \frac{1}{2}$ UCS with c_b acting over the two end walls with area wH, means the net vertical stress at the base is (γHwL - $2c_b wH$)/wL and if this net vertical stress is equal to the UCS then one obtains UCS = $\gamma H - 2c_b H/L = \gamma H$ - UCS H/L and this is straightforward to rearrange to yield Equation 1.

There are two restrictions noted by Mitchell et al. to this design equation. First, the exposed length L should not exceed 1.5 × H, otherwise the net weight will be too high and the cement bonds could be crushed (according to correlations between UCS and bond crushing observed in oedometer tests on sands). This is probably rarely a practical concern as most often exposed backfill faces are taller than they are wide (e.g., for long hole and Alimak stoping methods). The second restriction is that H needs to be no smaller than w tan(45°+φ/2) where φ is the angle of internal friction. This restriction is motivated by the notion that at failure a sliding plane may develop starting at the base of the exposed sidewall and extending up at an angle (45°+φ/2) to the contact wall opposite the exposed wall. This is a curious restriction for two reasons: first, the assumption of internal failure at the critical angle (45°+φ/2) is contradictory to the assumption $c_b = \frac{1}{2}$ UCS which inherently implies φ = 0; second, the development of Equation 1 as explained above on the basis of net weight does not require any assumption of a kinematically admissible sliding failure plane through the backfill. Furthermore, the essential simplicity of determining UCS without also determining the corresponding c and φ is lost if one wishes to strictly adhere to the geometric restriction.

Regardless of these restrictions (which, as noted, are probably moot for many practical backfill design applications), the origins of the $c_b = \frac{1}{2}$ UCS assumption need to be explored. In their 1982 paper the authors present a traditional Mohr's stress space diagram with test results from direct shear and UCS tests, reproduced here as Figure 1 with Mohr's circles at failure superposed for the UCS tests. There are two materials, one with UCS just over 100 kPa and one with UCS just under 200 kPa. The relatively weak UCSs were intentional because weak materials were required to appropriately scale with the model geometry used in the laboratory backfill simulations. For the stronger material (UCS ~200 kPa) the Mohr-Coulomb failure envelope for the direct shear tests at "large strain" appears to be consistent

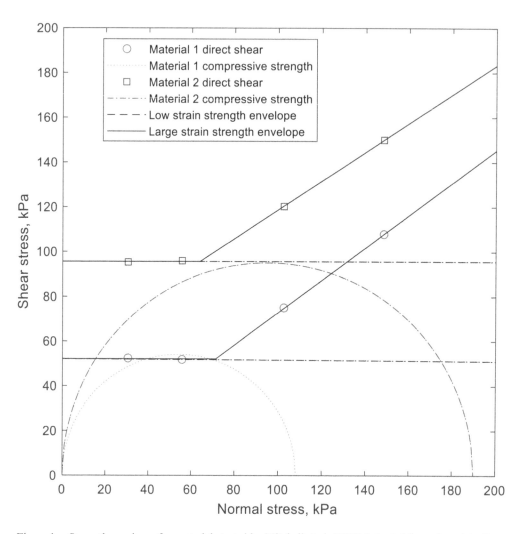

Figure 1.   Strength envelopes for materials tested by Mitchell et al. (1982) (adapted from the original).

with (i.e., tangential to) the Mohr's failure circle for the UCS; however, for the weaker material it is clearly not. The direct shear test results at "small strain" and for normal stresses less than about ½ UCS do indeed seem to show a constant shear strength equal to about ½ UCS, but stress-displacement results are never shown to clearly identify what is meant by "small strain" (probably because a very non-standard direct shear test method was used, for which obtaining shear displacements might have been difficult).

The way the researchers obtained relatively weak UCS values was by testing at very early ages, well before 28 days, and so it is not clear how well the hydration products had formed within the modelled backfill materials. Given the uncertainties about the materials, the strength testing methods, and the very unusual failure envelopes, it should be surprising that the resulting design formula (Equation 1) is so widely used without further material testing being undertaken to ascertain the validity of the fundamental assumption, c_b = ½ UCS.

To address this lack of material testing issue, a comprehensive set of quality laboratory tests was undertaken on a particular CPB from Williams mine. Conventional direct shear, specialty UCS for paste backfill, and a novel direct tension test method were used and will be described in the next section. It is important that these tests be carried out at appropriate normal stress levels. From the previous discussion of how Equation 1 can be rationalized based on net self-weight, the vertical stress through the fill mass will always be less than the UCS obtained from Equation 1. If some reasonable in situ stress ratio is assumed for localized areas within the backfill, then the confining stress should also be less than the UCS. Note that the point of tangency to Mohr's failure circle for the UCS test defines the normal and shear stresses acting on the inclined failure plane through the UCS sample: therefore, a UCS test sample has a non-zero compressive stress on its failure plane, and so does *not* represent failure in a truly "unconfined" condition. This is further addressed in the next section.

3 STRENGTH TESTING

This section describes the materials tested, the test methods, and the strength envelopes determined from the tests. Additional details and test results can be found in Pan (2019).

3.1 *Materials tested and sample preparation prior to testing*

The CPB materials tested come from Barrick Gold Corp.'s Williams mine, located on the north shore of Lake Superior near the town of Marathon, Ontario. The mineral deposit is finely disseminated gold in low-sulfide Canadian Shield host rock, and the resulting tailings are classified as non-plastic silt. The maximum particle size is less than 1 mm; the D_{60} is 0.030 mm and the D_{10} is 0.003 mm, giving $C_U = 10$; and 45% by weight pass 0.020 mm which is greater than the generally accepted 15% minimum as one of the qualifying paste backfill criteria. CPB is made by preparing the tailings with Ordinary (or Normal, or General Use, or Type 10) Portland Cement and mine process water (the strongest ionic concentrations of which are Na and S at about 9 and 7 mmol/l respectively). Quoted binder contents are based on weight of binder divided by weight of dry tailings. Water content is consistent at 38% on a mass of water divided by mass of solids (tailings + binder) basis (the common definition used in geotechnical engineering), which is equivalent to 27.5% on a mass of water divided by total mass basis (the common definition used in backfill design).

CPB samples are prepared by mixing the constituent materials using a hand blender for 10 minutes, then placed into molds specific to each test method (Figure 2) being careful to avoid air entrainment. The samples are submerged for 1 day to prevent oxidation, then removed from the molds and returned to the water to continue curing until testing time. This effectively eliminates oxidation of the sulfur-bearing minerals, keeps the samples saturated, and eliminates potential suction development due to "self-desiccation" effects during hydration (Simms and Grabinsky, 2009). The tested samples were observed to be homogeneous and had bulk properties consistent with field samples (Grabinsky et al., 2013, 2014).

3.2 *Uniaxial compression testing*

The sample mold helps ensure ideal cylindrical geometry. The ends are "polished" on glass to remove irregularities prior to testing. Uniaxial compression testing was conducted with the sample under water, compressed at constant vertical strain of 1% per minute (average 0.65 mm/min). The 1% per minute test rate was determined in many previous research projects on the same material to be sufficiently slow to neutralize internal water pressures/suctions uniformly within the sample when tested under water.

3.3 *Direct shear testing*

The sample molds result in cured blocks that fit tightly into the standard direct shear split box for the equipment used, and these were then tested according to ASTM

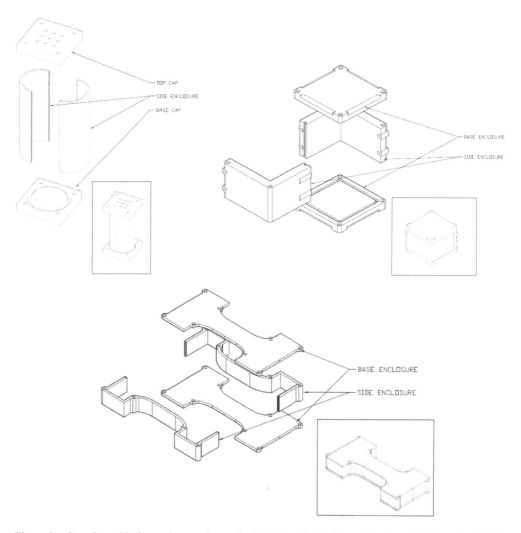

Figure 2. Sample molds for curing specimens for UCS (top left), direct shear (top right), and uniaxial tension (bottom).

Standard D3080: Standard Test Method for Direct Shear Test of Soils Under Consolidated Drained Conditions. The normal stresses ranged from 1 to 210 kPa which covered the desired range compared to the UCS, as discussed previously. The sample enclosure was filled with water during the test, similar to the UCS testing. The horizontal displacement rate was 0.5 mm/min which based on previous tests (Veenstra, 2013) was deemed to be sufficiently slow to neutralize pore pressures/suctions uniformly within the sample when tested under water.

3.4 *Uniaxial tension testing*

The uniaxial tension testing employed a novel test device illustrated in Figure 3. The sample mold for the "dog-bone" shaped specimen is shown in Figure 2. The upper portion of the apparatus bears on the lower protrusions of the specimen, and the lower (base) portion bears on the top protrusions. In this way, when the assembly is placed in a loading frame the applied compression is converted into tension within the specimen. The load

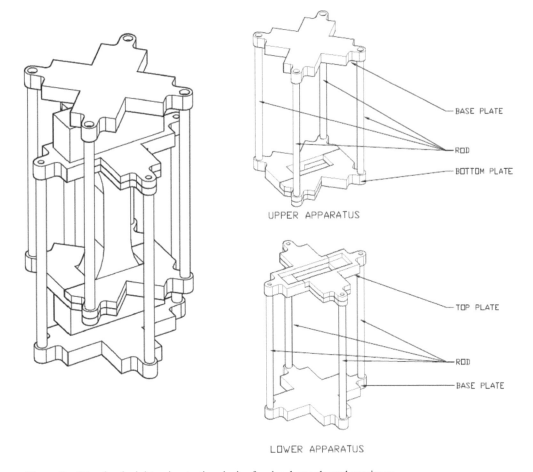

BASE PLATE

ROD

BOTTOM PLATE

UPPER APPARATUS

TOP PLATE

ROD

BASE PLATE

LOWER APPARATUS

Figure 3.   Novel uniaxial tension testing device for dog-bone shaped specimen.

frame is run in displacement control at 0.5 mm/min and the resulting tensile stress development to failure is very stable. The sample breaks horizontally at the specimen's mid-section, such that the failure surface is normal to the tension direction. These tests were not conducted under water, but this is thought to be not as important as for the other test methods due to the comparatively discrete nature of the failure surface and relatively small displacement to failure.

3.5   Test results and strength envelopes

Figure 4 shows testing results for 28-day cure times at 6.9% and 9.7% binder contents. The direct shear results (dot markers) exhibit good linearity and a Mohr-Coulomb envelope is fit to this data. The Mohr's failure circles for the UCS tests are closely tangent to this envelope, indicating good agreement between the two test methods. The Mohr's failure circles for the uniaxial tension tests are also generally tangent to the envelope, however, the actual failure surfaces are perpendicular to the direction of tension loading and so the corresponding envelope is better represented by a tensile cut-off. Regardless, it appears reasonable to extend the Mohr-Coulomb linear failure surface into the tensile range, as shown. Note that the point of tangency of the UCS circle occurs at a normal stress ~ ¼ UCS, and so the UCS does not represent failure in a truly "unconfined" condition. The direct shear tests, however, provide the

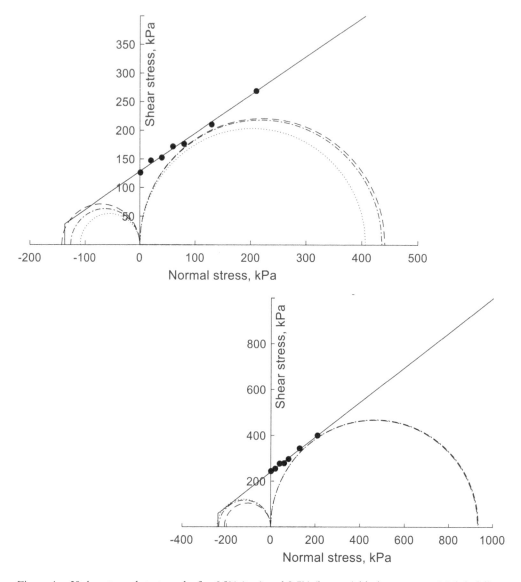

Figure 4. 28-day strength test results for 6.9% (top) and 9.7% (bottom) binder contents. Mohr's failure circles for uniaxial compression (3 tests each) and uniaxial tension (3 tests each); and dot markers for direct shear (7 test each).

additional evidence for actual cohesion (shear stress at zero confining stress) and failure shear stresses at normal stress up to the UCS tangent, and beyond.

4 IMPLICATIONS FOR EXPOSED SIDEWALL STRENGTH DETERMINATION

The failure envelopes just presented contrast starkly with the $c_b = \frac{1}{2}$ UCS criterion suggest by Mitchell et al. (1982). Had the materials not been tested under water, and had they been tested quickly, it might be that any dilation tendency would lead to suction and larger apparent shear resistance, but this would not be indicative of actual material behaviour.

Given that $c_b = \frac{1}{2}$ UCS is not an appropriate strength assumption for the Williams CPB material tested in this study (and, very likely not appropriate for similar CPBs made from non-plastic silty tailings) the obvious question arises: what are the implications for exposed backfill sidewalls designed based on Equation 1? This section explores an alternate solution based on the identified Mohr-Coulomb failure criterion and considers generalized design examples.

4.1 Alternative strength determination procedure

For a Mohr-Coulomb (c-φ) material with no pore water pressure, the shear stresses generated on the sidewalls are no longer constant but instead depend on the normal stress and the strength parameters, and so the vertical stress with depth z below the top backfill surface (again assuming only the two end walls remain in contact) can be written.

$$\sigma_v(z) = \sigma_{vmax}\left(1 - \exp\left(-\frac{z}{\frac{1}{2}L}K\tan\phi\right)\right) \tag{2}$$

where $\sigma_{vmax} = \frac{\gamma\frac{1}{2}L - c}{K\tan\phi}$ is the asymptotic value at large z and K is the horizontal stress ratio. Both σ_{vmax} and the decay exponent depend on the term $K\tan\varphi$. This term is reasonably insensitive to the assumed value of φ over the range expected for backfills. In the active condition $K_a\tan\varphi \approx 0.2$ and in the at-rest condition $K_o\tan\varphi \approx 0.3$. These are the likely limiting cases for preliminary analyses.

For practical cases the height of backfill is finite and so z = H. From the construction of the Mohr's diagram, UCS can be related to c and φ using UCS = c $2\cos\varphi/(1-\sin\varphi)$. Following the process outlined in section 1 for the net weight analysis, UCS = $\sigma_v(z=H)$ and making the substitutions into Equation (2) and simplifying results in.

$$UCS = \gamma\frac{1}{2}L\left[\frac{\tan\phi}{1 - \exp\left(-\frac{H}{\frac{1}{2}L}K\tan\phi\right)} + \frac{1 - \sin\phi}{2\cos\phi}\right]^{-1} \tag{3}$$

For the material test results presented in the previous section φ = 37°; and for the active and at-rest states (K_a and K_o) the UCS expressions reduce to the following cases:
Active, φ = 37°

$$UCS = \frac{1}{2}\gamma L\left[\frac{0.19}{1 - \exp\left(-0.38\frac{H}{L}\right)} + 0.25\right]^{-1} \tag{4}$$

At-rest, φ = 37°

$$UCS = \frac{1}{2}\gamma L\left[\frac{0.30}{1 - \exp\left(-0.60\frac{H}{L}\right)} + 0.25\right]^{-1} \tag{5}$$

For very tall stopes, H ≫ L, the above expressions evaluate to UCS = 1.02γL, 0.86γL, 1.14γL, and 0.91γL, respectively, so the range agrees with the Mitchell et al. solution (i.e., the second form presented in Equation 1) to within ±14% for the H ≫ L case. More general cases are considered in the next section.

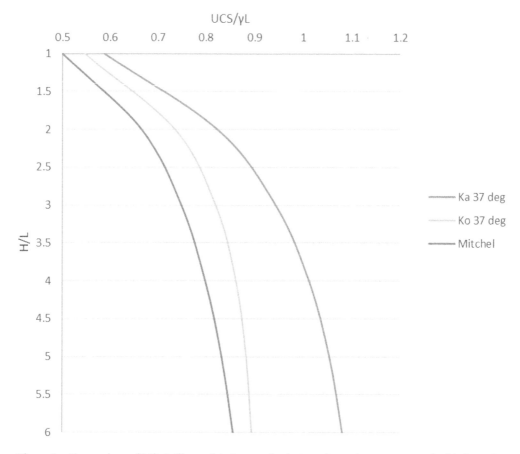

Figure 5. Comparison of Mitchell's predicted normalized strength requirement versus the friction solutions presented in Equations 4 and 5.

4.2 *General comparisons*

The second form of Mitchell's solution (Equation 1) can be normalized in terms of $UCS/\gamma L$ and H/L (in fact, this was the normalization used in their paper for comparing analytic predictions with laboratory model data). Figure 5 uses this normalization to compare the predicted normalized strength based on Mitchell's solution (i.e., $c_b = \frac{1}{2}$ UCS), with the predictions based on Equations 4 and 5 for the material tested in this work. In both cases the frictional solutions result in higher predicted UCS, because the assessed shear resistance in the upper portion of the stope has shear resistance less than $\frac{1}{2}$ UCS and so the net weight is greater at the base than would be the case for the constant c_b assumption. The differences are greater for the Ka assumption than for the Ko assumption, because the horizontal stress reaction is less for Ka than for Ko. The largest relative difference occurs for the Ka condition and for H/L between about 3 and 5, where the UCS based on Equation 4 is about 25% higher than the Mitchell et al. solution. The largest relative difference for the Ko condition occurs for H/L between about 1 and 2.5, where the UCS based on Equation 5 is about 10% higher than the Mitchell et al. solution.

4 CONCLUSION

The detailed laboratory test results presented in this work provide compelling evidence that the Williams CPB, and by extension probably most other CPBs made from non-plastic silty tailings, follow a Mohr-Coulomb (c-φ) failure envelope. This result contrasts starkly with the constant strength failure envelope ($c_b = \frac{1}{2}$ UCS) assumed in the generally accepted Mitchell et al. (1982) solution for the required UCS of a backfill with exposed sidewall.

Following the Mitchell et al. net weight approach, a generalized solution was developed to find the required UCS based on the c-φ failure envelope. Compared to the Mitchell solution, the required UCS determined using the frictional solutions could be up to 10% to 25% higher. However, this difference likely lies well within the range of Safety Factors used at most mines. Therefore, it should not be surprising that backfill designs based on the Mitchell's solution generally perform adequately well; however, the actual Safety Factors at these mines might be somewhat less generous than assumed. These currently "subtle" differences should nevertheless be kept in mind, as mines continue to strive for greater efficiencies and use less generous Safety Factors.

The quantified strength envelope could also have significance to the design of sill mats in undercut mining, where the flexure mode dominates for relatively thin mats (span/depth > 1; Pakalnis et al., 2005) and the design criterion is based on tensile strength which is notoriously difficult to determine for cemented paste backfills (Johnson et al., 2015).

ACKNOWLEDGEMENTS

The material testing shown in this paper was carried out by the second author during his MASc Thesis research at the University of Toronto, funded through a Natural Sciences and Engineering Research Council (NSERC) Discovery Grant to the first author.

BIBLIOGRAPHY

ASTM Standard D3080: Standard Test Method for Direct Shear Test of Soils Under Consolidated Drained Conditions.

Grabinsky, M. W., Bawden, W. F., Simon, D., Thompson, B. T., & Veenstra, R. L. 2013. In situ properties of cemented paste backfill from three mines. *Proceedings of the 2013 Canadian Geotechnical Conference, GeoMontreal 2013*. Montreal. Canadian Geotechnical Society.

Grabinsky, M. W., Simon, D., Thompson, B. D., Bawden, W. F., & Veenstra, R. L. 2014. Interpretation of as-placed cemented paste backfill properties from three mines. *MineFill 2014* (pp. 351–364). Perth: Australian Centre for Geomechanics

Handy, R. L. 1985. The arch in soil arching. *Journal of Geotechnical Engineering*, 111(3), 302–318.

Johnson, J. C., Seymour, J. B., Martin, L. A., Stepan, M., Arkoosh, A., & Emery, T. (2015, November). Strength and elastic properties of paste backfill at the Lucky Friday Mine, Mullan, Idaho. In 49th US Rock Mechanics/Geomechanics Symposium. American Rock Mechanics Association.

Mitchell, R. J., Olsen, R. S., & Smith, J. D. 1982. Model studies on cemented tailings used in mine backfill. *Canadian Geotechnical Journal*, 19(1), 14–28

Pakalnis, R., Caceres, C., Clapp, K., Morin, M., Brady, T., Williams, T., Blake, W. & MacLaughlin, M. (2005). Design spans – underhand cut and fill mining. 107th Annual General Meeting, Canadian Institute of Mining

Pan, A. 2019. *Mechanical properties of Cemented Paste Backfill under low confining stress*. MASc Thesis, University of Toronto. Institutional electronic repository: https://tspace.library.utoronto.ca/

Simms, P., & Grabinsky, M. 2009. Direct measurement of matric suction in triaxial tests on early-age cemented paste backfill. *Canadian Geotechnical Journal*, 46(1), 93–101.

Terzaghi, K. 1967. *Soil Mechanics in Engineering Practice*, John Wiley & Sons. Inc., New York.

Veenstra, R.L., 2013. *A design procedure for determining the in situ stresses of early age cemented paste backfill*. PhD Thesis, University of Toronto. Institutional electronic repository: https://tspace.library.utoronto.ca/handle/1807/36027

Minefill 2020-2021 – Hassani et al (eds)
© *2021 Taylor & Francis Group, London, ISBN 978-1-032-07203-6*

Cemented paste backfill response to isotropic compression

Murray Grabinsky & Mohammadamin Jafari
University of Toronto, Toronto, Canada

SUMMARY: CPB pressures up to 3.3 MPa have been measured arising from rock closure on the backfill. However, backfill has never been tested at pressures this high and therefore its response to high stress needs investigating. A high-pressure compression test device is described. Test results are presented for isotropic compression tests up to 6.5 MPa. The key parameters defining compression response are related to quality UCS test results. The general behaviour is very stable and extrapolating the results to several times the pressures used seems feasible, indicating CPB's suitability for deep and high stress mining.

Keywords: wall closure, backfill pressure, isotropic compression testing

1 INTRODUCTION

Backfill is the only form of regional ground support, and its use in deep and high stress mining is essential to maintain host rock stability and limit stress redistributions associated with ongoing mining that could otherwise lead to concentrated stresses and dangerous rockbursting conditions. Thompson et al. (2011) show field stress monitoring results from Kidd mine where initial stresses of a few hundred kPa jump to over 800 kPa with continued mining, and to over 1 MPa resulting from a Mn 3.8 rockburst that occurred a few hundred meters from the monitored stope. Subsequent work at Kidd mine employed higher capacity stress cells (Counter, 2014), and ultimately stresses in excess of 3.3 MPa were recorded using cells with 10 MPa capacity[1]. Although testing under these kinds of stresses is relatively common for rock materials, it is unheard of for backfills. The purpose of this work is therefore to generate new data in order to better understand CPB's response to isotropic compression under high stress regimes. The isotropic stress path is also fundamentally important to calibrating constitutive models where pressure stress is one of the stress invariants and yielding and hardening along this path forms an essential part of a more general "cap" that then describes the material's volumetric response to loading.

Isotropic compression testing is a comparatively rare test technique for *any* geotechnical material. Jafari (2019) provides a review of test results for naturally and artificially cemented soils. Naturally cemented soils have been tested by Airey (1993), Burland (1990), Cuccovillo and Coop (1999), Cuccovillo and Coop (1997). Artificially cemented soils have been tested by Arroyo et al. (2013), Clough et al. (1989), Coop and Atkinson (1993), Horpibulsuk et al. (2004), Huang and Airey (1998), Lo and Wardani (2002), Malandraki and Toll (2000), Malandraki and Toll (2001), Miura et al. (2001), Rios et al. (2012), Rotta et al. (2003), Santos et al. (2010), Xiao et al. (2014), and Xiao and Lee (2014).

---

1. Personal communications with D. Counter, November 6, 2019

DOI: 10.1201/9781003205906-12

In general, it is understood that the cement bonds create a structure within the solid particle's network. In isotropic compression loading, this will enhance the initial yield strength as compared to the uncemented material. The distinction between weak bonds and strong ones is based on the location of the intrinsic isotropic normal compression line in specific volume versus log-pressure space (Figure 1): for weakly cemented materials the bonds crush relatively quickly and the material soon returns to its unstructured state and behaves accordingly at large compressive stresses; for strongly cemented materials the initial nearly-linear response extends beyond the intrinsic isotropic normal compression line and much higher stress is required to degrade the bond structure and accelerate volume change with increasing stress (point s in Figure 1).

Exactly how CPB responds in this volume-stress space has yet to be determined. To investigate this, and to assess whether the material response corresponds to "weakly" or "strongly" cemented for the generally used binder contents (addressed subsequently) a new device had to be constructed. The device is considered first. Then the tested materials are described, and sample test results are shown and discussed within the above framework. Implications for practical mine design are then considered.

2   ISOTROPIC COMPRESSION TEST DEVICE

Samples were prepared to replicate the bulk properties determined from field testing (Grabinsky et al., 2013, 2014). A custom split mold (Figure 2) allows the material to be placed and cured under water for 24 hours, then removed without damaging the specimen when it is then returned to continue curing under water until testing. This avoids oxidation of sulfur-bearing minerals. The final sample's properties are homogeneous. The ends are polished on glass just before testing to remove surface irregularities where paste extruded into the top cap's vents.

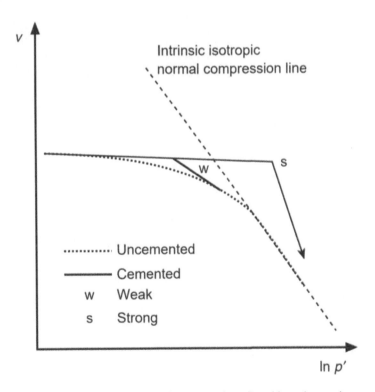

Figure 1.   Schematic comparison of the isotropic compression of weakly and strongly cemented carbonate sand (after Cuccovillo and Coop (1999)).

119

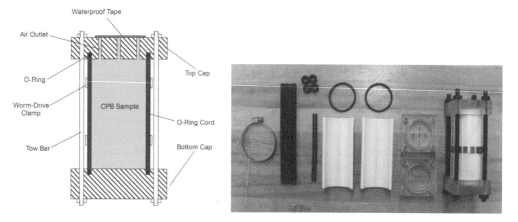

Figure 2. Sample split mold. left: schematic; right: accessories.

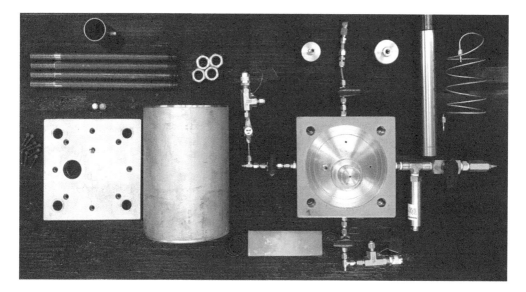

Figure 3. High-pressure isotropic compression cell components.

A cell was designed to test under isotropic compression at pressures up to 20 MPa (Figure 3), although the results shown here will be limited to 6.5 MPa range. Cell pressure was controlled according to feedback from a pressure transducer with 1000 kPa limits for cell pressure less than 1 MPa, and a pressure transducer with 20 MPa limits for cell pressure higher than 1 MPa. The combination of these sensors enhanced cell pressure accuracy in the high range of pressure. Two electrical pressure transducers with 1000 kPa limits were installed at the top and bottom of the specimen to measure pore water pressure at both ends of the specimen. Also, a frictionless rolling diaphragm type volume change device (VCD) with 0.01 ml resolution was used to measure the volumetric change of specimens. All sensors were calibrated before testing, and the calibration factors for each sensor were frequently checked during the testing program.

Care must be taken to check for apparatus compliances when such high pressures are used. It was found that two of the most significant sources of error when determining volume changes are 1) compliance of the porous disks and filter papers at each of the specimen's ends, and 2) membrane compliance. Methods for correcting these errors are addressed in Jafari et al. (2019).

Later in this work the isotropic compression test results will be compared with UCS results. It should be mentioned that although UCS testing is one of the (apparently) easiest strength tests to carry out, there are numerous details that must require attention if the results are to be reliable. If mining operations carry out routine UCS testing as part of a larger QA/QC protocol and find large scatter in the UCS results, then more attention to detail could well resolve many of the apparent testing issues. In particular, all of the UCS tests in this work were carried out under water and at an empirically determined loading rate found low enough not to introduce rate effects. If the testing is done at ambient laboratory atmospheric conditions, and if the loading rate is too high, then suctions can be induced by either the self-desiccation effect associated with hydration (Simms and Grabinsky, 2009), and/or pore pressure changes can be associated with sample contraction (at low strains) or dilation (at large strains), or the water pressures within the sample will not be uniform. To illustrate the effect of testing under water, Figure 4 shows test results for 3% and one-day cure time uniaxial tests, 3 tests under submerged conditions and 3 tests at ambient (unsubmerged) conditions. The three submerged test results are consistent and indicate a UCS just less than 30 kPa, where as the three unsubmerged tests all show significant strain hardening beyond 30 kPa and peak stresses up to four times the actual UCS. These additional apparent "strengths" are due to post-peak sample dilation and induced suctions, which actually increase the effective stress within the sample and lead to apparent strength when interpreted in a total stress context.

To further demonstrate that the apparent strengths in Figure 4 for unsubmerged tests are not real, Figure 5 shows uniaxial tests on 7.5% binder content and 3-day cure time specimens, all in unsubmerged conditions, for which apparent stresses are artificially high. The third sample was loaded to a level similar to the failure condition for the first two samples, and then water was splashed on the sample's exposed surfaces. This water was absorbed into the sample and reduced the suctions, thereby reducing the effective stress within the sample and so reducing the sample's load-bearing capacity. Over a few seconds the sample loses load-bearing capacity and ultimately fails. (The sample is tested in a displacement-controlled frame, so when the initial testing stops and water is splashed onto the sample, the displacement does not change but the reduction in load is registered by the frame's load cell.)

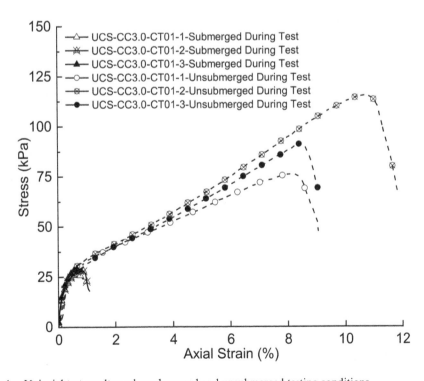

Figure 4.   Uniaxial test results under submerged and unsubmerged testing conditions.

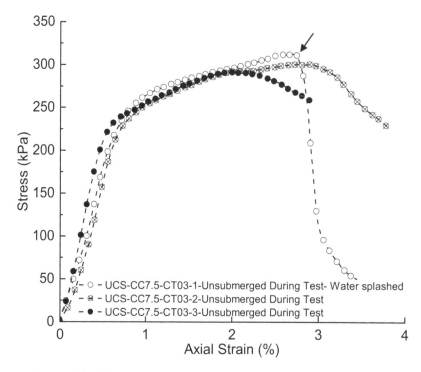

Figure 5. UCS tests with dilation-induced suction, and the effect of splashing water on the exposed surfaces of one of these specimens (at location of arrow).

3 MATERIALS TESTED

The CPB materials come from Barrick Gold Corp's Williams mine near Marathon, Ontario (Canada). The tailings are classified as non-plastic silt with 40% passing 0.020 mm which satisfies one of the generally accepted criteria for paste (minimum 15% passing 0.020 mm). Binder contents between 3% and 11.1% were tested because these correspond to the minimum and maximum generally used at the mine, depending on the detailed mine design scenario. Water content is consistent at 38% on a mass of water divided by mass of solids (tailings + binder) basis (the common definition used in geotechnical engineering), which is equivalent to 27.5% on a mass of water divided by total mass basis (the common definition used in backfill design). Tests were also carried out at varying cure times up to 28 days to better understand how the properties and behavior evolve with different hydration stages, however only the 28-day results will be presented here. The tailings contain relatively low amounts of sulfur-bearing minerals compared to other gold mines (in Canada in particular) and probably as a result do not suffer from long term strength degradation.

4 SAMPLE TEST RESULTS

Results are presented here for 3% and 11.1% binder content, at 1-day to 28-day cure times (Figures 6 and 7).

The 1-day cure times are thought to correspond to the "weakly cemented" material response characterized in Section 1, whereas at 3-day cure and longer the material strength increases and the "strongly cemented" material response is enhanced. For each test the initial response was linear and therefore the bulk modulus could be determined, and the onset of nonlinearity, YS_{ISO}, was readily identified (Figure 8) and this value was found to be only marginally larger (by about 7%) than the corresponding onset of nonlinearity from the UCS tests.

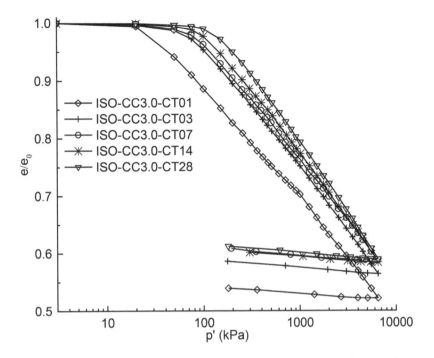

Figure 6.   Isotropic compression test results for 3% binder content, 1-day to 28-day cure times.

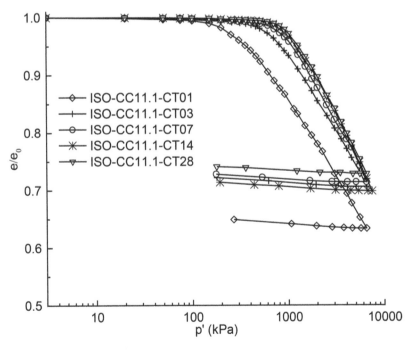

Figure 7.   Isotropic compression test results for 11.1% binder content, 1-day to 28-day cure times.

The overall response of each isotropic compression test can be generalized as shown for a particular test result in Figure 9. Initial loading (starting at A) is linear up to YS_{ISO} (point B) and deformation is controlled by the bulk modulus. Beyond YS_{ISO} the nonlinear response increases and approaches a new tangent (point C) to a new linear response characterized as the

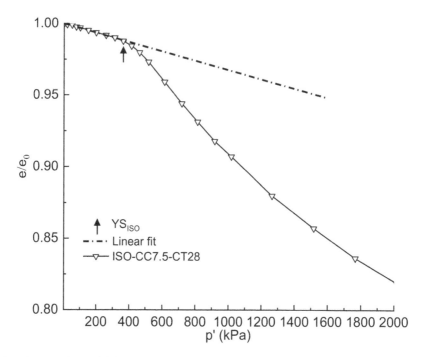

Figure 8. Identification of initial non-linearity (YS_{ISO}) in isotropic compression.

Post-yield Compression Line (C to D). Upon unloading (point D) the response is again almost linear with slope very similar (in log space) to the initial bulk modulus (A to B). Note that the overall response has many similarities to the one-dimensional compression (oedometer) results for overconsolidated clays.

The initial tangent for bulk modulus (line A to B) intersects with the Post-yield Compression Line at a point P_{ISO} in void ratio – log-pressure space (Figure 9). For all of the test results, over the range 3% to 11.1% binder content and 1-day to 28-day cure times, Jafari (2019) found that P_{ISO} = 1.14 UCS. Therefore, both YS_{ISO} and P_{ISO} can be strongly correlated to high-quality UCS test results, which is extremely useful when conducting parametric analyses to examine the design implications of intermediate binder contents.

Finally, it is instructive to consider extrapolating the available test results into higher compression ranges. If the 1-day test results at high stress is indicative of an "uncemented" response (i.e., all bonds destroyed) and thus used as a tangent for the inherent material's Post-yield Compression Line; and if in the same lower void ratio range the trend of the Post-yield Compression Lines is examined to estimate the tangent to the 28-day curve; then extrapolating these two tangents to lower void ratios and their intersection point, one finds a pressure stress approaching 100 MPa at void ratios around 0.3. While this approach is admittedly speculative, it is interesting that these values (100 MPa at 0.3 void ratio) would still lie completely in the plausible range for stable backfill remaining within the contained void space. As such, the backfill would continue to provide extremely effective regional ground support even under the most extreme loading conditions.

5   CONCLUSIONS AND IMPLICATIONS FOR PRACTICAL MINE DESIGN

More complete background, test results, and interpretations can be found in Jafari (2019) and these are also currently in peer review for anticipated journal publication. In the current article the isotropic test device has briefly been described, and the importance of high-quality UCS testing and straightforward steps that can be taken to achieve good results have been

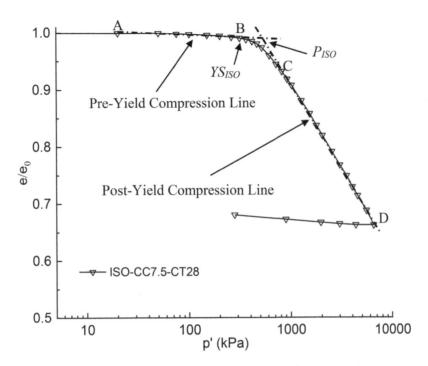

Figure 9. General characterization of response to isotropic compression.

highlighted. Even the bounding test results, shown for 3% and 11.1% binder contents, offer the first glimpse into the properties and behavior of CPB under multiple-MPa range isotropic compressive stresses.

The high-quality UCS test results available in this work allowed strong correlations between properties in isotropic compression loading and uniaxial loading. In particular, the onset of nonlinearity in isotropic compression was only marginally higher (factor 1.07) than the corresponding onset of nonlinearity in UCS tests. As well, the characteristic point for transitioning from initial elastic behavior (controlled by bulk modulus) to Post-Yield Compression behavior, P_{ISO}, was strongly correlated to UCS (factor 1.14). These correlations allow for parametric analysis where intermediate values of binder content and cure time might be considered for specific mine design scenarios.

The admittedly speculative extrapolation of available test results to their assumed "critical end point" suggests the tested backfill could stably sustain a 100 MPa isotropic stress at a void ratio around 0.3, at which point the backfill would likely be transformed into a something similar to siltstone.

More importantly, the test techniques and testing framework provide an approach that can be used to test other mine's CPB materials to determine if similar, or other correlations can be derived. Further such testing will lead to a much fuller understanding of CPB's effectiveness in providing regional ground support under the most extreme loading conditions.

ACKNOWLEDGEMENTS

This work was financially supported by Barrick Gold Corp and the Natural Sciences and Engineering Research Council Canada (NSERC) through a Collaborative Development and Research Grant to the first author.

BIBLIOGRAPHY

Airey, D. W. 1993. Triaxial Testing of Naturally Cemented Carbonate Soil. *Journal of Geotechnical Engineering* 119(9):1379–1398.

Arroyo, M., Amaral, M. F., Romero, E. & Fonseca, A. V. D. 2013. Isotropic yielding of unsaturated cemented silty sand. *Canadian Geotechnical Journal* 50: 807+.

Burland, J. B. 1990. On the compressibility and shear strength of natural clays. *Géotechnique* 40(3):329–378.

Consoli, N. C., Casagrande, M. D. & Coop, M. R. 2005. Effect of Fiber Reinforcement on the Isotropic Compression Behaviour of a Sand. *Journal of Geotechnical and Geoenvironmental Engineering* 131(11):1434–1436.

Consoli, N. C., Rotta, G. V. & Prietto, P. D. M. 2006. Yielding–compressibility–strength relationship for an artificially cemented soil cured under stress. *Géotechnique* 56(1):69–72.

Coop, M. R. & Atkinson, J. H. 1993. The mechanics of cemented carbonate sands. *Géotechnique* 43 (1):53–67.

Counter, D. B. 2014. Kidd Mine–dealing with the issues of deep and high stress mining–past, present and future. In *DeepMining 2014—Proc. Of the 7th Int. Conf. on Deep and High Stress Mining* (pp. 16–18).

Grabinsky, M. W., Bawden, W. F., Simon, D., Thompson, B. T., & Veenstra, R. L. 2013. In situ properties of cemented paste backfill from three mines. *Proceedings of the 2013 Canadian Geotechnical Conference, GeoMontreal 2013*. Montreal. Canadian Geotechnical Society.

Grabinsky, M. W., Simon, D., Thompson, B. D., Bawden, W. F., & Veenstra, R. L. 2014. Interpretation of as-placed cemented paste backfill properties from three mines. *MineFill 2014* (pp. 351–364). Perth: Australian Centre for Geomechanics

Horpibulsuk, S., Bergado, D. T. & Lorenzo, G. A. 2004. Compressibility of cement-admixed clays at high water content. *Géotechnique* 54(2):151–154.

Huang, J. T. & Airey, D. W. 1998. Properties of Artificially Cemented Carbonate Sand. *Journal of Geotechnical and Geoenvironmental Engineering* 124(6):492–499.

Jafari, M. 2019. *Experimental study on physical and mechanical properties of cemented tailings*. PhD Thesis, University of Toronto

Jafari, M., Shahsavari, M. & Grabinsky, M. 2019. Volumetric change measurement for Cemented Paste Backfill under high isotropic compression: laboratory challenges and tips. In *Proceedings of the Canadian Geotechnical Conference Geo St John's 2019*, Canadian Geotechnical Society.

Lo, S. R. & Wardani, S. P. 2002. Strength and dilatancy of a silt stabilized by a cement and fly ash mixture. *Canadian Geotechnical Journal* 39(1):77–89.

Malandraki, V. & Toll, D. 2000. Drained probing triaxial tests on a weakly bonded artificial soil. *Géotechnique* 50(2):141–151.

Malandraki, V. & Toll, D. G. 2001. Triaxial Tests on Weakly Bonded Soil with Changes in Stress Path. *Journal of Geotechnical and Geoenvironmental Engineering* 127(3):282–291.

Miura, N., Horpibulsuk, S. & Nagaraj, T. S. 2001. Engineering behaviour of cement stabilized clay at high water content. *Soils and Foundations* 41(5):33–45.

Rios, S., Fonseca, A. V. D. & Baudet, B. A. 2012. Effect of the Porosity/Cement Ratio on the Compression of Cemented Soil. *Journal of Geotechnical and Geoenvironmental Engineering* 138(11):1422–1426.

Rotta, G. V., Consoli, N. C., Prietto, P. D. M., Coop, M. R. & Graham, J. 2003. Isotropic yielding in an artificially cemented soil cured under stress. *Géotechnique* 53(5):493–501.

Santos, A. P. S. D., Consoli, N. C., Heineck, K. S. & Coop, M. R. 2010. High-Pressure Isotropic Compression Tests on Fiber-Reinforced Cemented Sand. *Journal of Geotechnical and Geoenvironmental Engineering* 136(6):885–890.

Simms, P., & Grabinsky, M. 2009. Direct measurement of matric suction in triaxial tests on early-age cemented paste backfill. *Canadian Geotechnical Journal*, 46(1),93–101.

Thompson, B. D., Grabinsky, M. W., Veenstra, R., & Bawden, W. F. 2011. In situ pressures in cemented paste backfill—a review of fieldwork from three mines. In *Proceedings of the 14th International Seminar on Paste and Thickened Tailings* (pp. 491–503). Australian Centre for Geomechanics.

Xiao, H.-W. & Lee, F.-H. 2014. An energy-based isotropic compression relation for cement-admixed soft clay. *Géotechnique* 64(5):412–418.

Xiao, H., Lee, F. H. & Chin, K. G. 2014. Yielding of cement-treated marine clay. *Soils and Foundations* 54(3):488–501.

Minefill 2020-2021 – Hassani et al (eds)
© 2021 Taylor & Francis Group, London, ISBN 978-1-032-07203-6

Evaluating cemented paste backfill plug strength and the potential for continuous pouring[1]

Murray Grabinsky
University of Toronto, Toronto, Ontario, Canada

Ben Thompson
RockEngineering Inc., Kingston, Ontario, Canada

Will Bawden
Mine Design Technologies, Kingston, Ontario, Canada

ABSTRACT: Currently there appears to be no rational method of defining plug strength in a typical "plug and cure" backfilling strategy, which is a significant problem when many engineered backfill containment structures include assumptions that the backfill plug isolates the structure from pressure applied by the subsequent "main" backfill volume. The proposed method for plug strength determination is presented as an analytical solution supported by numerical modelling and case study data. By extension, this method enables definition of required backfill strength for continuous backfilling.

Keywords: Cemented Paste Backfill, Continuous Pouring, Plug Strength

1 INTRODUCTION

Deep and high stress mining requires fast and effective backfilling to support the exposed side-walls surrounding an open stope, and thereby limit the stress concentrations that would otherwise build up and potentially lead to dangerous rockbursting conditions. For relatively tall open stopes (e.g., longhole and Alimak stoping methods) a containment structure, i.e. barricade or rock berm is constructed in the undercut access to contain the initially placed backfill "plug". The plug extends a few meters above the undercut's brow and must cure (gain strength) before starting the overlying "main" backfilling. The purpose of the plug is to isolate the barricade from induced stresses applied by this main backfill volume. The ideal situation is to engineer backfilling such that the plug will cure as it is being placed and will have sufficiently gained strength, on average, that backfilling can switch from plug to main backfills without stopping for a cure period. However, backfill plugs are typically placed and allowed to cure for between 24 hours and seven days prior to the onset of backfilling the main volume such that an adequate plug strength is achieved. There are exceptions, in which mines have optimized this process through understanding of barricade capacity (which requires an engineered barricade design and application of adequate quality control) and routine measurement of pressures applied by the backfill on the barricade, such that continuous pouring can be

---

[1]This is an abbreviated version of a manuscript submitted for peer review; the expected final version contains additional background information, full analytical solution development, and additional case studies.

DOI: 10.1201/9781003205906-13

safely achieved (e.g., Li et al., 2014). Continuous pouring provided potentially significant savings through reduced stope cycle times. It should be noted that the potentially fatal consequences of barricade failure, in terms of a high energy inrush of fluid backfill, justifies an extremely cautious approach to backfilling. Cemented Paste Backfill (CPB) is the only method that offers the potential for continuous backfilling; unlike CPB, hydraulic backfills require drainage periods during which no filling occurs to ensure safe backfilling.

Previously, the design of backfill containment structures has been considered by various authors, with most recently Oke et al. (2018) demonstrating that barricades must be considered as a system along with rock abutments as significant reduction in capacity can be expected in soft rock conditions. Design or definition of adequate plug strength should also be included in the backfill containment system design, as frequently the backfill barricade assumes a design load applied by the plug height of fluid backfill. This implies the plug will isolate the barricade from the effects of the "main" pour.

To our knowledge, there is limited published information that justifies plug strength requirements. For instance, it has been suggested that the backfill in the plug must generally achieve an Unconfined Compressive Strength (UCS) of 150 kPa (Henderson et al., 2005), however this recommendation is offered without justification in terms of explanation or engineering foundation. Clearly, the plug must have strength such that the backfill is liquefaction resistant, and it is frequently considered based on empirical data that 100 kPa constitutes an acceptable minimum strength to achieve this. A discussion of liquefaction resistance of paste is outside the scope of this discussion; readers are referred to Been et al. (2002), le Roux et al. (2005), Shahsavari et al. (2014). It is worth noting however, that liquefaction of paste has not to the authors knowledge occurred to cause barricade failure events. An important clarification should be made that barricade failures involving fluid paste do not constitute liquefaction events if the poured paste never transitioned from a fluid to a solid state.

In order to develop a rational engineering approach to backfill plug design it is first necessary to anticipate the potential failure mode. Forensic analyses of backfill and barricade failures (e.g., Grice, 1998; Torlach, 2000; Mangan, 2001; Grice, 2001; Yumlu and Guresci, 2007; Revel and Sainsbury, 2007) have identified some causal mechanisms (e.g., internal erosion and piping in hydraulic fills, rockfalls and groundwater intrusion in fresh fills, quality control issues with construction, over-pressurization in blind fills) but the actual backfill failure mechanism behind the barricade cannot be determined because the ensuing outrush of failed material erodes the original failure surfaces within the backfill plug.

The purpose of this paper is therefore to rationalize the potential failure mechanism occurring through the plug backfill in order to determine the limiting required plug strength to resist such failure. The original motivation for conducting this analysis was to determine adequate plug strength from a consulting perspective in order to provide a complete backfill barricade design package. Given the extent of efforts to produce engineered barricade designs, it would seem curious that determination of adequate plug strength typically receives relatively scant attention. In addition to the definition of plug strength for a plug pour, cure and main pour strategy, the analysis presented would logically extend to defining the plug backfill strength for which no plug cure is required – i.e., a continuous pour. While this design approach has not been verified in new field trials, a suggested method (subject to site specific safeguards) is provided based on case histories.

Two published case studies are used to motivate the plug strength analysis and identify critical design assumptions. A numerical analysis is carried out from which the failure mechanism is identified. An analytical solution for this mechanism is developed and compared with the numerical analyses. A general design approach is then described and applied to the back-analyses of the case studies. A critical limiting assumption, based on the case studies, is that the plug backfill is in a state of zero effective stress and therefore only material cohesion can be mobilized. Some consideration is therefore given to other mines where early effective stress development may occur, and how the suggested design approach would have to be adapted for these design scenarios.

2 CASE STUDIES

Two of the most thoroughly documented mine backfill in situ pressure case studies come from Cayeli mine (Thompson et al., 2012; this is an Editor's Choice article for "particularly high caliber and topical importance" and is freely available from the Canadian Geotechnical Journal web site at http://www.nrcresearchpress.com/doi/abs/10.1139/t2012-040). The first case study, stope 715, involved high barricade pressures which necessitated stopping the pour for a plug cure period. The second case study, stope 685, featured sufficiently low barricade pressures that a successful continuously poured stope was achieved. These are therefore useful bounding case studies for validating potential design methods against actual field conditions. Both stopes used flat reinforced shotcrete barricades with limited setback from the undercut's brow. The main differences were in the backfill rise rates, backfill binder contents and in the types of tailings used in the backfill mix designs.

These backfills were monitored using clusters of total earth pressure cells (TEPCs) and piezometers attached to cages for in-stope installation (Figure 1a). Several cages were placed between the barricade and the middle of the stope just above floor level, as well as suspended on cables up the stope's height (Figure 1b). Additional TEPCs and piezometers were attached to the barricade's fill side, and the displacements were measured on the barricade's free side. The mine's experience was that a barricade pressure limit of 100 kPa was the maximum, at which point backfilling should be halted.

Figure 2 shows the stress monitoring results for the two case studies. For each case, two cages are considered, both location in the approximate center of the stopes. Cages 3 are located within the plug while cages 5 are located within the main fills. The binder content within the 685 plug was 8.5%, with a 6.5% binder content deployed above 9 m height in the 18.6 m high stope. Paste reached the 6.4 m elevation after 24 hours of backfilling and the measured pressures were hydrostatic (i.e., vertical as well as the two horizontal total earth pressures are equal to pore water pressure) and so the paste was in a zero-effective-stress state at the Cage 3 location (at 2 m elevation) until the plug height was achieved. For reference (but

a)   b)

Figure 1.   a) A cage with instruments for monitoring total earth pressures and water pressures; b) cage installations in a stope.

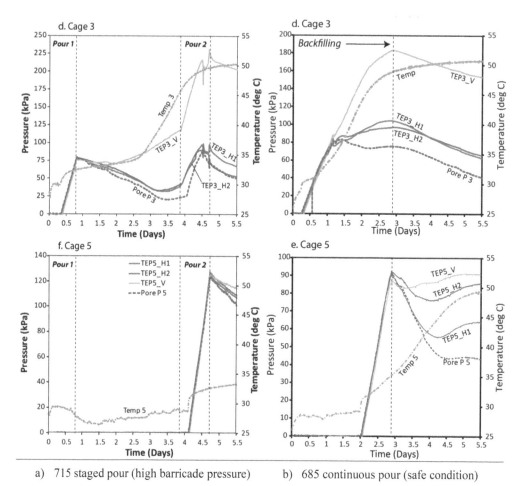

a) 715 staged pour (high barricade pressure)   b) 685 continuous pour (safe condition)

Figure 2.   Cayeli stress monitoring results (cages 3 are in plug; cages 5 are above plug), after Thompson et al. 2012.

not shown), non-hydrostatic loading was measured near the barricade 14 hours after paste reached the Cage 3 location. Barricade pressures did not increase linearly, and indeed peaked at 40 kPa (after 0.6 days of backfilling) and reduced before the plug pour was complete.

The plug height (6.4 m) for the 715 Stope was achieved within 20 hours of backfilling, during which time zero effective stress was measured throughout its filling. Unlike the 685 stope, 715 barricade pressures (not shown) increased linearly throughout the duration of the plug pour. The 715 stope featured 6.5% binder throughout; and indeed featured backfill made using a separate tailings stream to the 685 stope.

That there was significant difference in the rate of strength gain in the backfill for the respective plug volumes is indicated by temperature measurements, with the exothermic cement hydration responsible for temperature increase for 685 being very strong throughout the entire pour, while for 715 there is a slow temperature increase until a notable acceleration about 2½ days into the pour. Given that the backfill is a frictional material, zero effective stress represents a worse-case design condition in that it implies only the backfill cohesion can be relied upon.

For each stope, cages 5 are located within the main pours. In both cases the backfills undergo zero effective stress throughout the main pour period (i.e. > 20 hours of hydrostatic loading was measured within the 685 stope "main" volume). This implies a fluid loading condition was applied on the plug's top surface.

The 715 pour was stopped about 0.8 days into backfilling when the pressure cell located at about 1/3 barricade height reached the 100 kPa assumed barricade pressure limit. The barricade at this point was extensively cracked and leaking backfill water. Sophisticated reinforced concrete numerical analysis codes from the University of Toronto's structural analysis research group were used to back-analyze the barricade's pressure-displacement response and cracking patterns. These results indicated that the barricade was loaded close to its ultimate capacity (Grabinsky et al., 2014), and as such the plug cure backfill strategy was a requirement for safe backfilling.

In contrast, the 685 pour exhibited less pressure at the barricade than at the cage 3 location, suggesting strength gain of the backfill at the brow location backfill (regardless of the zero effective stress state measured at the center stope Cage 3 location) which allowed a safe continuous pour with no obvious signs of barricade distress and smaller barricade displacements. Readers are referred to Thompson et al. (2012) for fuller discussion of these results.

The conclusions arising from these case studies specific to the plug strength analysis are therefore: 1) as a worst-case scenario the plug can exist in a state of zero effective stress and therefore only cohesion will contribute to plug strength; and 2) the subsequent main backfill pour can be assumed to exert a fluid pressure on the plug's top surface. The next section seeks to identify the potential failure mechanism through the backfill plug that could arise under these circumstances.

3 MODELED FAILURE MECHANISM AND ANALYTIC SOLUTION

As a starting point for analysis the undercut is assumed to be square in cross-section and the plug is assumed to extend only to the brow height. The backfill is assumed to have constant cohesion and unit weight. A total stress is applied to the plug's top surface to represent the fluid loading arising from the main pour.

Analyses of this type were carried out using Flac3D which showed a failure mechanism developing with similarities to the Prandtl wedge solution for failure under a footing on undrained clay (Figure 3). The lateral extent is essentially limited by the undercut's sidewalls. A triangular wedge forms at the backfill's surface and mobilizes a radial fan (with a 22.5° subtended angle) which in turn pushes on the block of backfill within the undercut. The analyses were subsequently extended to include the plug portion above the brow, and also to consider a linear distribution of cohesion from zero at the plug's top surface in order to represent the plug's varying strength properties. Based on these analyses the equation for limiting plug stability can be written

$$q + \gamma(d + 0.55H) = c(3 + 4d/H + 4L/H) \tag{1}$$

The fluid loading q arising from the main backfill is simply the backfill's unit weight multiplied by the height of main backfill above the plug at any given time. The cohesion c is the average cohesion at the undercut's mid-height. The above solution does not incorporate barricade strength: it assumes the barricade is designed for initial plug fluid pressures only, and that the effect of the main pour should not be transmitted to the barricade. It is considered that this solution constitutes the primary contribution of this paper in terms of a means of a rationally engineered determination of adequate plug strength for backfill system design. In order to achieve this for a mine with no calculated value of c, the relationship c = ¼ UCS has been used, and appropriate QA/QC and safety factor assessments are required.

Another contribution of this paper is to extend this analysis to provide framework showing how the minimum plug strength to allow a continuous pour can be determined. To this end, the case study data is assessed and back-analysis is conducted.

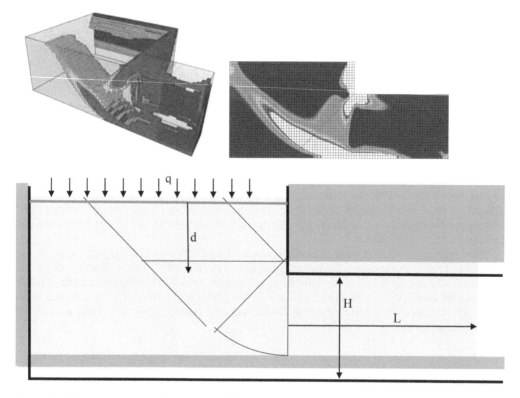

Figure 3. Numerical analysis showing plug failure mechanism (top) and analogy to Prandtl mechanism (bottom).

For a continuous pour analysis it is important to have comprehensive data for the development of cohesion with time during the entire filling process. Veenstra (2013) showed that this could be achieved using direct shear tests. Figure 4 shows results for plug CPBs from four mines, with cohesion determined starting at 4 hours (after paste mixing) up to about 14 days. For the two Cayeli case studies, stope 715 uses clastic tailings with 6.5% binder, while stope 685 uses non-clastic tailings with 8.5% binder plug and 6.5% binder main volume. For the two stopes, the early plug strength developments are dramatically different: cohesion for the stope 715 clastic tailings only increases significantly after about 2½ days (60 hours), which roughly corresponds to when the temperature dramatically increases in the stope.

The delayed strength development for the 715 backfill plug is likely responsible for the relatively high pressure barricade conditions discussed previously. It should also be noted that every CPB mix design exhibits unique cohesion development with time, due to the different curing stages associated with each combination of tailings mineralogy and binder chemistry. A plug strength design procedure is suggested next, and then applied to the back-analysis of Cayeli 715 and 685 stopes.

4 DESIGN STEPS AND BACK-ANALYSES

In order to determine plug strength requirements for a specific stope geometry, it is necessary to calculate the imposed fluid backfill head acting on the plug surface and apply Equation 1 and then determine adequate time (based on the specific paste recipe) required for the plug to cure. A rigorous QA/QC process is required to determine adequate plug strength has been achieved based for specific stopes using at minimum, Unconfined Compressive Strength testing and in-situ instrumentation (i.e. to measure barricade pressures).

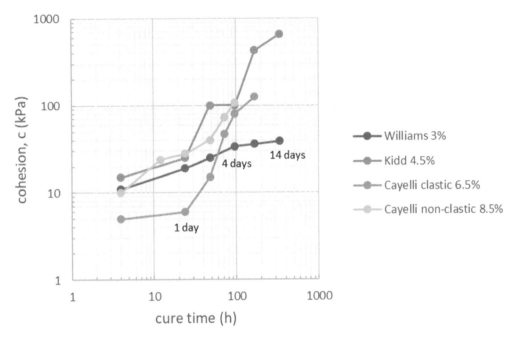

Figure 4. Laboratory determined cohesion development with time (adapted from Veenstra, 2013).

This paper does not attempt to review the complete nature of such a process, and caution is required to ensure safe backfilling practices are maintained. In the following description an idealized process by which an operation could determine plug strength requirements are suggested and indeed, the theoretical possibility of a continuous pour could be assessed. Where applicable, these steps have been used to back-analyze the case study data and calibrate the plug strength determination method presented herein.

The recommended plug strength design steps are:

1. Design the backfill barricade to resist the head of fluid backfill to the full plug height, in case of binder hydration in the plug occurring under a prolonged zero effective stress state (or confirm an existing design satisfies this condition).
2. Determine cohesion as a function of cure time as shown in Figure 4 for the plug paste.
3. Conduct a cavity monitoring survey of the open stope prior to filling. Based on cavity geometry and on the backfill plant's delivery rate, determine the expected fill rise rate(s) (or determine a representative worst-case rise rate for typical stopes).
4. Determine when the backfill will reach barricade mid-height elevation and establish this as the reference time (t_{ref}) with respect to a representative average curing time (and therefore cohesion developed) for the backfill plug. If possible, adjust (increase) this time for the mixing time in the backfill plant, the transport time in the pipe distribution network, and the travel time from the deposition point to the final placement location (however, this adjustment is not as important for rapidly hydrating backfills).
5. Determine critical times (with respect to the above-determined reference time) when the paste covers the barricade, when plug filling is complete, when different stages of binder hydration take place, when significant changes in stope geometry (and therefore fill rise-rates) occur, etc. and calculate the expected backfill cohesion for these reference cure times.

Calculate the plug's ultimate resistance to imposed surcharges using Equation 1 and convert this into an equivalent head of fluid backfill, i.e., $h_{max,eq} = q/\gamma$. Alternatively, do the same using Equation 2:

$$h_{max,eq} = (c/\gamma)\ (3 + 4d/H + 4L/H) - (d + 0.55H) \qquad (2)$$

7. Consider the result of Step #6 when the plug fill completes. A negative value means the plug does not have sufficient strength to support its own self-weight and therefore the plug requires additional cure time before the main pour starts.
8. Consider subsequent fill stages. If the calculated $h_{max,eq}$ is less than the actual expected main pour height above the plug, h_{ac}, then the plug needs strengthening through higher binder content and/or longer cure time, or else through pore water pressure dissipation and effective stress development (which implies employing a more sophisticated analysis technique).

Note that if a mining operation implements the above, then the time when backfilling reaches the undercut's mid-height (Step 4) is the time to collect multiple control samples (probably taken at the paste plant) for subsequent strength testing when the plug pour completes (Step 7). If the control samples' strengths satisfy the specified design criterion, then continuous pouring may theoretically proceed; otherwise, starting the main pour must wait until sufficient strength develops in the control samples (and, presumably in the plug backfill). These specific control samples would be in addition to samples collected routinely during the entire pour for more general QA/QC purposes. The following two case studies illustrate the above design steps.

It is emphasized that the above presents a theoretical approach, and any attempt to determine an adequate plug strength, or accelerate backfilling should deploy all means of verification that safe backfilling (i.e. at a minimum, barricade instrumentation) is occuring.

4.1 *Cayeli 715 case study: Plug Cure Required due to High Barricade Pressures*

Cayeli 715 is a long-hole stope approximately 15 m long, 8.5 m wide and 15 m backfilled height, with the barricade constructed very close (L = 1.8 m) to the brow. The barricade height is 4.5 m but the brow angles up into the stope with an average height H = 5.0 m, so that L/H = 0.36. The backfill unit weight is $\gamma = 22.4$ kN/m^3. Based on the backfill plant's delivery rate the average rise rate is expected to be 0.37 m/h but more detailed geometry is used to obtain better values for expected elevation with time in this analysis. Table 1 shows the (abbreviated) predicted response and the plug cannot even support its own self-weight. This analysis result is consistent with the observation that the barricade suffered significantly high pressure, and as such verifies the proposed design method for a "poorly performing plug" case. The analysis might seem trivial except for the fact that the 715 pour was eventually completed successfully and so the actual filling strategy requires further investigation. As an aside, this

Table 1. Key backfill times and plug capacities for 715 stope (assuming continuous pour).

Note	Pour time (h)	Fill el. above floor (m)	Cure time (h)	Backfill c (kPa, Figure 4)	q_{max} (kPa, Eq. 1)	h_{eq} (m, Eq. 2)	Fill el. above H (m)
start	0	0	n/a	n/a	n/a	n/a	n/a
mid-baricade	8 (t_{ref})	2.2	0	0	n/a	n/a	n/a
initial plugset	12	3.3	4	5	n/a	n/a	n/a
top-barricade	16	4.5	8	5.2	n/a	n/a	n/a
fill undercut	17.5	5.0	9.5	5.3	-38	n/a	0
top of brow	19	5.6	11	5.4	-48	n/a	0.6

result shows the importance of employing barricade pressure measurements to verify safe backfilling.

In actuality the pour was stopped when the maximum barricade pressure reached close to the threshold value at the completion of the plug height, and as planned, a plug cure period was enforced. This plug cure time resulted in sufficient strength to support the main pour. A similar back-analysis of the actual filling history is presented in tabular form in Table 2, and graphically in Figure 5. The cohesion increased significantly by the time the pour resumed (to 65 kPa) and at that time could theoretically support an imposed main volume backfill height 12.2 m which exceeded the required 8.4 m additional height for the main pour. The main pour therefore continued successfully and by the time it completed the plug had 87 kPa cohesion and could have supported almost double the imposed main backfill height. In fact, with the benefit of hindsight offered by this retrospective analysis, theoretically the main pour could have safely resumed even earlier. Finally, it should be noted that the delayed hydration for this backfill and its impact on early strength development was not fully appreciated at the time of the fieldwork (however, the mine no longer uses this particular mix design).

4.2 Cayeli 685 case study: Successful continuous pour

The overall geometry of the 685 stope was slightly larger than the 715 stope, with length 23 m, width 8.5 m and backfill height up to 17 m. Because of the larger volume, the average backfill rise rate was lower: 0.23 m/h. The flat barricade was 5.2 m high and was set back from the brow a distance L = 2.5 m, and the average undercut height was H = 6.5 m so that L/H = 0.38. The depth of plug above the brow was 3.1 m, giving d/H = 0.48. The performance of this filling strategy is evaluated using the proposed design procedure, with results tabulated in Table 3 and shown graphically in Figure 6. Again, detailed geometry was used to assess the relevant heights and times during the filling process, and cohesions were interpolated based on the results presented in Figure 4.

The relatively rapid strength development for this stope's CPB was clearly important to the success of this continuous pour. With reference to the field results presented in Figure 2, Cage 3 near the plug's base indicates zero effective stress up until backfilling reaches the top of brow (33 hours) yet the analysis indicates the plug has already gained significant strength (28

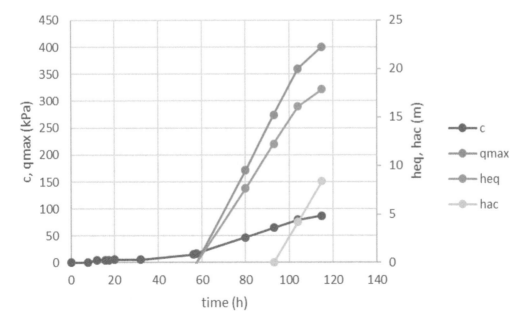

Figure 5.   Actual performance of 715 stope (with plug cure period).

Table 2. Key backfill times and plug capacities for 715 stope (based on actual backfilling history).

Note	Pour time (h)	Fill el. above floor (m)	Cure time (h)	Backfill c (kPa, Figure 4)	q_{max} (kPa, Eq. 1)	h_{eq} (m, Eq. 2)	Fill el. above H (m)
start pour 1	0	0	n/a	n/a	n/a	n/a	n/a
mid-baricade	8 (t_{ref})	2.2	0	0	n/a	n/a	n/a
initial plugset	12	3.3	4	5	n/a	n/a	n/a
top-barricade	16	4.5	8	5.2	n/a	n/a	n/a
fill undercut	17.5	5.0	9.5	5.3	n/a	n/a	n/a
end pour 1	20	6.6	12	5.4	n/a	n/a	n/a
24 h cure	32	6.6	24	6	n/a	n/a	n/a
48 h cure	56	6.6	48	15	n/a	n/a	n/a
self support	57.5	6.6	49.5	17	0	0	0
72 h cure	80	6.6	72	47	171	7.6	0
start pour 2	93	6.6	85	65	274	12.2	0
96 h cure	104	10.8	96	80	360	16.1	4.2
end of filling	115	15.0	107	87	400	17.9	8.4
168 h cure	176		168	125			

Table 3. Key backfill times and plug capacities for 685 stope (based on actual continuous pour).

Note	Pour time (h)	Fill el. wrt bar. base (m)	Cure time (h)	Backfill c (kPa, Figure 4)	q_{max} (kPa, Eq. 1)	h_{eq} (m, Eq. 2)	Fill el. above plug (m)
start	0	0	n/a	n/a	n/a	n/a	n/a
mid-baricade	9 (t_{ref})	2.8	0	0	n/a	n/a	n/a
initial plugset	13	3.6	4	10	n/a	n/a	n/a
top-barricade & 12 h cure	21	5.2	12	24	n/a	n/a	n/a
fill undercut	27	6.5	18	26	40.5	1.9	n/a
top of brow & 24 h cure	33	7.8	24	28	44.1	2.1	n/a
plug filled	39	9.6	30	31	56.3	2.6	0
48 h cure	57	13.5	48	40	114	5.3	3.9
end of filling	70	16.4	61	58	230	10.7	6.8
72 h cure	81		72	73			

kPa cohesion, enough to support 2.1 m main fill). By the time the plug pour is completed the refence cohesion has reached 31 kPa and could support 5.3 m of main fill. During the main pour the plug cohesion continues to increase. Potentially the lowest "strength factor" occurred at about 57 hours, when the plug cohesion was 40 kPa and could support 5.3 m of main fill, while the main fill was actually 3.9 m above the plug. At the end of filling the "strength factor" increased somewhat to 10.7 m of supportable fill versus 6.8 m actual.

In summary, analysis of the first case study showed that a continuous pour would not be possible, and indeed barricade pressure data demonstrated safe loading conditions would be exceeded if the standard plug cure strategy was exceeded. For this same stope, the analysis showed that the extra cure period afforded by the time delay to main pour was sufficient to allow the late hydrating CPB to gain strength and so successfully support the resumed main pour. The second case study shows that the back analyzed strength gain with time was sufficient to support a continuous pour, although possibly just minimally at 57 hours into the pour.

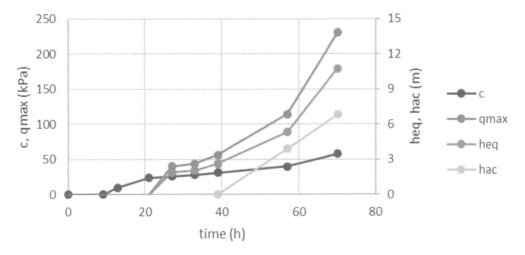

Figure 6.   Actual performance of 685 stope (continuous pour).

5   DESIGN FOR EFFECTIVE STRESS CASES

Not all the field trials previously studied showed prolonged periods of zero effective stress. Thompson et al. (2011) show field results from Williams mine where the initial pour starts to develop effective stress before the backfill reaches the brow. Indeed, when the analysis approach presented here is applied to that case study, an unstable plug is predicted whereas the field trial resulted in a successful continuous pour. It is very likely the conservative assumption of cohesion only (Williams cohesion development is also shown in Figure 4) is not appropriate for such backfilling situations. However, attempting a full effective stress analysis is a comparatively non-trivial task. The effective stress analyses demonstrated by Helinski et al. (2010), Witteman and Simms (2010), Veenstra et al. (2011), El Mkadmi et al. (2013), Doherty et al. (2015), Shahsavari and Grabinsky (2015), and Cui and Fall (2017), demonstrate that numerical analysis techniques are well developed to simulate these situations, however the time and effort to obtain the input parameters required for these simulations is likely to be prohibitively expensive for many mines.

6   CONCLUSION AND FUTURE WORK

A method has been proposed to define plug strength requirements appropriate for typical plug pour, plug cure and main pour backfilling strategies. This paper has not attempted to define an adequate Factor of Safety (FS), which should include site specific factors such as expected variability of the strength properties of as-placed material. Use of instrumentation is recommended to verify barricades remain in a safe pressure range during backfilling. Indeed, many operating mines with CPB systems do not use backfill instrumentation and likely do not know how quickly effective stresses develop in their backfill plugs. These operations should implement simplified field instrumentation programs to determine their drainage conditions. Even if drainage is not very substantial (i.e., effective stresses are low or are not consistently achieved) the comparatively conservative design approach recommended here offers a good first step rational analysis approach, as compared to empirical design methods.

Mines that find their plugs consistently and reliably achieve early effective stress development can still use the recommended design procedures, realizing that they will be conservative. If the design results constrain backfill design unnecessarily, then the next step is full effective stress analysis but this step must be undertaken recognizing the significant resource requirements associated with determining a much wider range of material parameters, and with the additional numerical analyses.

The industry standard UCS testing is not appropriate for determining cohesion, because the obtained UCS is a function of friction as well as cohesion. The test method suggested by Veenstra (2013) is a viable alternative, but new test methods should be explored to determine if faster and easier techniques can also be used.

Mines that are currently production-limited due to stope cycle time are the most likely to realize significant benefits from implementing the above approach to investigate the feasibility of continuously backfilling using rationally designed backfills. It is our future intention to back-analyze existing continuous pour data from other mines to further calibrate this method. Any mines seeking to increase backfilling efficiency through accelerating backfilling should proceed with great caution and utilize instrumentation to verify safe barricade pressure conditions are maintained. The entire backfill community would benefit greatly if these mines collect and report data in a systematic way, so that additional mines might also undertake these recommended procedures.

ACKNOWLEDGEMENTS

This work relies heavily on the research program Geomechanical Design of Cemented Paste Backfill Systems carried out between 2007-2013 (Phase I) and 2017-2019 (Phase II), jointly sponsored by the Natural Science and Engineering Research Council Canada (NSERC) and the parent companies and personnel from Williams, Kidd, and Cayeli mines (Barrick, Xstrata, and Inmet respectively at the time of the field studies). While the research involved many graduate students and research personnel, this work has made particular use of testing results by Ryan Veenstra. The three-dimensional Flac numerical modeling was carried out by Jeff Oke while at Mine Design Engineering (MDEng), Kinston, Ontario.

BIBLIOGRAPHY

Been, K., Brown, E. T., & Hepworth, N. 2002. Liquefaction potential of paste fill at Neves Corvo mine, Portugal. *Mining Technology*, 111(1), 47–58.

Cui, L. and Fall, M., 2017. Multiphysics modeling of arching effects in fill mass. *Computers and Geotechnics*, 83, pp. 114–131.

Doherty, J.P., Hasan, A., Suazo, G.H. and Fourie, A., 2015. Investigation of some controllable factors that impact the stress state in cemented paste backfill. *Canadian Geotechnical Journal*, 52(12), pp. 1901–1912.

El Mkadmi, N., Aubertin, M. and Li, L., 2013. Effect of drainage and sequential filling on the behavior of backfill in mine stopes. *Canadian Geotechnical Journal*, 51(1), pp. 1–15.

Grabinsky, M.W., Cheung, D., Bentz, E., Thompson, B.D., and Bawden, W.F., 2014. Advanced structural analysis of reinforced shotcrete barricades. In *11th International Symposium on Mining with Backfill*, Australian Center for Geomechanics

Grice, A.G., 1998. Stability of hydraulic backfill barricades. In *Proceedings of the 6th International Symposium on Mining with Backfill*, AusIMM, pp. 117–120.

Grice, A.G., 2001. Recent minefill developments in Australia. In *Proceedings 7th International Symposium on Mining with Backfill*, SMW (Seattle – canceled), pp. 351–357.

Helinski, M., Fahey, M. and Fourie, A., 2010. Behavior of cemented paste backfill in two mine stopes: measurements and modeling. *Journal of geotechnical and geoenvironmental engineering*, 137(2), pp. 171–182.

Henderson, A., Revell, M.B., Landriault, D. and Coxon, J., 2005. Chapter 6: Paste Fill. In Potvin, Y., Thomas, E. and Fourie, A. (eds) *Handbook on Mine Fill*. Australian Center for Geomechanics. ISBN 0-9756756-2-1.

le Roux, K., Bawden, W.F. & Grabinsky, M.W. 2005. Field Properties of Cemented Paste Backfill at the Golden Giant Mine, *Institution of Mining and Metallurgy. Section A: Mining Technology*, 114(2), pp.65–80

Li, J., Ferreira, J.V., and Le Lievre, T., 2014. Transition from discontinuous to continuous paste filling at Cannington Mine. In *11th International Symposium on Mining with Backfill*, Australian Center for Geomechanics.

Mangan, S., 2001. Coroner critical of Normandy over Bronzewing deaths. *MiningNews.net*, 2 August 2001. (www.miningnews.net/archive/news/1192079/coroner-critical-normandy-bronzewing-deaths, accessed 2019 05 15).

Oke, J., Thompson, B. D., Bawden, W. F., Lausch, P., & Grabinsky, M. W. 2018. Backfill Barricade Design: Practical Experiences and Recommendations. In *52nd US Rock Mechanics/Geomechanics Symposium*. American Rock Mechanics Association.

Revell, M.B. and Sainsbury, D.P., 2007. Paste bulkhead failures. In *9th International Symposium on Mining with Backfill, Montreal*, Paper, 10 p., Canadian Institute of Mining, Montreal: Proceedings on CD-ROM.

Shahsavari, M. and Grabinsky, M., 2015, September. Mine backfill porewater pressure dissipation: numerical predictions and field measurements. In *Proceedings of the 68th Canadian Geotechnical Conference*, Quebec City, QC, Canada (pp. 20–23).

Shahsavari, M., Moghaddam, R., & Grabinsky, M. 2014. Liquefaction screening assessment for as-placed cemented paste backfill. In *Proceedings of the Canadian Geotechnical Conference GeoRegina 2014*. Canadian Geotechnical Society.

Thompson, B.D., Grabnsky, M.W., and Bawden, W.F., 2011. In situ monitoring of cemented paste backfill pressure to increase backfilling efficiency. *CIM Journal*, 2(4):199–209.

Thompson, B.D., Bawden, W.F., and Grabinsky, M.W. 2012. In-situ measurements of cemented paste backfill at the Cayeli Mine. *Canadian Geotechnical Journal*, 49(7):755–772.

Torlach, J.M., 2000. *Potential hazards associated with minefill*. Safety Bulletin No. 55, Mining Operations Division, Department of Minerals and Energy, East Perth WA.

Veenstra, R.L., Bawden, W.F., Grabinsky, M.W. and Thompson, B.D., 2011. An approach to stope scale numerical modelling of early age cemented paste backfill. In *45th US Rock Mechanics/Geomechanics Symposium*. American Rock Mechanics Association.

Veenstra, R.L., 2013. *A design procedure for determining the in situ stresses of early age cemented paste backfill*. PhD Thesis, University of Toronto. Institutional electronic repository: https://tspace.library.utoronto.ca/handle/1807/36027

Witteman, M. and Simms, P., 2010. Hydraulic response in cemented paste backfill during and after hydration. In *Proceedings of the Thirteenth International Seminar on Paste and Thickened Tailings*, pp. 199–207. Australian Centre for Geomechanics.

Yumlu, M. and Guresci, M. 2007. Paste backfill bulkhead failures and pressure monitoring at Cayeli mine. *CIM Bulletin*, vol. 100, pp. 1001–1010.

Binders, admixtures, and other chemicals

to improve fill performance

Minefill 2020-2021 – Hassani et al (eds)
© 2021 Taylor & Francis Group, London, ISBN 978-1-032-07203-6

The influence of the flocculant on the process of thickening and depositing of copper ore flotation tailings

Maciej Gruszczyński & Stanisław Czaban
Institute of Environmental Engineering, Wrocław University of Environmental and Life Sciences, Poland

Robert Pratkowiecki, Zbigniew Skrzypczak & Paweł Stefanek
KGHM Polska Miedź S.A., Hydrotechnical Unit, Poland

SUMMARY: In the mining industry both segregation and thickening of flotation tailings are usually used to obtain backfill material. The use of flocculants in the process of thickening tailings causes a change in the rheological parameters of the obtained mixture which directly affects the flow conditions and segregation of the solid phase. The paper presents the results of semi-technical tests of thickening, deposition and segregation of the copper ore flotation tailings. Thickening tests with and without flocculants were carried out. The material obtained at the underflow was deposited in prepared measuring ditches. During the experiments were monitored such parameters as the flow rate, particle size distribution of the solid phase and the density of the feed, overflow and underflow. The results of the experiment were analyzed with regard to the use of the underflow stream as a backfill material. After exceeding critical concentration of the solid phase, the movement of non-Newtonian fluid was observed in measuring ditches whose role was to simulate the spread of backfill in the post-mining void.

Keywords: backfill material, paste, segregation, thickening, non-Newtonian fluid

NOTATION

c_u coefficient of uniformity [-]
d_{10} Grain size indicator, particle diameter corresponding 10% cumulative undersize particle size distribution [μm]
d_{50} Grain size indicator, particle diameter corresponding 50% cumulative undersize particle size distribution [μm]
d_{60} Grain size indicator, particle diameter corresponding 60% cumulative undersize particle size distribution [μm]
x Distance [m]
y Elevation [m]
ρ Solids density [kg·m^{-3}]

1 INTRODUCTION

Tailings are fine-grained materials that remain after valuable metals and minerals have been extracted from ore (Lee et al. 2017). One of the most important directions of flotation tailings management - as far as the impact of the mining industry on the natural environment is concerned - is the deposit of tailings at the mine in the form of paste (Stefanek & Serwicki,

DOI: 10.1201/9781003205906-14

2014). The primary advantage of paste fill is: the fill has little or no bleed water thus simplifying mine dewatering (Fehrsen & Cooke 2010), This technology requires the production of paste from tailings in the dewatering process (Li et al. 2017) such as hydrocyclonage or filtration so as to reduce the water content (Stone 2014). We can define paste as a mixture with a reduced amount of water (Annor. 1999) which can be pumped and does not separate during flow and depositing (Belem & Benzaazoua 2008, Theriault et al. 2003, Yilmaz et al. 2011).

In copper ore mines in Poland there is space to be filled with tailings. The estimated volume of voids in the goafs amounts to respectively: O/ZG "Lubin"– 2.2 million m^3, O/ZG "Rudna"- 3.9 million m^3, O/ZG, "Polkowice – Sieroszowice" – 1.9 million m^3, the total capacity of the goafs is approximately 8 million m^3 (Mazurkiewicz et al. 2015).

The technological process of producing paste requires reducing the amount of water in the tailings, which generates significant costs (Gruszczyński 2019). The paper presents the results of thickening using the thickener on a semi-technical scale. The flocculant was used in the thickening process, which may affect the geotechnical parameters of the generated underflow (ICOLD, 2001). During the conducted experiments, the obtained thickened tailings were deposited in specially prepared experimental fields so that the flow angles could be assessed.

2 PURPOSE AND CONDUCT OF THE RESEARCH

The aim of the study was to determine the impact of the flocculant on the course of the process of thickening, depositing and segregating of tailings (Palkovits 2011). The Mobile Thickening Unit (MTU) (Figure 1) was used for the tests. While preparing the experiments, the prototype equipment destined for the MTU was designed and built. The research was carried out for tailings from copper Ore Enrichment Plant (OEP). In view of the anticipated measurement difficulties, the MTU was located on a wheeled platform (Figure 1).

Thickening tests were carried out without and with the addition of the flocculant for 5 measurement series. The parameters of feed, underflow and overflow from the MTU were determined and the process of depositing on the Tailing Storage Facility (TSF) beach was assessed. During the tests the MTU operating parameters were monitored "on-line" using the installed equipment. Samples of the deposited underflow from the beach were collected to determine segregation as a function of distance from the discharge point. Tailings beach slope, deposition process, erosion and accumulation areas where determined using the laser scanning technique and surface numerical model. During the tests, samples of the mixture were taken and their physical properties were determined. The impact of the flocculant on the segregation process of the deposited material on the beach was indicated as well as longitudinal slopes of the deposited material. It was observed that the addition of the flocculant significantly changes the regime of tailings flow on the beach.

3 MOBILE THICKENING UNIT

The main element of the installation was a cone thickener DN900. The thickener had the height of 100 cm (Figure 2). The thickener was equipped with: feed stream confusor, rake arm, overflow weir, underflow stream diffuser. The thickener was additionally equipped with a sight glass enabling observation of the flocculation process and bed height. All the elements of the device in contact with the feed were made of stainless steel. Figure 2 presents the technological scheme of the MTU.

Due to the expected distribution of density and grain size of the solid phase in horizontal distributing pipelines, it was decided to supply the thickening station with 3 spigots of different heights. The spigots were installed on section N3 of Φ800 distributing pipeline carrying tailings of copper ore flotation from OEP. They were installed at three heights in relation to the axis of the pipeline, i.e. above the axis, in the axis and below the axis. Each spigot was equipped with a DN100 gate valve with a Perot type quick coupler.

Figure 1.   Mobile Thickening Unit during measurements.
Source: Own source.

The installation was equipped with two U-bend density meters. The density meters were made in the form of loops from pipes DN63 SDR17 PEHD100 (Figure 2). The density meters consisted of two measuring sections 2.27 m long located vertically. In both measuring sections, the medium flowed in opposite directions, i.e. up and down. Pressure differences were measured on each measuring section using an Aplisens electronic pressure transmitter. Impulse tubes supplying pressure to the measuring membrane were rinsed and fully filled with tap water. The accuracy of the pressure difference measurement was 0.1% of the full measuring range of the device, i.e. 0.025 kPa. Before testing, the density meters were checked and calibrated on fresh water for the full range of flow rate measurement, i.e. from 0 to 10 $m^3 \cdot h^{-1}$. Impulse tubes were made of transparent polyurethane DN10, which allowed for ongoing assessment of patency, purity and air in the pressure difference measurement installation.

The installation was equipped with two electromagnetic flow meters dedicated to slurry flow from Techmag DN 50 and DN 32. The first flow meter was installed on the feed line, between the density meter and the supply pump and the other on the underflow line. The flow meters were equipped with an electronic display showing the current flow and the sum of the flow. The electromagnetic flow meters used were not sensitive to possible stationary bed in the measuring head cross section (Baker 2000).

The installations were equipped with sampling points for laboratory tests. The points were located on vertical sections of pipelines and equipped with DN 6.5 ball valves. This solution allowed the collection of the most reliable samples of tailings from three streams, i.e. from the feed, underflow and from overflow. The samples were sealed in plastic containers after collection. The secured samples were sent to the geotechnical laboratory. Particle size distribution measurements were made with a Mastersizer 2000 instrument.

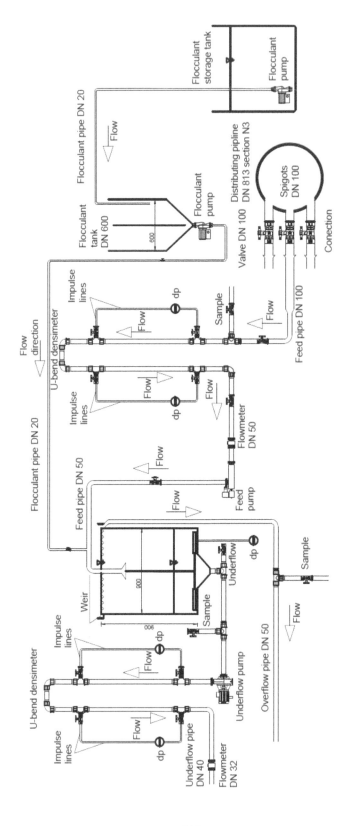

Figure 2. Technological scheme of the Mobile Thickening Unit.
Source: Own study.

The AN 820VHM flocculant was prepared in two phases. In the first phase, the flocculant was prepared at a 1 % concentration in a 100-liter plastic container. According to the manufacturer's instructions, the flocculant matured for 48 h. Then the flocculant was diluted to a concentration of 0.1 %. The flocculant prepared in this way was pumped from the container to the 280 dm^3 flocculant tank with which the thickener was equipped. It was assumed that the initial dose of the flocculant would be 50 g·Mg^{-1} of the solid phase of the feed stream. As the actual solid phase mass flow was determined only after laboratory tests, the solid phase amount was assumed at 2.0 Mg·h^{-1} on the basis of those preliminary tests. Thus the initial flocculant amount was assumed to be 100 g·h^{-1}. During the tests, the flocculant dose was being increased until clean water was obtained in the overflow.

Thickened tailings were pumped to the previously prepared test stands to assess beach slope and particle size distribution along the deposited material. Seven measuring fields were prepared in the form of trapezoidal channels (ditches) from R1 to R7 (Figure 3.), 150 m long and 0.30 m wide at the bottom.

In order to assess the sedimentation processes in the prepared fields, the technique of laser scanning and mathematical modelling of the terrain surface was used. After filling the experimental ditches, laser scans of the study area were made. In order to achieve high accuracy the scanning operation was performed three times.

4 RESULTS OF THE CONDUCTED EXPERIMENTS

The research was carried out in difficult field conditions in the TSF. Supplying the MTU from the tailings transfer installation enabled the station to operate in conditions as close as possible to a real installation. Figure 4a shows the efficiency of individual streams entering and leaving the MTU. Figure 4b presents the measured densities of feed, underflow and overflow of the MTU.

Figures 4a and 4b show four MTU modes. The start-up mode lasted from 13:25 to 13:50. The MTU worked without the addition of the flocculant from 13:50 to 15:10. The flocculant dosing started at 15:10 and lasted for an hour to 16:10. Turning off the MTU lasted from 16:10 to 16:40.

Figure 3.   Photography and scan of prepared measuring channels.
Source: Own study

147

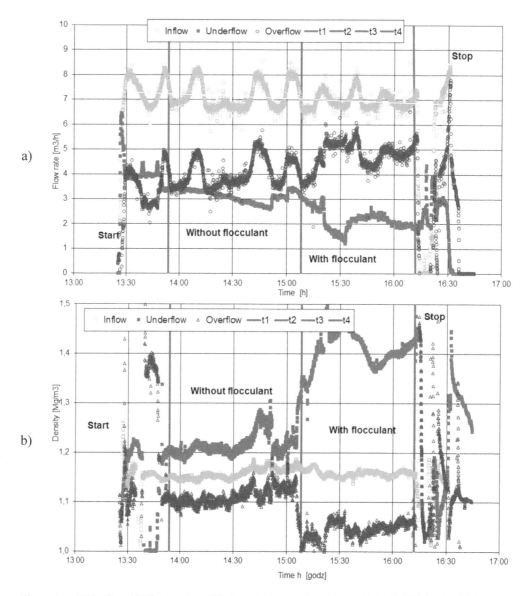

Figure 4. a) The flow- b) The density- of feed, underflow and overflow of Mobile Thickening Unit.
Source: Own study.

The presented data refer to the case with the use of a flocculant dose 50 g·Mg⁻¹. It can be observed that in the presented case the full purity of the overflow was not achieved.

During the tests, samples were taken from the feed stream, overflow, underflow and from the experimental ditches in which the thickened tailings were deposited. The grain size distribution of the samples taken was determined. The measurements can be divided into two groups, i.e. without flocculant and with flocculant. Figure 5a shows the particle size distributions of individual streams and tailings deposited in the experimental channels for the case of the MTU working without the flocculant. In the thickener, the coarse particles whose grain size indicator d_{50} is 70 μm are segregated and flow in underflow. Fine fraction with the grain size indicator d_{50} 10 μm flows to overflow. Feed flow d_{50} was 20 μm and for tailings disposited in the dich was 100 μm.

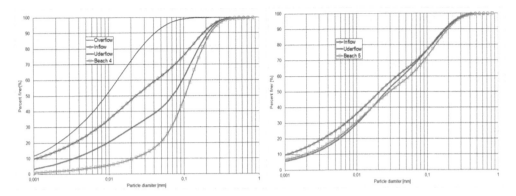

Figure 5. Average grain size composition in feed, underflow, overflow and in experimental channel; a) for the tests without flocculant; b) tests with flocculant.

Source: Own study.

Another test discussed in the work is the use of the flocculant in the process of thickening tailings in the MTU. When the MTU was operating with the flocculant, clean water was achieved in the overflow in which no solid particles were observed (Figure 5b). Grain size indicator d_{50} of tailings in feed, underflow and experimental channel is almost the same and amounts to 20 µm.

After filling the experimental channels, the slope was assessed. Before starting the experiment, the shape of the entire deposition area and of the ditches was measured using laser scanning techniques. Figure 6 presents a numerical model of the deposition area generated on the basis of laser scanner measurements. Comparison of numerical terrain models was used to determine the shape of the deposited thickened tailings.

Table 1 presents the deposited underflow length and parameters for all seven ditches. In three cases, the MTU worked without the flocculant. In four cases, the tailings going to the ditches were thickened with the addition of the flocculant. In three cases, clean water was

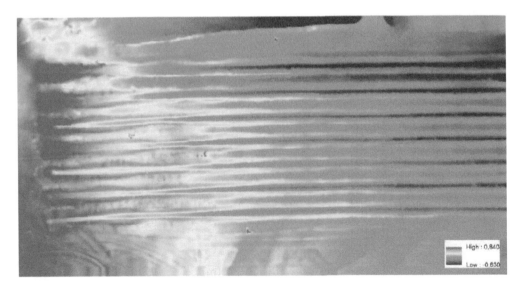

Figure 6. Numerical Terrain Model of the measuring ditch region.

Source: Own study.

Table 1. Types of deposits in experimental ditches.

Experimental ditch	Addiction of flocculant	Overflow transparency	Deposit length [m]
R1	No	No	18
R2	Yes	No	16
R3	Yes	Yes	7
R4	No	No	16
R5	Yes	Yes	40
R6	No	No	12
R7	Yes	Yes	35

Source: Own study

achieved on the overflow. The maximum measured length of deposited overflow was 40 meters, while the minimum length was 7 meters.

Figure 7 shows the longitudinal profile of ditches R1 to R7. Longitudinal profiles were made for all the ditches on the basis of numerical model assessment. The measurements showed the influence of distance from the discharge point on the slope of disposed material. For measuring series in which no flocculant was used, the slopes of the beach in the initial section were from 3 to 6%, and with the increase in the distance they decreased to 2-3 % (R1, R4 and R6). The average slope of deposited underflow ranged from 2.6 to 3.6 %. For the experiments in which the flocculant was used, the slopes were much lower. At the beginning the slope was within a range of 2.4-3.5 %, and at a further distance about 1%. The average slope of deposited underflow ranged from 1,1 to 3,6 %.

Parameter coefficient of uniformity is defined as (1):

$$c_u = \frac{d_{60}}{d_{10}} \tag{1}$$

Figure 8 shows the grain size indicator coefficient of uniformity c_u distribution along the length of the ditches. The parameter distribution in ditches R3 R5 and R7 oscillates within the

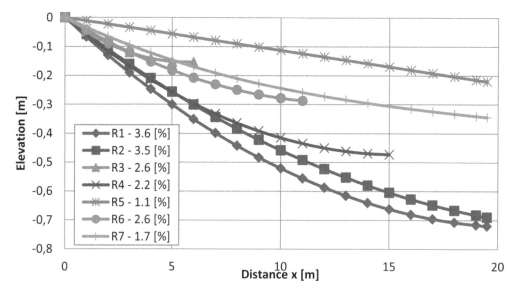

Figure 7. Slope of ditches R1 to R7.
Source: Own study.

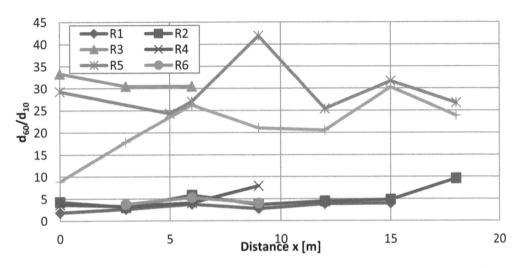

Figure 8.   Dependence of the grain size indicator – coefficient of uniformity d_{60}/d_{10} on the distance from the discharge point. Ditches R1 – R7.

Source: Own study.

value of 25. The parameter is more or less horizontally arranged. The discussed case is the use of the flocculant in a dose ensuring a clean overflow. In other cases, the parameter d_{60}/d_{10} is almost horizontal and does not deviate from the value of 5. A strong dependence between the coefficient of uniformity d_{60}/d_{10} and the flocculant use is observed.

5   CONCLUSIONS

– The proposed measurement methodology has received good practical confirmation. This applies both to the apparatus installed on the Mobile Thickening Unit and also to the depositing process evaluation,
– Underflow from the thickener with the flocculent applied can be used as a material for the preparation of backfill paste or for sealing goafs,
– As a result of the observation of the process of depositing tailings with the flocculant applied, it was found that the tested slurry, in almost the entire length, flows with the movable bottom sediment regime. The basic type of particle displacement on the beach is slow dragging and rolling. The suspended fine particles are dust and clay fraction. Studies have shown that in the place where the mixture stream falls on the layers of embedded soil, erosion funnel is formed,
– The depositing process for which the flocculant was used in the MTU was different. The slurry with the flocculant moved homogeneously and not heterogeneously,
– For the experiments in which the flocculant was used, the tailings deposited on the beach were practically not segregated, whatever the distance from the place of discharge, there was no reduction in granulation or reduction of the sand fraction content,
– Macroscopic observations have shown that the deposited material with the flocculant has greater porosity. Dewatering process occurs much more slowly. This may mean the possibility of liquefying deposited tailings under the influence of rockburst as well as much larger volumes of water retained compared to the current state.

BIBLIOGRAPHY

Annor, A. B. 1999. *A Study of the Characteristics and Behaviour of Composite Bacldill Material*, McGill University Department of Mining and Metallurgical Engineering, Thesis for degree of Doctor of Philosophy.

Baker, R. C. 2000 *Flow Measurment Handbook*, Cambridge University Press.

Belem, T. and Benzaazoua, M. 2008. *Design and application of underground mine paste backfill technology*, Geotechnical and Geological Engineering., vol. 26, no. 2, pp. 147–174.

Fehrsen, M. and Cooke R. 2010. *Paste fill pipeline distribution systems - current status*, pp. 1–13.

Gruszczyński, M. 2019. The assessment of the impact of the rheological parameters of fine-grained copper ore flotation tailings on friction head losses in pressure pipes. Wrocław University of Environmental and Life Sciences

ICOLD, 2001. *Tailing Dams risk of dangerous occurrences*, International Commission on Large Dams, vol. XXXIII, no. 2, p. 144.

Lee et al. 2017 – Lee, J. K., Ko, J. and Kim, Y. S. *Rheology of Fly Ash Mixed Tailings Slurries and Applicability of Prediction Models*, vol. 7, no. 9, p. 165.

Li et al. 2017 - Li, H., Zhang, J., Zhang, D, Li, Z and Hu, G. *Paste pipeline transportation for Chambishi Copper Mine*, [in:] A Wu & R Jewell (eds), Proceedings of the 20th International Seminar on Paste and Thickened Tailings, University of Science and Technology Beijing, Beijing, pp. 393–401, https://doi.org/10.36487/ACG_rep/1752_44_Li

Mazurkiewicz et al. 2015 - Mazurkiewicz, M., Popiołek, E., Niedojadło, Z., Sopata, P. and Stoch, T. *Some Aspects Of Using Goafs For Locating Post-Flotation Waste In LGOM Mines*, Archives of Mining Sciences, vol. 60, no. 4, pp. 941–954.

Palkovits, F. 2011. *Paste Thickening: Considerations for Backfill vs. Tailings Management*, Engineering and Mining Journal. no. November, pp. 34–40

Stefanek, P. and Serwicki, A. 2014. *Ograniczenie oddziaływania OUOW Żelazny Most na środowisko poprzez zmianę technologii składowania odpadów*, Warsztaty 2014 z cyklu Górnictwo – człowiek – środowisko zrównoważony rozwój, pp. 394–406.

Stone. D. 2014. *The evolution of paste for backfill*, pp. 31–38, 2014

Theriault et al. 2003 - Theriault, J. A., Frostiak, J., Welch, D. and Corporation, B. G. *Surface Disposal of Past Tailings at the Bulyanhulu Gold Mine, Tanzania*, Mining Environmental III Conf., pp. 1–8.

Yilmaz et al. 2011 – Yilmaz, E., Belem, T., Benzaazoua, M., Kesimal, A., Ercikdi, B. and Cihangir, F. *Use of high-density paste bacfill for safe disposal of copper/zinc mine tailings*, Gospodarka Surowcami Mineralnymi, vol. T. 27, z., pp. 81–94.

Minefill 2020-2021 – Hassani et al (eds)
© 2021 Taylor & Francis Group, London, ISBN 978-1-032-07203-6

Determining the required underground grout pack production profile for narrow tabular mining operations

Bernardt Van Der Spuy

Paterson and Cooke Consulting Engineers (Pty) Ltd, South Africa

SUMMARY: South African gold and platinum mines commonly utilise narrow tabular mining for ore extraction. The removal of the ore body leaves open voids with stoping widths that can exceed 2 m in some mines. Props and cementitious grout packs are used extensively as stope support systems to stabilize the rock mass of the voids and to reduce the risks associated with rockfalls and rockbursts. The construction of grout packs for support follows the stope face advance and mining production is inter alia dependant on the production profile of the grout packs. The grout pack production system of a mine generally consists of a surface mixing plant and a reticulation system to the underground workings. A cementitious grout slurry is prepared at the mixing plant and is pumped to the underground stopes for filling of grout packs positioned between the hanging wall and footwall. Construction of grout packs in stopes is labour intensive and requires planning to ensure that the grout pack production profile meets the mining production profile. To determine the planned monthly grout volumes of a mining operation a variety of parameters need to be considered. These parameters should be accurately estimated at the start of a production month and are based inter alia on historical data and the mining plan of the operation. This paper presents a method to assist mining staff in estimating the required grout pack production profile, determining the required capacity of grout plants and reconciliation of the actual and planned grout pack production profile for a production month.

Keywords: backfill, mining, grout, optimisation, cemented hydraulic fill

1 INTRODUCTION

South African gold and platinum mines commonly utilise narrow tabular mining for ore extraction. The removal of the ore body leaves open voids with stoping widths of some mines in excess of 2 m. Props and cementitious grout packs are used extensively as stope support systems to stabilize the rock mass of the voids and to reduce the risks associated with rockfalls and rockbursts (Daehnke, van Zyl & Roberts, 2001).

The construction of grout packs for support follows the panel face advance and mining production is inter alia dependant on the production profile of the grout packs. Grout plants utilized in the South African mines generally consist of a batch mixing system with a continuous feed to underground operations.

In order to estimate the required grout plant/s production capacity for a mining operation a variety of parameters need to be considered. These parameters should be accurately estimated at the start of a production month and are based inter alia on historical data and the mining plan of the operation.

DOI: 10.1201/9781003205906-15

The actual grout plant/s production capacity of a mining operation is determined by the possible grout flow rate that can be delivered to the underground operation. If the required volumes exceed the grout plant production capacity, the grout pack production profile will not be able to satisfy mining requirements. This paper presents a methodology to determine the required and actual plant/s production capacity in order to identify possible bottlenecks of the system and provides possible solutions to increase the capacity of the grout system.

2 GROUT PACK FILL VOLUMES

The pack types used to support the hanging wall depend on a variety of factors including the required in situ load bearing characteristics of the packs to support the hanging wall, in-stope space constraints and the stope standard.

Packs can have a square (see Figure 1) or circular (see Figure 2) plan area. Generally, square packs contain a centrally mounted wooden support pole while circular packs are reinforced and confined by placing the pack in a steel cage with support poles before filling and pressurizing the pack.

The grout volume required to fill a pack is determined by the volume of the bag used to contain the grout during filling. After the first fill of the pack is complete it is pressurized (commonly referred to as "stressing") by adding additional material to ensure a bond is created between the hanging and foot wall of the stope. This pressurization of the pack causes an increase in filling volume as the pack bulges, material consolidates and water seeps through the bag material.

2.1 Theoretical fill volume (non-pressurized pack/non-stressed pack)

A non-stressed, theoretical fill volume of a pack can be determined based on the pack plan area and the stoping width of a panel. This is the volume of the bag used to contain the grout material during filling. Panels mined in South African tabular mines generally have stoping widths of between 1.0 m and 1.8 m. However, the development of gulleys and raises around

Figure 1. Square grout pack.
Source: own study

154

Figure 2. Circular grout pack with steel reinforced cage.
Source: own study

panels may require the installation of ledging packs on the panel edges which can have stoping widths exceeding 2 m. The theoretical fill volume of a circular pack is determined as follows:

$$V_{ns-r} = \frac{\pi \times D^2}{4} \times H \qquad (1)$$

where:
V_{ns-r} = non-stressed fill volume of round pack (m³)
D = outside diameter of pack (m)
H = stoping width (m)
The theoretical fill volume of a square pack is determined as follows:

$$V_{ns-s} = (L^2 - \left(\frac{\pi \times d^2}{4}\right)) \times H \qquad (2)$$

where:
V_{ns-s} = non-stressed fill volume of square pack (m³)
L = plan width of pack (m)
d = outside diameter of wooden pole (m)
H = stoping width (m)

2.2 *Theoretical fill volume (pressurized pack/stressed pack)*

The theoretical, pressurized/stressed fill volume of a pack is dependent on the following factors:

– Theoretical fill volume (Non-pressurized pack).
– Bulging of the pack during filling and pressurization.
– Seepage and consolidation of the grout material during filling and stressing.

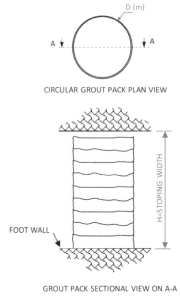

CIRCULAR GROUT PACK PLAN VIEW

GROUT PACK SECTIONAL VIEW ON A-A

Figure 3.   Theoretical fill volume - unpressurized circular grout pack with steel reinforced cage.
Source: own study

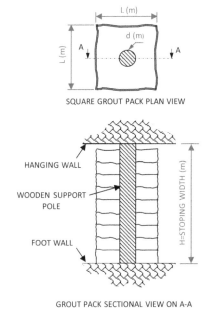

SQUARE GROUT PACK PLAN VIEW

GROUT PACK SECTIONAL VIEW ON A-A

Figure 4.   Theoretical fill volume – unpressurized square grout pack with wooden support pole.
Source: own study

The theoretical fill volume of a stressed pack is calculated as follows:

$$V_s = V_{ns} + V_b + V_{sc} \qquad (3)$$

where:
V_s = stressed fill volume of round or square pack (m³)
V_{ns} = non-pressurized fill volume of round or square pack (m³) – Refer to Section 2
V_b = additional fill volume due to bulging (m³)
V_{sc} = additional fill volume due to seepage from pack and consolidation of grout
material (m³)

2.2.1 *Additional fill volume due to bulging (V_b)*
During pack filling, the grout material introduced into the grout pack causes an internal pressure to be applied on the pack wall segments. This internal pressure causes the woven pack material to stretch, resulting in a "bulging" effect and the whole pack experiences an increase in filling volume.

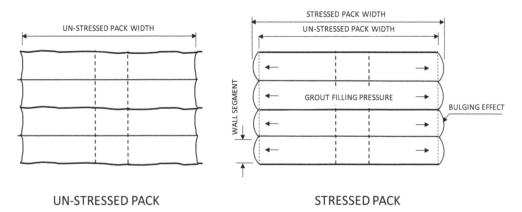

Figure 5. "Bulging" effect on a single square bag during pack filling and stressing.
Source: own study

2.2.2 *Additional fill volume due to seepage from pack and consolidation of material (V_{sc})*
Consolidation is a time-related process of dewatering the grout and increasing the density by seepage of interstitial water from voids between particles. As grout material is introduced into packs, the internal pressure induced causes water and fine particles in the grout mixture to seep through the woven grout pack.

The lost water and fines causes the consolidated density of the grout material in the packs to be higher than the pumped grout mixture density which in turn causes an increase in the pumping volume required to fill each pack.

The consolidation and seepage effect of the grout material is based on the mass concentration of grout material being introduced into the pack as well as the maximum bed packing concentration of the material. Material property test work is required to accurately calculate the increased grout volume required due this effect.

3 STOPE STANDARD

When constructing in-stope supports for the hanging wall, grout packs are spaced at fixed intervals to limit spans and provide the hanging wall support required to ensure a safe working environment for underground operations. The mining operation will generally have a stope standard (or distribution profile) to dictate the spacing of packs. The stope standard is generally determined by the mining operation's rock engineers and is dependent on the grout pack sizes, mining method employed, required accessibility for effective sweeping of panels and the ground conditions. As mining reaches deeper levels and ground conditions deteriorate, additional stope standards may be adopted.

Figure 6 presents a typical stope standard to be used by grout pack preparation crews for a breast mining operation. Crews will distribute the empty bags and relevant reinforcement at the specified intervals before pack filling commences.

In order to provide access space between the last row of packs and the mining face for mucking and crew access while still providing sufficient support for the hanging wall, the stoping standard requires a maximum first pillar row spacing of 'e' m from the mining face. Ground conditions require in-stope packs to be spaced 'b' m apart on strike and 'a' m apart on dip.

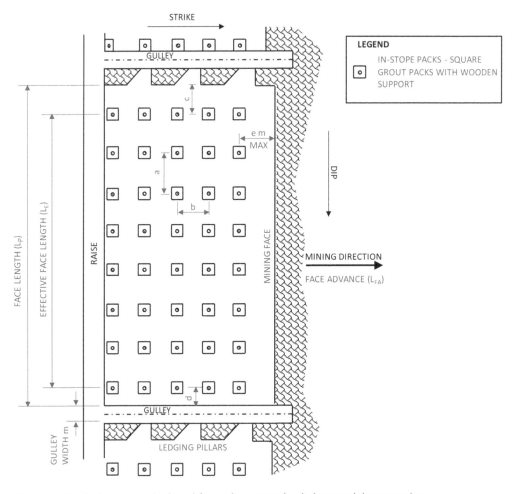

Figure 6.   Typical stope standard used for pack construction in breast mining operation.
Source: own study

4   TOTAL VOLUME OF GROUT REQUIRED PER MONTH

The total volume of grout required per month is based on various mine production parameters presented in the following section.

4.1   *Mine production parameters*

4.1.1   *Total stoping area to be mined for month (A_T)*
The total planned underground stoping area to be mined per month for the mine is projected at the start of a production month and is determined by the mining plan. Typically expressed in m²/month.

4.1.2   *Average panel face length (L_P)*
The panel face lengths of underground panels are measured by surveying at the end of a production month. A planned, average panel face length can be estimated at the start of a production month based on historical surveyed data. Typically expressed in metres.

4.1.3   *Average stoping face advance (L_{FA})*
The face advance of a mine is dependent on a variety of factors including the blasting schedule, stoping method and production efficiency of the mine. A planned, average stoping face advance can be estimated at the start of a production month based on historical data and the mining plan. Typically expressed in metres/month.

4.1.4   *Average stoping mining area per panel per month (A_P)*
An average planned mined area per panel/month is based on the planned, average stoping face advance as well as the planned, average panel face length:

$$A_P = L_P \times L_{FA} \tag{4}$$

where:
A_P = average stoping mining area per panel per month (m²/month)
L_P = average panel face length (m)
L_{FA} = average stoping face advance (m/month)

4.1.5   *Number of panels mined per month (N_P)*
The planned number of panels to be mined per month is based on the planned, total stoping area to be mined for month and the calculated average mining area per panel/month:

$$N_P = \frac{A_T}{A_P} \tag{5}$$

where:
N_P = number of panels mined per month
A_T = total stoping area to be mined for month (m²/month)
A_P = average stoping mining area per panel/month (m²/month)

4.2   *Total volume of grout required to satisfy planned mining parameters*

The mine production parameters (discussed in section 4.1) and the stope standard (discussed in section 3) are used to determine the required number of grout packs for the month. The required number of grout packs are subsequently used in conjunction with the grout pack fill volumes (discussed in section 2) to estimate the total volume of grout required for the production month.

4.2.1 *Required number of grout packs for month (N_{GP-TOT})*

The required pack spacing on the stope standard is used in conjunction with the number of panels mined per month (N_P) to determine the planned number of grout packs to be filled per month.

The number of packs required on dip is determined by dividing the effective face length by the pack spacing on dip:

$$N_{GP-DIP} = \frac{L_E}{a} + 1 \qquad (6)$$

where:

N_{GP-DIP} = number of packs required per panel on dip
L_E = effective face length (m)
a = pack spacing on dip (m)

The effective face length can be described as the total length between the first pack on dip and the last pack on dip. The effective face length is calculated from the average panel face length (L_P) and the stope standard. The effective face length for the stope standard presented is calculated as follows:

$$L_E = L_P - c - d \qquad (7)$$

where:

L_E = effective face length (m)
c = first pack spacing from ledging pillar in panel (m)
d = last pack spacing from gulley edge (m)

The number of packs rows required on strike is determined from the average stoping face advance for the month (L_{FA}) and the pack spacing on strike as determined by the stope standard:

$$N_{GP-STRIKE} = \frac{L_{FA}}{b} \qquad (8)$$

where:

$N_{GP-STRIKE}$ = number of packs required per panel on strike
L_{FA} = average stoping face advance (m/month)
b = pack spacing on strike (m)

The number of required packs per panel is then determined as follows:

$$N_{GP-PANEL} = N_{GP-DIP} \times N_{GP-STRIKE} \qquad (9)$$

where:

$N_{GP-PANEL}$ = number of packs required per panel
$N_{GP-STRIKE}$ = number of packs required per panel on strike
N_{GP-DIP} = number of packs required per panel on dip

Finally, the total number of required grout packs for the month is determined from the number of panels mined for the month:

$$N_{GP-TOT} = N_{GP-PANEL} \times N_P \qquad (10)$$

where:

N_{GP-TOT} = required number of packs for month
$N_{GP-PANEL}$ = number of packs required per panel
N_P = number of panels mined per month

4.2.2 *Total volume of grout required for month (V_G)*

The planned, total volume of grout required per month is determined from the number of required packs for the month and the theoretical pressurized pack volume discussed in Section 2.

$$V_G = N_{GP-TOTAL} \times V_S \tag{11}$$

where:
V_G = total volume of grout required for month
N_{GP-TOT} = required number of grout packs for month
V_S = pressurized volume of round or square pack (m³)

5 REQUIRED SURFACE GROUT PLANT OUTPUT CAPACITY

After the total volume of grout required for month is determined it can be related back to a required hourly grout flow rate to satisfy the planned mine production parameters. The flow rate is dependent on the possible daily crew production and the grout plant utilisation as discussed in the following sections.

5.1 *Possible daily crew production*

A production day on a South African platinum mining operation can typically be divided into three shifts:

– 1 x 8 hour blasting and cleaning shift
– 2 x 8 hour grout filling shifts including preparatory work such as drill, sweep, stope support, services installation etc.

5.1.1 *Blasting and cleaning shifts*

Only blasting crews are present in stopes to prepare panels for blasting during the first part of the shift. After blasting is completed, panel cleaning crews muck blasted ore into gulleys and raises for transportation to surface. During the blasting and cleaning shift no pack filling can be done as the safety of filling crews are of concern.

5.1.2 *Grout filling shifts*

During the remaining two production shifts in the mining operation, packs are constructed and filled in panels. Panels that will be mined in the following blasting and cleaning shift take preference to satisfy the required mining production profile of the mine. Backlogged packs are filled after the panels to be blasted are prepared.

A typical 8-hour pack filling shift consists of a number of required activities in addition to the actual filling of packs. The actual time available for the filling of packs during a shift is influenced by these additional daily shift activities which can be planned, 'normal shift activities' or unforeseen. These additional activities reduce the time available for pack construction and filling during a shift and consequently influences pack production.

In order to maximise the pack production of the mine, the time allocated to filling of packs should be kept at a maximum by optimising the time spent on other activities.

Planned, 'normal shift activities': Table 1 presents the activities that are identified as 'normal shift activities' for grout pack production crews. The activities listed will vary from mine to mine and the table provides typical items witnessed by the author.

Table 1. Typical planned, 'normal shift activities' for grout pack production crews.

Activity	Description
Travel	**Start of shift**: The time required from entering shaft cage on surface and travel to the underground safety meeting locations and workplaces. **During shift**: The time required to travel between different workplaces where grout production is required. **End of shift**: The time required for travel from the workplace back to shaft cage and finally up to surface.
Meetings	Grouting crews may be required to attend daily safety and planning meetings prior to commencing work.
Panel inspection	A safety inspection of the panel by a designated safety representative is conducted before any worker is allowed to enter the panel. If the ground conditions in the panel are deemed to be unsafe, the installation of temporary supports is required before any work may be conducted in the relevant panel.
Grouting preparation	This is the time allocated to the preparation of a workplace before grout pumping commences. The grouting preparation includes the preparation of filling hoses and may include the construction of packs to be filled.
Pack filling	Time available for grout pumping during the grouting shift. During this time the actual filling of packs takes place.
Line flushing/ filling	Grout pipelines are flushed with mine water before and after grouting to prevent the setting of material in pipelines. At the start of a grouting shift time is also allocated to the filling of pipelines with grout material from surface, through the relay station and to the underground discharge point.
Relay station cleaning	The relay stations are flushed and cleaned after each shift.

Source: own study

Figure 7 presents a typical timesheet of a pack filling crew during an 8-hour shift. The following should be noted:

– The timesheet assumes that the pack filling operation utilises relay stations for the grout system.
– The table makes provision for one grouting crew to consecutively fill packs in two panels by allowing additional time spent on travel and grouting preparation activities between panels.
– Packs were constructed in the preceding shift and are ready for filling upon arrival of grouting crews.
– The timesheet excludes unforeseen activities which should not form part of a normal pack filling shift.

Based on the timesheet a total of 3 h 30 min was available for pack filling during the 8-hour shift. Historical grout plant audits conducted by P&C suggests that this available time can vary between 1h 30 min and 4h 00 min.

Unforeseen shift activities: A number of 'unforeseen shift activities' that should not form part of the 'normal shift activities' were identified during historical grout plant audits conducted by P&C. These activities further reduce the available time for pack filling and consequently influences pack production. Some of the identified activities cannot be avoided and provision should be made during planning stages to address and mitigate the time spent on the relevant activity.

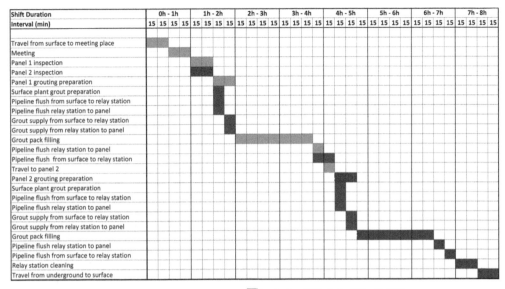

Grouting Crew Activities for Panel 1 Grout Pack Filling
Grouting Crew Activities for Panel 2 Grout Pack Filling
Surface Plant Crew Activities

Figure 7. Typical time sheet of pack filling crew over 8 hour shift.
Source: own study

Table 2 presents the 'unforeseen shift activities' that were identified during previous studies. Previous audits have shown that the time spent on unforeseen activities can significantly reduce the time available for pack filling and even result in a loss of a complete grouting shift. All unforeseen activities should be mitigated during planning stages in order to maximise the pack production of the mining operation.

5.2 Grout plant utilisation

5.2.1 Number of available production days per month ($N_{PRODUCTION}$)
The following factors influence the number of available production days:

– Annual shutdown for maintenance
– Work schedule of grouting crews

The average number of available production days per month can be calculated as follows:

$$N_{PRODUCTION} = \frac{(365 - N_{MAINTENANCE}) \times \eta}{12} \qquad (12)$$

where:
$N_{PRODUCTION}$ = number of available grout pack production days per month
$N_{MAINTENANCE}$ = number of scheduled shut-down days per annum for maintenance, etc.
η = availability factor of grout plant and underground staff
Platinum mining operations in South Africa generally work on an 11-days per fortnight schedule. In this case an availability factor of 78.6% (11/14) will be applied.

Table 2. Typical planned, 'unforeseen shift activities' for grout pack production crews.

Activity	Description
Additional traveling time	Time spent on traveling in addition to the normal travel activities listed in Table 1.
Grouting preparations that should not form of "normal" grouting preparation	Operations where packs are prepared by grouting staff require the grout materials (bags, jackpots and wooden poles) to be present upon arrival of the crew. This may not always have been completed during preceding shifts due to a variety of reasons. The fetching of these materials should not form part of the 'normal' grouting shift. Operations where packs are prepared by mining staff require this activity to be completed in preceding shifts but may not always be completed due to a variety of reasons. These grouting preparations that should not form part of the 'normal' grouting preparations includes transporting of pack support materials, measuring the required length of the wooden pole used to provide initial bag support before filling, mounting of packs, wooden pole and hydraulic jackpod and pressurizing of the jackpod.
Non- or malfunctioning equipment	Non- or malfunctioning equipment used for pack construction and filling requires pack filling crews to repair or replace the item during the shift.
Absence or insufficient number of prepared packs	The absence or insufficient number of prepared packs to be pumped upon arrival in a panel require crews spent additional time on travel to another panel where packs may be prepared or even lose an entire shift.
Panel cleaning	In order to ensure adequate support of panel hanging wall, panel areas are cleaned of blasted rock and debris prior to the construction of packs. The cleaning of panels is usually done during the preceding blasting shift. If a panel is not cleaned upon the arrival of grouting crews, packs cannot be installed. Uncleaned panels can cause a reduction in time available for filling or even the loss of an entire shift.
Installing temporary supports	Panel inspections may require the installation of temporary supports if ground conditions are unsafe.
Grout piping problems	Grout piping problems can include blocked pipes, grout pipes bursting, etc.
Relay station problems	Relay station problems can include blockages, equipment failures, etc.
Additional meetings/discussions	Additional time spent on meetings or discussions when required.
Unavoidable mining activities	Mining activities occurring in conjunction with the grouting shift can also influence the time available for pack filling. These mining activities includes injuries, mucking of raises and gulleys during grouting shifts and skips receiving material from nearby orepasses are not emptied before grouting shift begins. (Grout pack filling cannot occur if a skip at a nearby orepass is full as this can cause the grout material to run into the skip and setting).

Source: own study

164

5.2.2 Lost production days per month ($N_{LOST\ DAYS}$)

A number of factors can cause the available grout pack filling production days per month to be "lost" and consequently decrease the utilisation of the filling operation, namely:

– Labour (staff strikes, staff shortages, etc.)
– Unscheduled maintenance
– Mining stoppages (accidents, etc.)

A direct result of these "lost days" is an increase in the required grout pack production profile for the remaining production days.

5.2.3 Effective number production days per month ($N_{EFFECTIVE}$)

The effective number of production days available per month is the determined as follows:

$$N_{EFFECTIVE} = N_{PRODUCTION} - N_{LOST\ DAYS} \qquad (13)$$

where:
$N_{EFFECTIVE}$ = effective number of available grout pack production days per month
$N_{PRODUCTION}$ = number of available grout pack production days per month.
$N_{LOST\ DAYS}$ = number of un-scheduled lost days per month
Allowance for a number of lost production days should made during planning stages to ensure that the filling operation will still meet the required production profile of the mine.

5.3 Required grout plant/s output capacity

Based on the total volume of grout required as discussed in section 4, the time available for pack filling during the day- and nightshifts and the effective number of production days available for the month, the required grout plant output flow rate can be calculated as follows:

$$Q_{REQUIRED} = \frac{V_G}{N_{EFFECTIVE} \times (h_{DAYSHIFT} + h_{NIGHTSHIFT})} \qquad (14)$$

where:
$Q_{REQUIRED}$ = required grout flow rate to underground operations (m³/h)
V_G = total volume of grout required for month (m³)
$N_{EFFECTIVE}$ = effective number of available grout pack production days per month
$h_{DAYSHIFT}$ = hours available for pack filling during dayshift
$h_{NIGHTSHIFT}$ = hours available for pack filling during nightshift
The surface grout plant/s actual output capacity discussed in section 6 should exceed or, as a minimum requirement, match the surface grout plant/s required output capacity ($Q_{REQUIRED}$) to satisfy the production profile of the mine.

It is crucial that mine staff understands the hours available during a shift for pack filling. Grout plants should be sized to provide the required instantaneous flow rate to the underground operations. Increases in mining production will increase the required grout plant/s output capacity and sufficient provision should be made during planning stages to address planned future increases in the production profile pf the mine.

6 ACTUAL GROUT PLANT/S PRODUCTION CAPACITY

Grout plants utilized in the South African mines generally consist of a batch mixing system with a continuous feed to underground operations. Tailings, fly ash, cement and water is blended in a mixing tank to produce a grout material at a design mixture density. This grout material is then fed to a range of feed pumps which deliver the material to underground operations via a number of grout ranges.

The maximum grout output flow rate that can be delivered from surface to underground is determined by the lowest capacity of the following:

– Plant mixing system
– Grout plant pumps
– Surface grout piping
– Shaft grout piping

The surface grout plant/s actual output capacity is determined by the plant mixing system.

6.1 Plant mixing system output flow rate

The capacities of the surface grout plants' individual batch mixing systems determine the maximum volume of grout that can be produced per hour. Each grout blend component is fed consecutively into the mixing tank at a specific mass flow rate, after which the blend is mixed and transferred to grout storage tanks. The total mass flow rate of the various grout blend components in conjunction with the blend mixing and transfer time determine the continuous output flow rate of the mixing system.

$$Q_{AVAILABLE} = \left(\frac{m_t}{\rho_t} + \frac{m_f}{\rho_f} + \frac{m_c}{\rho_c} + \frac{m_w}{\rho_w}\right) \times \frac{3600}{(t_t + t_f + t_c + t_w + t_m)} \tag{15}$$

where:
$Q_{AVAILABLE}$ = available grout flow rate to underground operations (m³/h)
m = mass added to mixing tank per batch (kg/batch)
ρ = density of each component into the mixing tank (kg/m³)
t = feed time of each component into the mixing tank (s/batch)
t_m = the time dedicated to mixing the components and transferring the blend into the grout pumps feed tank (s/batch)
Subscripts:
t = tailings
f = usually fly ash, but can be any supplementary cementitious material
c = cement
w = water

6.2 Grout plant pumps capacities

The combined grout pump capacities should be sufficient to deliver the total mixing system output flow rate to underground operations at the required pumping pressure.

Standby pumps are advised to ensure 100% pumping capacity during maintenance or pump failures.

6.3 Piping ranges

The quantity of grout piping ranges from the plant to the underground operations should be sufficient to ensure that 100% of the full mixing system output flow rate can always be delivered to underground operations.

Standby pipe ranges are advised to ensure 100% grout delivery capacity during pipe maintenance, failures or blockages.

7 REQUIRED VS ACTUAL GROUT PLANT/S PRODUCTION CAPACITY

After the surface grout plant/s required output capacity ($Q_{REQUIRED}$) calculated, it should be compared with the surface grout plant/s actual output capacity ($Q_{AVAILABLE}$) to ensure that the plant/s will be able to satisfy the required production profile of the mining operation.

If the required output capacity exceeds the actual output capacity of the surface grout plant/s, the actual grout pack production will not be able to satisfy the required production profile. The actual output capacity of the surface grout plant/s can be increased as discussed in the following sections.

7.1 *Increase the shift hours dedicated to pack filling*

The number of shift hours dedicated to the actual filling of packs can be increased by mitigating the time spend on unforeseen activities as well as by optimising the time spend on 'normal' shift activities. 'Normal' shift activities which can be streamlined should be identified and procedures be adopted to reduce unnecessary time spend on these activities.

7.2 *Mitigate the number of lost production days for month*

By decreasing the number of 'lost' production days per month, the surface grout plants required output capacity will decrease. 'Lost' days as a result of labour difficulties and mining stoppages are however generally difficult to predict and control. Unscheduled maintenance can be mitigated by ensuring that regular inspections and preventative maintenance is conducted on grout equipment and piping to identify possible failures in the near future.

7.3 *Overtime production*

Overtime production will increase the effective number of production days available and hence decrease the surface grout plant/s required output capacity. Overtime production however increases labour costs and should be seen as a temporary solution.

7.4 *Utilise other existing grout plants to supply grout to the operation*

If other grout plants exist in the vicinity of the mine operation they may be utilised as an additional supply to the operation. The utilisation of additional existing grout plants will require the installation or rerouting of additional grout piping. If the grouting operation utilises underground relay stations to distribute grout to panels, the relay stations may also require additional capacity to manage the increase in grout supply from the surface.

7.5 *Construction of additional grout plants to supply grout to the operation*

The construction of additional grout plants to supply grout to the mining operation will require capital investment as well as the employment of additional operations staff. Again, if the grouting operation utilises underground relay stations to distribute grout to panels, the relay stations may also require additional capacity to manage the increase in grout supply from the surface.

8 GROUT PLANT AND RETICULATION SYSTEM AUDITS

If a mining operation is experiencing difficulties relating to the grout production, it is advised that a grout plant and reticulation system audit is completed. Audits should include establishing the current grout cycle, establishing current grout reticulation to determine maximum possible grout plant throughput. The audit should then determine the required grout production profile of the mine for current and future operations.

Based on the information gathered during the audit as well as historical data analysis, possible areas of concerns can be identified. These areas are then analysed in order to make suitable recommendations to rectify the problems experienced.

9 CONCLUSION

To determine the required monthly grout volumes of a mining operation a variety of parameters needs to be considered. These parameters should be accurately estimated at the start of a production month and are based inter alia on historical data and the mining plan of the operation.

Grout plants should be sized to provide the required flow rate to the underground operations for the hours available during a shift for pack filling. The actual grout plant/s production capacity of a mining operation should exceed the required grout plant/s production capacity to satisfy mining requirements. The actual output capacity of the surface grout plant/s can be increased by optimisation of the existing system as discussed in this paper.

If a mining operation is experiencing difficulties relating to the grout production it is advised that a grout plant and reticulation system audit is completed to determine the mine requirements as well as the parameters influencing the grout pack production profile of the mine.

BIBLIOGRAPHY

Daehnke, A, van Zyl, M and Roberts, M.K.C. 2017. Review and application of stope support design criteria. *Journal of The South African Institute of Mining and Metallurgy, vol. 101, no 3,* pp. 135–164.

Rheological yield stress measurement of paste fill: New technical approaches

Maria Silva
Somincor, Lundin Mining Corporation, Portugal

Martin Hansson
Sika Tunnelling and Mining, Sweden

Matilde Costa e Silva
Cerena, Técnico, University of Lisbon, Portugal

SUMMARY: Paste backfill technology has become increasingly relevant to the mining industry, providing not only a safe way for the underground disposal of tailings but also essential advantages for ground stabilization and optimized ore recovery. Yield stress is among the most important rheological properties of paste backfill, which determines its transportability during long distances. Measurement of yield stress is a challenging and complex task, given the high number of variable factors. So far, no standard procedures and methods have been established for measuring the rheological properties of paste backfill, in particular yield stress.

This experimental work consists of the development of an accurate laboratory testing programme that will allow for the evaluation, measurement and understanding of rheological yield stress of paste fill. For this study, tailings from Zinkgruvan (Sweden) and Neves-Corvo (Portugal) mines were analysed. A series of laboratorial tests were conducted including the following test procedures: slump, flow table spread, fall cone and the vane technique (applied using a viscometer and a rheometer).

The correlations between the yield stress, measured by vane technique and other test methods, were obtained from the test results. Additionally, preliminary conclusions were drawn regarding the influence of physical properties of tailings (particle size distribution, dry content, uniformity coefficient and coefficient of gradation) on yield stress by a statistical study using multiple linear regression models.

The fall cone test has resulted in the best correlation measurement of dry content and of yield stress measurements using the viscometer and rheometer. Being a simple, inexpensive, and expedited method for paste yield stress measurements, it is considered effective for quality control and/or rapid on-site measurements of paste fill.

Keywords: Paste Fill, Yield Stress, Rheology, Fall Cone

1 INTRODUCTION

With increasing public pressure to handle mine waste material more carefully, paste backfilling has proven to be the most important creation of the last 30 years, because it significantly reduces the disposal of surface tailings, by placing them safely in underground stopes (Yilmaz and Fall, 2017). Aside from providing a safe way for storing tailings in underground mined-out stopes,

DOI: 10.1201/9781003205906-16

paste backfill also has essential advantages for ground stabilization, such as: providing a safer work environment for miners, working platform, and enabling the exploitation of adjacent stopes (increasing ore recovery).

The major challenges associated with this increasingly popular technology in underground mines are related to the massive consumption of binder and its associated costs, including the efficient/reliable transport of freshly mixed paste to the mined-out stopes. In order to optimize the paste mix design, a maximized solid content is desired while maintaining a pumpable consistency. However, high solid mixes often create increased densities during transport, in particular as the mine advances into greater depth and the reticulation system increases in length.

Technological advancements in thickening, filtration, centrifugation, mixing and pumping ability are all dependent upon an increased understanding of paste flow behaviour/workability and strength characteristics (Sofra, 2017).

1.1 *Paste fill rheology*

Rheology is among the most important properties of paste fill materials. Understanding the rheological properties of a paste fill material is essential to predict its behaviour during mixing, pumping, transport and curing.

Paste fill normally exhibits pseudoplastic behaviour with a yield stress, and in some instances, thixotropic behaviour.

Yield stress is the critical shear stress that must be exceeded before irreversible deformation and flow can occur. This property is considered by many authors as the most important rheological parameter for the design and operation of a paste system. Physical and chemical/mineralogical properties of paste fill have a significant influence on yield stress, such as: solids content, percentage of cement, particle size distribution (PSD), particle shape, mineralogy, surface properties of tailings, among others.

In terms of PSD, the quantity of fine particles present is very important for transportation because fine particles help to float the coarse grains in the paste and provide a non-settling paste flow within the network of pipes (Ercikdi et al, 2017). As a rule of thumb, a minimum of 15% of particles having a diameter lower than 20 µm (fines fraction) is considered sufficient to generate water retention properties required for the transport of tailings paste through a borehole/pipeline system (Landriault, 1995).

Particle morphology or shape can have dramatic effects on rheology, particularly when plate or needle-like particles with high aspect ratios are present.

The mineralogy of tailings influences a number of paste backfill features, such as water retention, strength, settling characteristics and abrasive action (Ercikdi et al. 2017). However, it is not clear how it specifically influences the rheological behaviour of paste. The chemical composition of tailings is of utmost importance for the mechanical strength development of paste backfill.

The precise yield stress, as a true material constant, has turned out to be very difficult to measure, as different tests often give out different results: "depending on the measurement geometry and the detailed experimental protocol, very different values of the yield stress may be found" (James et al, 1987; Nguyen & Boger, 1992; Barnes, 1997,1999; Barnes & Nguyen 2001).

Consequently, there is a need to standardise procedures for yield stress measurements and for the conditioning/preparation of paste samples in order to achieve significant improvements in mine backfill (Silva, 2017).

2 EXPERIMENTAL WORK

The experimental work consists of the development of an accurate laboratorial testing program to evaluate, measure and understand the rheological behaviour and properties of fresh paste fill. To this effect, a series of experimental tests were conducted at GEOLAB - Instituto

Superior Técnico in Lisbon, using different paste fill mixtures. These were produced using Portland cement type II A/L 42.5R, tap water and tailings from Zinkgruvan (Sweden) and Neves-Corvo (Portugal) mines, subsidiaries of Lundin Mining Corporation. Both tailing samples were collected following the filtration process (filter cake) and were kept inside 45-litres barrels.

An optimal interval of solid content was set for each mine tailings, where the laboratorial tests could produce valid results and the material would behave like a real mix at the mine site.

To reduce the number of variables, the percentage of cement was set at 4% and, for each mine, three mix designs were established.

The Neves-Corvo (NC) tailings were used to produce mixtures of 78%, 80% and 82% of solids. In the case of Zinkgruvan (ZG) tailings, these presented different physical properties, visible on PSD curves. For barrels 1 and 2 the mixtures had 78%, 79% and 80% of solids, while for barrels 3 and 4, the mixtures made had 79%, 80% and 81% of solids.

2.1 *Physical and chemical characterization of mine tailings*

The physical and chemical characterization of tailings was conducted by the Corporate Analytic Service of Sika Technology in Zürich, Switzerland.

The particle size distribution (PSD) analysis was determined using a laser diffraction technique. Each sample was manually sieved with a mesh size of 1mm to separate big agglomerates. The cumulative and density distributions of NC and ZG tailings are shown in Figure 1. The curves correspond to each barrel of tailings.

ZG tailings are much coarser than NC tailings, showing an accentuated slope of the curve starting in 20/30 μm of particle size. In terms of the differences between barrels, the PSD curves for NC barrels are equals, which proves that they were collected at the same time. In contrast, the PSD curves of ZG barrels show a clear difference between barrels 1/2 and 3/4, suggesting that the tailings were collected on separate days. It is also possible that barrel 4 may have not been collected at the same time as barrel 3, due to a slight difference in the maximum values of the curves.

Given this, only NC barrel 1 (NC1) and ZG barrels 1 and 3 (ZG1 and ZG3) samples were considered for particle size, chemical composition and mineralogy. These were determined by scanning electron microscopy (SEM) with energy dispersive x-ray analysis (EDS).

Based on SEM images, it is noticeable that ZG tailings have more angular and elongated particles than NC tailings (Figure 2).

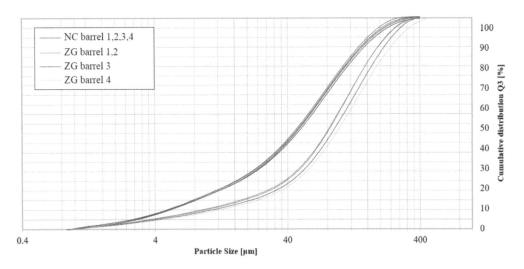

Figure 1.   Cumulative distributions of NC tailings (on top) and ZG tailings (on bottom).

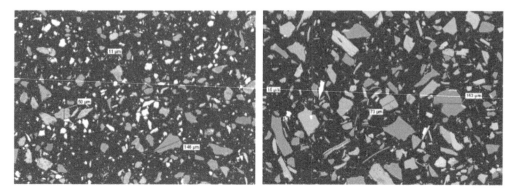

Figure 2.   SEM images of samples NC1 and ZG3.

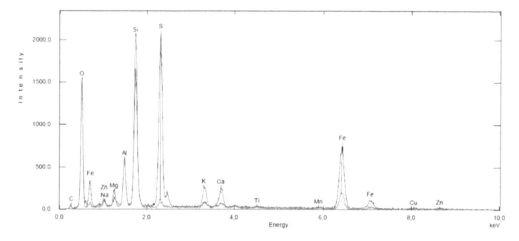

Figure 3.   EDS spectra comparison of the samples NC1 (black), ZG1 (red) and ZG3 (green).

In terms of chemical composition, the Bulk EDS spectra presented in Figure 3 clearly shows a much higher sulphur and iron content in NC tailings, as well as a higher silica and alkali content in ZG tailings.

2.2   Sample preparation

The tailings from each barrel were divided into samples of approximately 4 kg and sealed in plastic bags to prevent evaporation.

To measure the water content of each sample/bag, the tailings were mixed for homogenization by a Hobart mixer and then a small specimen, was taken, weighted and dried in the oven for 24 hours at 80°C.

Knowing the water content and total weight of the sample, it was allowed to calculate the quantity of cement and water to be added to produce the desired mix.

The mixing procedure combined different rotational speeds, agitators and mixing times. First, the tailings were placed inside the mixer bowl and mixed for 1 minute using the flat beater at speed 1. The whip attachment then replaced the flat beater and the water was slowly added at speed 1. After 1 minute of mixing, the cement was slowly added. At last, the final mixture was prepared at speed 2 for 3 minutes to ensure that the materials were well blended by the Hobart mixer.

2.3 Test procedure

Mixing the trials were then undertaken in the following order: dry content, bulk density, viscometer, rheometer, fall cone (with 30° and 60°), flow table and slump.

The procedure adopted for bulk density is similar to standards ASTM C138/C 138M-01a and BS EN 12350-6 2009. Standard SCAN-P 39:80 was followed for determining the dry content.

The slump test was carried out using a mini cone with a bottom diameter of 100 mm, a top diameter of 50 mm and a height of 150 mm. The procedure adopted was similar to standards EN 12350-2 and ASTM C143/C 143M-05a, used for concrete.

To measure the slump-flow, using a flow table for cement mortar, a cone with a bottom diameter of 100mm, a top diameter of 70mm and height of 50mm was used. The procedure adopted was similar to standard ASTM C230, used for cement mortar. The diameter of the paste was measured twice: the first measure was taken after the cone lifting and the second measure after the 15 strokes made with the up and down movement of the table.

The absolute values of yield stress were determined by vane technique, widely accepted by many authors as the best method for achieving the "true" yield stress: "the great advantage of the vane is the material yields on itself this technique is now used worldwide for measurement of yielding in a variety of materials and was motivated by the need to understand the true yielding behaviour under plant/site conditions" (Sofra, 2017). This technique was applied in a controlled shear rate mode (stress growth technique) by a viscometer and a rheometer with vane spindles.

The viscometer used was the DV1 Brookfield rotational viscometer, model RV, with four-blade vanes (V73 and V75) and the Wingather SQ software, V4.0.7. This equipment was programmed to a constant speed of 0.3 rpm and a running time of 2 minutes and 15 seconds.

The rheometer used was the Anton Paar RheolabQC instrument with the vane spindle ST22-6V-16 and the Rheoplus software V3.62. In this case, the rheometer was programmed to a constant speed of 0.1 rpm and a running time of 3 minutes and 20 seconds.

In an attempt to obtain a practical and inexpensive test method to measure the yield stress onsite, some experiments were carried out with a Controls S.R.L fall cone instrument, using a standard metal cone of 30 degrees. This method is well known in soil mechanics for measuring the liquid limit and the undrained shear strength of fine-grained soils.

Due to the weight of the cone, it became wholly immersed at every attempt. As such, the exact geometry of the metal cone and bar had to be recreated and attached using a lighter material. After some unsuccessful reproductions, two final prototypes were printed in a Formlabs 3D printer, at the maximum definition (0.025mm) with a special resin for top quality output, durability and toughness. The cones were printed with 30 and 60 degrees, weighing 12g and 18g, respectively (Figure 4). The procedure is quite simple. A container was filled with paste and the top surface levelled with a spatula. The cone tip resting on the paste surface is released and, following penetration, it is stopped as quickly as possible. The depth of penetration is then recorded. The containers used were 53 mm in diameter and a 68 mm in height for the 30°-cone, and 85 mm in diameter and 57 mm in height for the 60°-cone.

3 RESULTS

A total of 72 mixtures were produced and the number of values recorded were in excess of 500. The average of all results is presented in Tables 1 and 2.

The shear stress-time curves and the correlation graphs between the yield stress and the dry content of ZG and NC mixtures are shown in Figures 5, 6 and 7.

Throughout all figures, it is possible to identify the domains of each mix design, especially in NC charts. These domains are mainly due to the differences in solids concentration between mixtures, being the dry content interval higher in NC than in ZG mixtures. With exception to the ZG curves, the residual shear stresses also show very perceptible domains achieved by the rheometer. It is possible to observe that, after yielding, the curves recover some of the stress

Figure 4.   Printed cones with 30 and 60 degrees and respective containers.

Table 1.   Experimental result of ZG mixtures.

ZINKGRUVAN MIXTURES

| Solid content [%] | Dry Content [%] | Density [kg/m³] | Vane Technique | | Flow Table | | Fall Cone | | Slump [cm] |
			Yield stress (viscometer) [Pa]	Yield stress (rheometer) [Pa]	Initial D [cm]	Final D [cm]	30° Cone [mm]	60° Cone [mm]	
78%	73.47	2046.9	45.27	92.50	13.17	24.08	32.97	25.82	11.4
78%	73.61	2060.97	48.65	134.67	12.17	24.08	31.2	22.7	10.93
79%	74.88	2020.03	55.17	167.67	10.58	23.42	25.77	19.83	9.5
79%	74.66	2030.27	54.31	217.33	10.58	22.5	23.73	18.67	9.5
79%	75.07	2052.17	65.06	256.67	10.63	22.13	22.8	19.47	8.4
79%	74.7	2020.17	62.81	225.67	10.5	22.58	24	19.83	7.33
80%	75.59	1983.09	62.85	353.33	9.67	21.67	20.27	16	6.67
80%	75.71	2034.88	80.62	245.67	9.58	20.08	16.8	14.27	6.77
80%	76.77	2080.5	93.32	377	9.75	18	11.67	10.8	5.43
80%	75.64	2033.6	85.60	281.33	9.58	19.33	15.23	14.23	6.67
81%	76.61	2027.41	86.69	349	9.5	18.33	14.5	12.27	5.9
81%	76.95	1997.49	98.94	373.33	9.5	17.83	13.03	11.03	5.93

and, therefore, the residuals are not distributed by domains/intervals as shown in NC curves. This suggests that ZG paste fill could be time-dependent for shearing at low shear rates (most probably thixotropic behaviour). However, additional tests should be undertaken to prove and understand this rheological behaviour.

The test methods applied in this experimental work are well known in the mining industry for evaluating the rheology of concrete/paste backfill, particularly the slump test. There are

Table 2. Experimental result of NC mixtures.

NEVES-CORVO MIXTURES

Solid content [%]	Dry Content [%]	Density [kg/m³]	Vane Technique Yield stress (viscometer) [Pa]	Yield stress (rheometer) [Pa]	Flow Table Initial D [cm]	Final D [cm]	Fall Cone 30° Cone [mm]	60° Cone [mm]	Slump [cm]
78%	77.29	2374.51	114.15	184.33	14.33	23.25	31.7	22.13	12.17
78%	76.86	2396.11	78.16	177	17.83	24.92	35.43	25.07	12
78%	76.91	2369.63	84.03	187.33	16	23.83	33.67	22.37	11.83
78%	76.79	2378.35	68.61	190.33	17.75	25.17	34.87	24.27	12.5
80%	79.05	2598.43	177.32	381.67	10.25	20.75	23.2	16.33	9.5
80%	78.84	2416.48	170.12	491.33	10	21.17	25.23	18.87	8.83
80%	79	2422.93	167.48	449.33	9.92	20.58	23.73	16.4	8.33
80%	78.77	2433.95	160.87	480.33	10.75	21.67	26.6	16.6	10.33
82%	81.21	2515.84	245.93	1023.5	9.125	16.125	10.4	9.15	4.75
82%	80.95	2458.19	339.95	1360	9.17	17.75	12.3	9.3	5
82%	80.73	2483.49	364.48	1553.33	9.25	17.08	10.87	8.13	4.33
82%	80.84	2515.63	340.68	1680	9.25	18.67	13.6	9.8	4.5

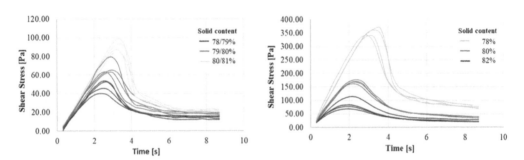

Figure 5. Shear stress-time curves for ZG (left) and NC (right) mixtures as recorded by the viscometer.

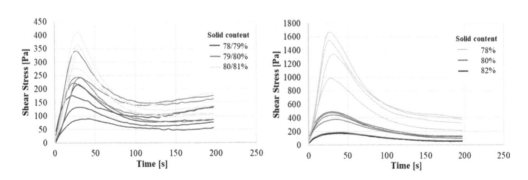

Figure 6. Shear stress-time curves for ZG (left) and NC (right) mixtures as recorded by the rheometer.

formulations that allow us to determine the yield stress of samples from the results of these tests. However, the purpose of this work was to evaluate the fitting capacity of these method-ologies to different paste fill mixtures, with different tailings and mix designs, and to deter-mine the veracity and quality of the results and their correlations with the yield stress values measured by a rheometer and a viscometer.

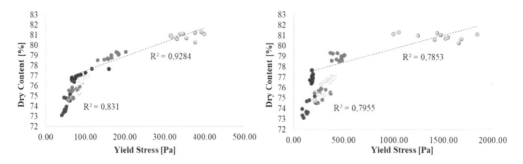

Figure 7. Correlation graphs of yield stress recorded by the viscometer (left) and rheometer (right) versus dry content of mixtures.

From the test method results, the measurement that achieved the worst performance was the initial diameter of flow table. It was not able to characterize mixtures with higher dry contents using this measurement, because the diameter of the cone mould is identical.

The correlation graphs between the results obtained by the slump, fall cone and flow table spread (final diameter) were correlated with the dry content and the absolute yield stress (measured by the viscometer and rheometer) of all mixtures are presented in Figures 8 and 9.

4 RESULTS DISCUSSION

The mix designs set for each mine tailings, during the initial run practice, were different in terms of workability range. This range was smaller for ZG mixtures (78–80% and 79–81% solids content) than for NC mixtures (78%–82% solids content), mainly due to the physical properties of both tailings.

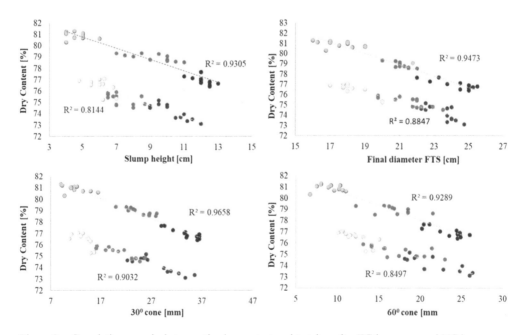

Figure 8. Correlations graphs between the dry content and tests' results. ZG in orange and NC in green.

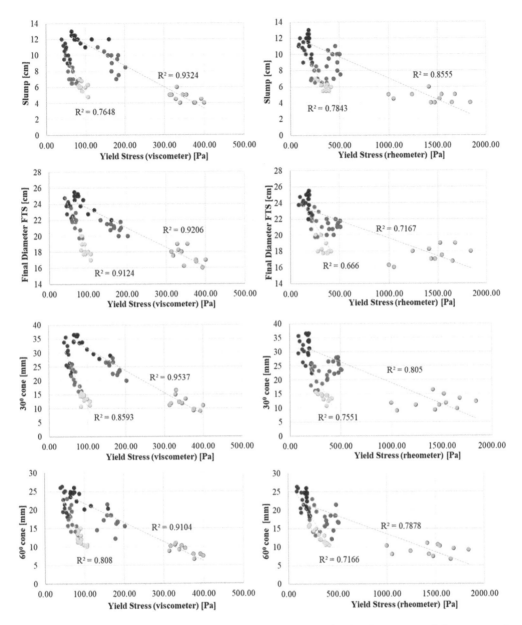

Figure 9. Correlations graphs between the yield stress recorded by the viscometer and rheometer and the tests' results. ZG in orange and NC in green.

In general, finer slurries generally display a more gradual increase in yield stress with increasing solids concentration, whereas coarse materials show a relatively sharp transition from liquid-like to solid-like behaviour (Sofra, 2017), which may justify the smaller interval of workability for ZG paste fill when compared with a finer material such as NC tailings, using the same cement percentage and water.

When observing Tables 1 and 2, the dry content values are consistent for each mix design. Another noticeable effect is the density difference of both paste mixtures, which can be explained by the increase in solids concentration of the mixtures but, moreover, because of the tailings' mineralogy. NC tailings contain denser elements (Fe_2O_3, SO_3, etc.) than ZG tailings (SiO_2, Al_2O_3, etc.).

The absolute yield stress values were measured using the viscometer and the rheometer. The rheometer results were less consistent than the values measured by the viscometer. This can be explained by the higher precision of the equipment and lower rotational speed, making it particularly susceptible to slight deviations of the vane spindle from the centre of the container, to existent air bubbles, small agglomerations of particles, wall slip and other factors. Another aspect is related to the dimensions of the container utilised for these measurements: the distance from the vane spindle to the edge of the container was too small. Previous projects that involved the use of a vane in a cup to minimise slip (Barnes and Carnalli, 1990) and infinite medium analysis (Kreiger and Maron, 1954) have proven the major advantages of using the vane in an infinite medium for this type of measurement, such as (Sofra, 2017): insertion of a vane causes less sample disturbance, minimizing thixotropic breakdown; the vane in cup is less susceptible to errors arising from large particle sizes; yielding occurs between layers of fluid; minimizing the effects of wall slip, among others. Since ZG tailings are coarser than NC material, the presence of larger particles and more air voids causes more disturbance during the rotation of the vane spindle. This could be the reason for the inconsistent results using ZG tailings.

Due to the higher rotational speed of the viscometer, the values of yield stress are smaller. The dimensions of the container were high enough compared to the diameter of the vane spindle, creating an infinite medium for measuring paste rheology. Both mine mixtures presented consistent results and behaviours.

In general, it is worth noting that NC mixtures present higher yield stress values in comparison to ZG mixtures for the same solids' concentration, suggesting that the dry content could not be the only variable responsible for the variance of yield stress. In this view, the physical and chemical differences of both tailings can also be influencing factors. In terms of the influence of tailings mineralogy on rheological properties of paste fill, only general information is available in academic works. Therefore, it is, in this case, difficult to conclusively assess which parameters most affect the rheology of paste backfill in terms of chemical composition. In general, as NC mixtures are denser than those from ZG, due to their mineralogy, the shear necessary to start flowing is naturally higher.

From the test method results, the measurement that presented the worst performance was the initial diameter of flow table. Apart from this, all methods were able to evaluate mix design differences and obtain a good correlation with yield stress values.

The final diameter proved to be a perfect measurement of paste fill rheology. The data was consistent and the correlations with yield stress values were even better in comparison to the slump test (Figures 10 and 11).

Finally, the adapted fall cone test turned out to be the methodology that performed the best, with very good correlation with the dry content of the mixtures and with the yield stress values achieved by the viscometer and rheometer (Figure 9 and 10). The weight and geometry of the cones combined with the time released proved to be a successful procedure (Figure 12) for measuring paste fill rheology.

As mentioned before, this test method is widely used in soil mechanics to obtain the undrained shear strength of fine-grained soils, based on the "cone factor", K. Reports from different authors show K values ranging from 0.4-1.33. Differences are mostly attributed to the cone surface roughness (Llano et al., 2018).

To compare the K factor of the 30°-cone used for this experimental work with other standard cones, this was calibrated following the procedure proposed by Llano et al., 2018. To this effect, the yield stress values measured by the viscometer that presented the best correlations with the fall cone were used. All the results obtained for both mixtures were applied.

A correlation graph with the yield stress and the Q/h^2 was drawn, where Q is the total cone weight and h is the cone penetration depth. The slope of the best fit straight lines corresponded to the K factor. This varies between 0.26-0.39.

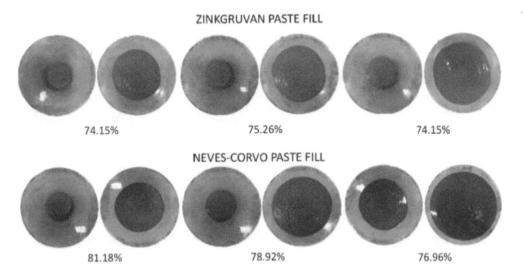

Figure 10. Photographic record of flow table tests.

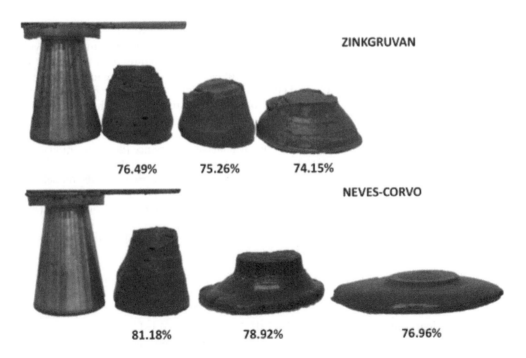

Figure 11. Photographic record of slump tests.

These K interval are admissible in the range values for the 30° standard cones. However, this is a conservative estimation and for greatest accuracy, further values should be taken.

4.1 Data modelling

As previously concluded, the dry content of both mixtures was not the only influencing property on yield stress results. In an attempt to mathematically prove the influence of the physical

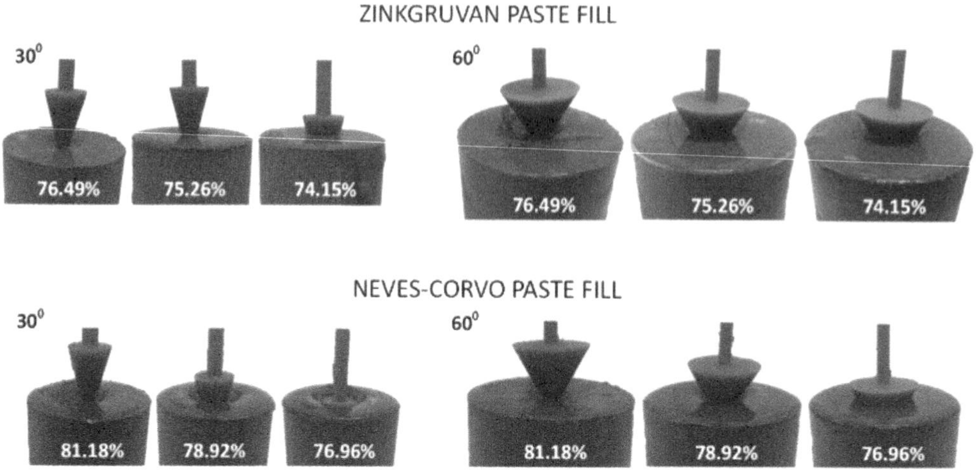

ZINKGRUVAN PASTE FILL

NEVES-CORVO PASTE FILL

Figure 12. Photographic record of fall cone tests.

properties tailings on yield stress, two detailed statistical studies were developed. They are based on the proposal and validation of multiple linear regression models to predict yield stresses measured by both equipment (dependent variable) in all mixtures, as a function of the following quantitative predictors: dry content, percentage of material passing under 20 micron ($<20\mu m$), uniformity coefficient (C_u) and coefficient of gradation (C_c). The remaining characteristic diameters of PSD were not considered because they are applied in the calculation of coefficients C_u and C_c (Silva, 2017).

The validation of each candidate model included the statistical significance of the estimated parameters, the suitability of the adjusted model by analysis of variance (decomposition of the total sum of squares of the residuals) and the statistical quality of the regression model through residual analysis (Paneiro et al., 2015). In this sequence, the quality of the models was assessed using the following criteria: coefficients of determination (R^2 and R^2 adjusted), analysis of variance (ANOVA table) and Akaike and Schwarz-Bayesian information criteria (AIC and SBIC).

Considering the laboratorial data and statistical analysis, the best multiple linear regression models can be written as follows (Equations 1 and 2):

$$\log(yield\ stress_{visc}) = 4.6571 + 0.7482\log(DryC) + 0.1724\log(C_C) - 0.2201\log(C_u)$$
$$+0.2374\log(<20\mu m) \tag{1}$$

$$\log(yield\ stress_{rheo}) = 5.8085 + 1.0157\log(DryC) + 0.1816\log(C_C)$$
$$-0.3180\log(C_u) - 0.5957\log(<20\mu m) \tag{2}$$

It is worth noting that *yield stress_{visc}*, *yield stress_{rheo}*, *DryC*, C_C, C_u and $<20\mu m$ correspond to normalized variables.

The models presented very high adjusted coefficients of determination (0.9570 and 0.8937, respectively) and rejected the null hypothesis by f-test statistics analysis, which shows a very good level of explanation of the laboratorial data variability (Silva, 2017).

Based on these models, it is possible to conclude that the dry content of mixtures as well as the particle size of tailings are strong influencing variables in yield stress values determined using a viscometer and rheometer. However, it is necessary to consider other influencing variables such as mineralogy/chemical composition that may have a significant influence on yield stress. Furthermore, the weight of each quantitative predictor is questionable. To validate these models, a higher number of results and other influencing parameters should be analysed.

5   FINAL CONSIDERATIONS AND RECOMMENDATIONS

Rheology can play an essential role in better understanding and dealing with thickened tailings. Due to the necessity of creating a standard experimental testing program to evaluate paste fill worldwide, a simple testing procedure using a drop cone measurement was developed.

Paste mixtures were produced using tailings from Zinkgruvan and Neves-Corvo mines, 4% of cement and tap water. To determine the absolute values of yield stress, the vane technique with a viscometer and a rheometer at different rotational speeds was applied. The results obtained showed that the range of yield stress for both mine mixtures was different. From the properties assessed, the dry content of mixtures was the main variable that affected this rheological property.

The laboratory program included the following test methods: dry content, bulk density, viscometer, rheometer, fall cone (with 30° and 60°), flow table spread and slump. Of all laboratory test methods, the least convincing measurement was the initial diameter of the flow table. Apart from this, all test methods enabled the assessment of the difference between mix designs and obtained a good correlation between the yield stress measured by both pieces of equipment. The flow table (final diameter) and the 30° fall cone presented the best performances.

To verify the influence of the physical properties of tailings on yield stress, two detailed studies based on the proposal and validation of multiple linear regression models were developed. The models obtained have shown a strong correlation of tailings' physical properties on yield stress. However, these must be validated with more results and other influencing factors such as mineralogy/chemical composition included in future works.

The fall cone test adopted for this experimental work turned out to be the methodology of choice that performed best, with a good correlation of dry content and the yield stress values achieved by the viscometer and rheometer. The weight and geometry of the cones combined with the release time have proven to be a successful procedure for measuring paste fill rheology, and thus represent one of this work's major contributions. This equipment can become a simple, inexpensive, convenient and fast method for determining yield stress, ideal for quality control and/or fast on-site measurements.

ACKNOWLEDGEMENTS

This research was sponsored by Sika Corporation (Switzerland R&D Department, Sweden, Portugal and Spain), and we would like to personally thank Fabian Erismann, Christophe Kurz and Rute Silva for their guidance and assistance during the project.

We are grateful to Zinkgruvan and Neves-Corvo mines for providing the material and equipment necessary for the laboratorial tests.

We gratefully acknowledge Rodolfo Machado and Hugo Brás for sharing their experience and knowledge, which greatly supported the elaboration of the laboratorial testing programme developed in this experimental work.

BIBLIOGRAPHY

Barnes, H. A. 1997. Thixotropy – a review. *Journal of Non-Newtonian Fluid Mechanics 70*, pp. 1–33.
Barnes, H.A. 1999. The yield stress – a review or 'παντα ρει' – everything flows? *Journal of Non-Newtonian Fluid Mechanics 81*, pp. 133–178.

Barnes, H.A. and Carnali J.O. 1990. The vane-in-cup as a novel rheometer geometry for shear thinning and thixotropic materials. *The Journal of Rheology 34(6)*, pp. 841–866.

Barnes, H. A. and Nguyen, Q. D. 2001. Rotating vane rheometry – a review. *Journal of Non-Newtonian Fluid Mechanics 98*, pp. 1–14.

Ercikdi et al. 2017 – Ercikdi, B., Cihangir, F., Kesimal, A. and Deveci, H. 2017. *Practical Importance of Tailings for Cemented Paste Backfill.* Yilmaz, E. and Fall, M. (eds) Paste Tailings Management, Springer, Cham, pp. 7–32.

James, A. E., Williams, D. J. A. and Williams, P. R. 1987. Direct measurement of static yield properties of cohesive suspensions. *Rheologica Acta 26*, pp. 437–446.

Kreiger, I.M. and Maron, S.H. 1954. Direct determination of the flow curves of non-Newtonian fluids, III. Standardised treatment of viscometric data. *Journal of Applied Physics 25 (1)*, pp.72–75.

Landriault, D. 1995. Paste backfill mix design for Canadian underground hard rock mining. *Proceedings of the 97th annual general meeting of the CIM rock mechanics and strata control session*, pp. 652–663.

Llano-Serna, M. A., Farias, M. M., Pedroso, D. M., Williams, D. J. and Sheng, D. 2018. Considerations on the Experimental Calibration of the Fall Cone Test. Geotechnical Testing Journal 41 (6), pp.1131–1138

Nguyen, Q.D. and Boger, D.V. 1992. Measuring the flow properties of yield stress fluids. *Annual Review of Fluid Mechanics 24*, pp. 47–88.

Paneiro, G., Durão, F., Costa e Silva, M. and Neves, P. 2015. Prediction of ground vibration amplitudes due to urban railway traffic using quantitative and qualitative field data. *Transportation Research D (40)*, pp. 1–13.

Silva, M.A. 2017. *Contribution to laboratorial determination of rheological properties of paste backfill.* Lisbon: IST Press, 82 pp.

Sofra, F. 2017. *Rheological Properties of Fresh Cemented Paste Tailings.* Yilmaz, E. and Fall, M. (eds) Paste Tailings Management, Springer, Cham, pp. 33–57.

Yilmaz, E. and Fall, M. 2017. *Introduction to Paste Tailings Management.* Yilmaz, E. and Fall, M. (eds) Paste Tailings Management, Springer, Cham, pp. 1–6.

Backfill reticulation: pumping, piping, hydraulic analyses

Minefill 2020-2021 – Hassani et al (eds)
© 2021 Taylor & Francis Group, London, ISBN 978-1-032-07203-6

Automated diverter valves at Kirkland Lake Gold Fosterville mine improve safety and efficiency

Russell Evans
Kirkland Lake Gold

Gary Trinker
Victaulic

SUMMARY: Traditional manual approaches to managing backfill systems are being replaced with improved and more efficient methods that are faster and safer. Automated diverter valves are among the technologies being used to effect efficiency and safety gains. Using diverter valves eliminates time-consuming manual operations that require special equipment and put workers in harm's way. A case study that assesses automated diverter valve operation and manual diversion at an Australian mine provides a comparison of functionality, time-savings, and safety advantages inherent to the diverter valve.

Keywords: Backfill Diverter Valve, Risk Reduction, Efficiency

1 INTRODUCTION

Many mine owners around the world are using manual backfill processes that rely on experienced workers investing considerable time and effort in risk-prone processes. These methods require significant manhours and are dependent upon having skilled laborers on site and expensive, specialized equipment on hand.

As mine owners employ different means of improving economics, many are using backfill systems to dispose of tailings and enhance ore recovery and are looking for safe and efficient ways to manage the backfill process. At Kirkland Lake Gold Fosterville mine in Victoria, Australia, automated diverter valves were installed at distribution switching points to improve this process, directing paste to the intended stopes without the need for underground crews and additional equipment.

Prior to the adoption of diverter valves, making changes in paste distribution was a manual operation that involved long sweep elbows to be disconnected from the piping network and reconnected to a different downstream pipe. This manual operation was time consuming, and when the necessary manpower and/or equipment was not available, resulted in considerable downtime.

The diverter valves address many of the issues encountered during manual operations. The valves function reliably in a broad temperature range, and valve operation does not require specialized equipment or skills. Because the valves can be operated remotely, they improve safety by reducing the risk of injury by removing people from harm's way.

This paper examines the use of automated diverter valves to replace manual diversion of paste in the Fosterville Gold Mine (FGM) reticulation system. Replacing the manual process with a remotely controlled system of valves resulted in dramatic reduction in safety risks and improved uptime by enabling safe and timely switching of the reticulation route.

DOI: 10.1201/9781003205906-17

1.1 Traditional approaches

Most medium and high grade mining operations use backfill. The backfill is comprised of sand, tailings, aggregate and binders and can be abrasive resulting in wear on the system. When maintenance is required in the reticulation system or when backfill is required in a different stope the system is shutdown and manual switching of the piping is needed.

Each time this is required, multiple workers and equipment are involved and the resulting downtime is significant. In addition, this manual process introduces safety risks for the workers and takes them from more productive activities.

1.2 Introducing innovation

Victaulic designed the Series 725S Diverter Valve specifically for use in backfill operations. Providing multidirectional service, the valve eliminates the need for manual manipulation of backfill piping systems.

Designed to withstand high pressures of underground systems, the valve is rated to 103.5 megapascals [MPa] (1500psi) and uses a 4D bend radius flow path to reduce wear from abrasive slurry and high-flow conditions. All wetted surfaces of the Series 725S valve are constructed from martensitic stainless steel, which provides both excellent abrasion resistance and corrosion resistance. Continuous flushing is not necessary because the smooth profile of the plug prevents clogging. This valve is customizable, accommodating electric, hydraulic or pneumatic actuation, and can be installed vertically, horizontally or in any intermediate position. The valve is offered with either grooved, double grooved or ringed ends, allowing the specifier to choose their preferred mechanical couplings for quick and simple installation and removal.

Installing automated diverter valves at switching points on paste piping allows backfill to flow to alternative stopes without requiring manual manipulation of the pipe/pipe fittings to redirect flow.

2 A NEW APPROACH

Efficiency gains are being achieved by replacing manual backfill systems with methods that are faster and safer. Automated diverter valves are among the technologies being used to effect efficiency and safety gains. Using diverter valves eliminates time-consuming manual operations that require special equipment and introduce risks for workers.

Installing automated diverter valves in the Kirkland Lake Fosterville gold mine illustrates how functionality and safety can be improved, leading to time savings and ultimately better economics.

2.1 Mine history

The FGM is approximately 20 km east of the city of Bendigo and 130 km north of Melbourne in the State of Victoria, Australia. The FGM and associated infrastructure are on Mining License 5404, which is 100% owned by Kirkland Lake Gold Ltd. (Figure 1).

Gold was discovered in the area in the mid-1800s, and mining activity was underway by 1894. Exploration activities by various owners of what is now the FMG ensued from 1973, with heap leaching operations from an oxide pit beginning in 1993. By 1998, Perseverance Corp., which operated the mine, was producing 40,000 oz/year from the oxide ore. In mid-2001, a sulfide resource was developed, and an open pit was initiated in 2004. The first gold pour at the site took place in 2005.

The owners started an underground decline in March 2006, and the first underground stope was mined in December 2006. By early 2008, underground ore had become the main mill feed. The 500,000 th ounce of gold was produced in March 2011. Over the next seven years, the mine produced gold at a rate of 150,000 oz/year.

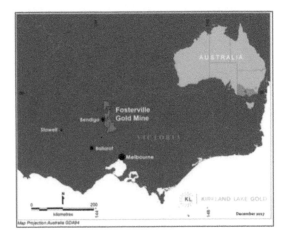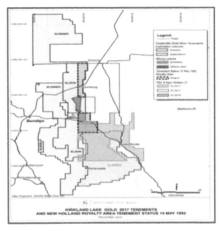

Figure 1.   The Kirkland Lake Fosterville Gold Mine is approximately 20 km east of the city of Bendigo and 130 km north of Melbourne in the State of Victoria, Australia, on Mining License 5404.

Source: Kirkland Lake Gold, Toronto, Ontario.

Kirkland Lake Gold Ltd., headquartered in Canada, purchased the mine in November 2016 and began an aggressive exploration program that yielded impressive results, including the Swan Zone, which contains 2.34 M oz at an average grade of 49.6 g/t gold. FGM produced approximately 600,000 ounces of gold in 2019.

2.2   Mining method

Current mining at FGM is undertaken predominantly as owner-miner. FGM uses an open stoping, retreat mining method with the use of backfill to extract gold ore from the Phoenix decline ore bodies. Figure 2 shows the actual and proposed mining layout at FGM.

Stoping widths vary as dictated by grade distribution in the block model and strike length is determined by rock mass and hanging wall stability assessments. Once a stope has been mined out, the void is generally backfilled to ensure stability in accordance with planned future vertical and horizontal exposures. The open stopes are relatively small, ranging from 500 m^3 to 2,700 m^3 with an average of 1,500 m^3. The current annual backfill requirement is 100,000 m^3 to 110,000 m^3 per year.

The selection of the specific mining method within the open stoping regime is based on previous experience at the Fosterville Mine and expectations of ore zone geometry and geotechnical conditions. A standard level interval of 20 vertical meters is usually applied across all mining areas. However, this can be varied to maximize the extraction of the economic material.

Underground mining is carried out using a conventional fleet, including twin boom development drills, production drills, loaders, trucks and ancillary equipment.

2.3   Backfilling

Historically, mining at FGM had been done top down without backfill or with limited quantities of cemented rock fill (CRF) using underground development waste mixed with cement slurry. With the discovery of the high-grade Swan Zone, a decision was made to design and construct a paste backfill system. In the interim, while awaiting the construction of the new paste plant, the stopes would all be filled with CRF. Because of the relatively flat dipping nature of the ore body, the hanging walls required topping up with flowable fill – a mixture of various sands, cement and water. The objective of this process was to reduce hanging wall failure. While it was effective, the backfill method was slow, and operating costs were high.

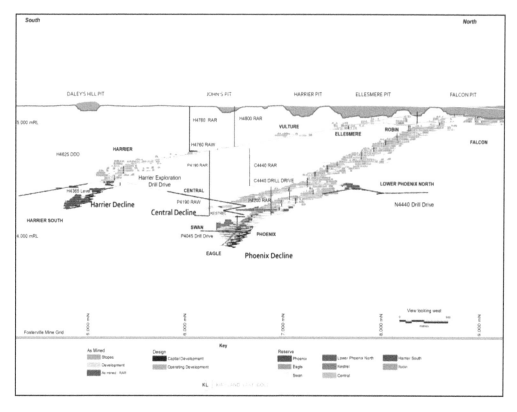

Figure 2.   Actual and proposed mining layout at the Kirkland Lake Fosterville Gold Mine.
Source: Kirkland Lake Gold, Toronto, Ontario.

The decision to switch to paste backfill was based on ensuring that the backfill achieved full confinement of hanging wall voids leading to higher productivity, lower dilution and reduced operating costs.

The plant process design combines full plant tailings and binder at a rate of 65 m³ per hour. The paste plant design includes a thickener and two large vacuum disc filters feeding a continuous mixer where binder and water are added to the filter cake to produce paste with the required slump or yield stress. Lime can also be added during this process.

The paste plant is located so the paste fill flows by gravity to the Phoenix ore zones. The paste is delivered through a 1,040 m borehole from the surface paste plant to the P4190L. The 7 5/8" steel casing in the borehole is lined with a 12 mm ceramic epoxy polymer coating. The borehole casing has an ID of 150 mm. Figure 3 shows the paste reticulation long section at FGM.

3   IMPLEMENTING AUTOMATION

The current paste reticulation system is made up of 150NB Sch80 A106B SMLS pipe connected with Victaulic HP70ES couplings. The pipe lengths include exact 6 m, 3 m, 1m and 300 mm straight lengths as well as Victaulic cast 3D bends of 90, 45, 22.5 and 11.25 degrees. The hanging and thrust support is galvanised steel.

The reticulation system instrumentation includes pressure sensors and flow meters as well as a dump valve at the bottom of the main borehole and pressure relief spools on each level. There are permanent closed circuit television (CCTV) cameras at two location on the P4190L.

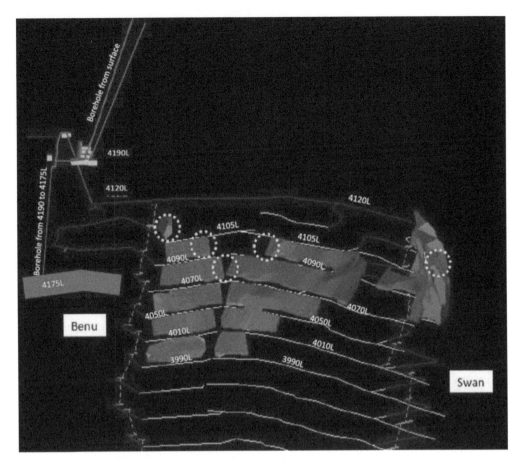

Figure 3.   The Paste Reticulation Long Section at Fosterville Gold Mine.
Source: Kirkland Lake Gold, Toronto, Ontario.

Portable CCTV cameras are positioned at the pour point for each stope and at the fibrecrete bulkheads on the undercut of each stope. Total earth pressure cells and piezometers are installed inside most bulkheads. Workers monitor the instrumentation and CCTV cameras in real time from the control room at the paste plant.

A tradeoff study was done to evaluate the benefits of using actuated diverter valves versus manually changing the direction of pipe spools when switching the pour locations.

The relatively small stope volumes averaging only 1,500 m^3 of paste was one of the factors that contributed to the decision to implement the use of diverter valves. Based on the stope volumes and the production rate of 65m^3 per hour, reticulation changes would be required at least once per day and possibly two times per day.

The decision was made to use remotely actuated diverter valves based on:

• Improved safety
• Higher productivity
• Lower operating costs.

The Victaulic 725S diverter valve was selected from among the various diverter valves available on the market based on design and price.

3.1 Fosterville diverter valve installation detail

The Victaulic 725S diverter valves at FGM are equipped with 415 V electric actuators, limit switches and Auma remote control units. Pneumatic actuators and 240V electric actuators were considered, but although pneumatic actuators could have reduced the capital cost, electric actuators were selected because electric power is required at the valve for the limit switches and solenoids and on each level for pressure sensors, flow meters and CCTV cameras. Underground power at FGM is 1,000V.

The underground instrumentation and diverter valves are controlled with a separate dedicated PLC and communications system. Rockwell Automation equipment was selected as this was the current standard at FGM. The surface Paste Reticulation Control Panel includes an Allen-Bradley ControlLogix 5571 PLC, FLEX I/O module and a Managed Ethernet switch.

This panel receives the inputs from the underground instrumentation and CCTV cameras and controls the diverter valves. This control panel communicates with the main Paste Plant PLC via ethernet which then enables the Paste Plant Operator to see the data from the instrumentation and control the diverter valves.

On each level underground there is a Form-4 Paste Board. These boards contain a 1000V to 415V transformer, a 1000 V to 240 V transformer, RT1100 Power Shield UPS and 24VDC power. These boards also contain FLEX I/O modules and ethernet switches.

The communication between the surface and underground control panels is through fiber optic cable. The fiber optic cable is connected to the ethernet switches in each panel and the FLEX I/O then connects the instruments and valves to the PLC.

Each diverter has a unique valve number in the PLC control logic, and each discrete paste line has a number. The inlet line and the A and B outlet lines from the diverter are numbered, with the numbers displayed on the valve as well as the underground pipes.

3.2 Diverter valve operation

The Paste Fill Note includes instruction for the setup of each diverter valve for each pour (Figure 4). The instructions indicate the required direction for each diverter using the diverter valve number and identifies the specified direction of flow to either the A or B outlet, including the paste line number for the outlet.

The Paste Fill Note is delivered to both the surface paste plant operator and the underground paste operator. The surface paste plant operator switches the valves to the required positions on the human machine interface (HMI) and advises the underground paste operator that the valves are in the correct position. The plant operator can see the route of the piping

Figure 4.   The Paste Fill Note includes instruction for setting up each diverter valve for each pour.
Source: Kirkland Lake Gold, Toronto, Ontario.

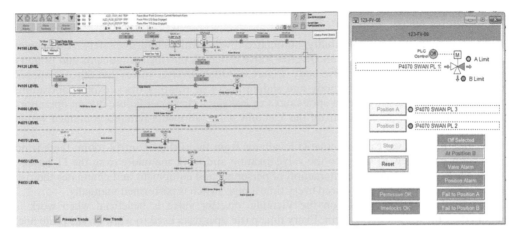

Figure 5.   The plant operator can see the route of the piping highlighted in blue on the HMI display.
Source: Kirkland Lake Gold, Toronto, Ontario.

highlighted in blue on the HMI display. Once there is pressure or flow in the line, the colour changes from blue to red (Figure 5).

Using the Paste Fill Note, the underground paste operator checks all the pipes along the route to the stope, including the diverters. By consulting the diverter valve instruction in the note, the underground paste operator ensures the yellow indicator on the valve is pointing towards the correct line number indicated in the note. Once that has been confirmed, the underground operator switches the diverter control panel to the local mode so the direction of the valve cannot be changed during pouring.

4   RESULTS

Replacing the traditional manual approach to backfilling at FGM has enabled more stream-lined operations, increased productivity, and reduced safety risks. The new process is allowing the mine to produce more ounces per day and is improving profitability.

4.1   *Higher productivity, reduced operating costs*

Once a stope is filled, the time required to change the piping location to the next stope is critical. In reticulation systems where there are no diverter valves, the steps required include traveling to the lockout location, performing the lockout (sometimes this takes place on the surface), traveling to the switching location, manually switching the pipe, and finally traveling back to the lockout location and removing the locks. This can take a team of three paste operators anywhere from two to five hours, depending on the layout of the mine and the lockout procedure in place. Using diverter valves eliminates these two- to five-hour delays, allowing paste filling to commence quickly. With the paste filling process expedited, mining can resume more rapidly, and ultimately, more ounces of gold can be mined each year.

Another benefit of this tremendous time savings is the substantial reduction of non-productive time. Adopting an automated process means the diverters can be used during shift changes when there is no access to the underground. Automation also frees the paste operators to perform other work during time that otherwise would have been spent manually managing the paste fill process.

The cumulative effect is reduced operating costs. In a typical lower cost mine such as FGM, the total cost to purchase and install a single diverter valve can be repaid in less than one month through increased ore production.

4.2 *Improved safety*

In addition to improving profitability, automated diverter valves improve safety. An inherent safety feature of the Victaulic 725S Diverter Valve is that all moving components and primary seals are contained within the valve body (Figure 6). Other diverter valves that rely on exposed compression seals between the body and sliding plates or rotating discs, expose operations personnel to high pressure media spray and seal leakage if there is a failure during operation. In the unlikely event of seal wear on the Victaulic valve, all media remains contained within the valve body.

Because these valves can be actuated locally or from the surface control room, there is no need for a paste crew to travel to the valve location and work manually from the IT basket. Using these valves reduces the interaction of personnel and equipment and eliminates the risk of personal injury that is introduced every time a pipe switch is performed manually.

The local actuation feature on the Victaulic diverter valve is beneficial when work is required on the reticulation system. Every time paste operators need to work on the paste line, some form of lockout is required. The operator can travel to the level where the work is to be performed, change the diverter control unit to local actuation, switch the valve away from the line to be worked on and lock out the actuator to isolate the line. This functionality provides the highest level of safety for the paste operators working on the line because they can manage the process without relying on other personnel to ensure isolation, further reducing risk.

4.3 *Additional benefits*

The FGM reticulation system includes a dedicated dump valve on the P4190L so the contents of the surface borehole can be dumped into a dedicated sump in case of emergency. The

Figure 6.   Series 725S cross section.
Source: Victaulic Co, Easton, Pennsylvania.

Victaulic 725S diverter valve is rated for use as a dump valve under full pressure, so installing these valves on each level provides the additional benefit of being able to use the diverters for an emergency dump.

The cost of production delays caused by borehole blockages in mines varies depending on the production rate and the All-In Sustaining Costs (AISC) of the mine versus the current gold price. In broad terms, a month of lost production amounts to a few million dollars at most mines. Also, there are additional costs for clearing borehole blockages, (or in extreme cases total borehole replacement), which can exceed $1 million. Therefore, a strategy of a reliable borehole dumping system and a backup surface borehole is used at most mines.

5 CONCLUSIONS

Replacing the traditional manual approach to backfilling at FGM has enabled more stream-lined operations, increased productivity, and reduced safety risks. The new process is allowing the mine to produce more ounces per day and is improving profitability. The outlook for the FGM is positive, with continuing exploration success in the Harrier Zone and downdip in the Phoenix Zone. Expansion of the underground paste reticulation system to these areas will begin in 2020 and 2021. If the stope geometries are similar to those in the Swan Zone, the current diverter valve usage model will be implemented on these levels based on successes to date.

As backfill technology continues to develop and evolve, so will diverter valves. To meet the evolving needs of mine operators, valves containing multiple outlets for paste diversion, paste evacuation and flush water diversion, valves of higher-pressure ratings, valves of larger diameters, valves of more abrasion resistant materials, and valves capable of full automation and integration into intelligent systems are all conceived or being developed.

Minefill 2020-2021 – Hassani et al (eds)
© *2021 Taylor & Francis Group, London, ISBN 978-1-032-07203-6*

Filter revamping, the economic way to get old filters in tailings dewatering back on track

Jürgen Hahn
BOKELA GmbH, Karlsruhe, Germany

SUMMARY: The filtration of tailings becomes more and more an indispensable part of the environmental management system of preparation plants. This requires powerful, reliable and economical filter technology. The BOKELA BoVac disc filter represents a new generation of high performance disc filters, which have set a new standard in the alumina industry, in the dewatering of coal slurries and which are operated in many applications of tailings dewatering such as gold/copper, zinc or gold/silver tailings (Hahn et al. 2015). This modern vacuum disc filter type is in approximately 80% of all applications the most economical technology for tailings dewatering – both in CAPEX and OPEX (Hahn et al. 2017 and 2014). However, running disc filters of older design are often not operated under optimal conditions and the results do not meet expectations because they are operated up to their capacity limit or beyond. In this case, it must be decided whether the operation target should be achieved with a new filters or by optimizing the existing filter system. To answer this question it is necessary to evaluate the potential for improvement and to have ideas as to which concrete measures are to be taken for an effective filter modification. The optimization of running filters with the BOKELA filter revamp program is a very economical alternative to new investments. With this revamping program inadequate performance, excessive maintenance and high operating costs of existing filters can be quickly and inexpensively eliminated by transferring advanced design features of the BOKELA disc filter design to disc filters of old design. Benefits include: a 30% to 135% increase in filtration capacity, improved cake moisture, improved filter operation and reduced maintenance. However, the costs amount only some 20% to 30% compared to a new investment.

Keywords: disc filter, filter optimisation, filter capacity, availability, cost reduction

1 NEW INVESTMENT OR REVAMPING?

The first step in deciding between a new filter installation and an existing filter plant optimisation is to verify the current performance and capability of the existing plant, and its improvement potential.

In many cases the revamping of running filtration plants improves the filter capability to such an extent that the required targets can be achieved as effectively and reliably as with new equipment. This requires that the revamping is informed by deep know-how and experience concerning the filtration process and filtration equipment. The upgrading of operating filter plants is realised much quicker and impairs the whole production process significantly less than the planning and implementation of new equipment. Capacity increases of 30 % up to 135 % can be achieved by revamping.

Investing in new equipment usually means a prolonged multistage procedure including;

DOI: 10.1201/9781003205906-18

- a time consuming pre-engineering phase to specify and pre-plan rebuilding measures, to work out a specification of the new technology, etc.
- technology screening to identify the best suited technology available on the market, which often demands the performance of tests or trials
- OEM screening, calling for bids, to compare and evaluate the competitive offers with respect to technical and economic criteria, and to carry out negotiations
- complex engineering, since new equipment often requires modifications to the filter building or even a new building, both of which means comprehensive modifications to, or installation of completely new, piping, wiring, instruments, etc.
- potential for schedule slip in the installation of the new equipment through unforeseen delays, e. g. delayed delivery of one or more components
- overcoming of acceptance barriers by the operators and maintenance staff who also need a training and familiarisation period (with increased risk of malfunctions) to learn how to operate the new equipment well
- costs for additional peripheral equipment
- Applications and authorisations for financing the new investment.

Compared to this comprehensive and administratively heavy procedure, revamping existing filter equipment proceeds much more simply and directly, as numerous revamping projects have proven. The modernisation and upgrading measures of a filter revamping project normally cause less or no changes to the building, and the repercussions on the periphery of the filter plant are significantly reduced. The existing equipment is upgraded at the site and stays in place, so, the effort of pre-engineering, logistical planning, inquiring and ordering of supplementary peripheral equipment etc. are minimal compared to the installation of new equipment (Hahn et al. 2016).

In summary, the main advantages of a revamping process are;

- reduced planning and engineering effort
- compressed-schedule, fast realisation
- step-by-step engineering
- involvement of the owner's know-how and plant technical knowledge
- use of well-understood and operator-accepted equipment
- minimal costs for peripheral & supplementary equipment
- coverage of costs by the maintenance budget

The optimisation of existing processes and equipment however, demands a fundamental understanding of the dependencies between the product to be filtered, the applied filtration process and the filter equipment used. Only substantial filtration know-how and expertise and the understanding of these dependencies allow the engineering of optimal solutions, and finally guarantee the improved performance. Against this background BOKELA has developed a successful concept to de-bottleneck and revamp existing filtration plants.

2 BOKELA FILTER REVAMPING PROGRAMME - FILTER OPTIMISATION IN THREE STEPS

On the basis of numerous filter revamping projects in nearly all industries, BOKELA has developed a special program for filter optimisation in three steps as shown in Figure 1. The three phases allows rigorous cost control, involvement of own plant technology and minimising risk. It is carried out with the know-how and the experience gained by upgrading of the drum, belt, disc and plate filters, filter presses, Niagara filters, Kelly filters etc. of nearly all OEMs.

2.1 Diagnostic step

In the first phase, the filtration behaviour of the product and the filter performance are examined in lab and field tests. This analysis of the existing state defines the real capacity of the

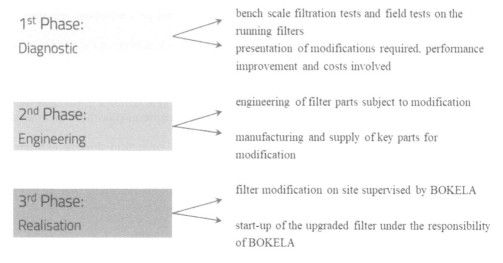

1st Phase: Diagnostic	→ bench scale filtration tests and field tests on the running filters
	→ presentation of modifications required, performance improvement and costs involved
2nd Phase: Engineering	→ engineering of filter parts subject to modification
	→ manufacturing and supply of key parts for modification
3rd Phase: Realisation	→ filter modification on site supervised by BOKELA
	→ start-up of the upgraded filter under the responsibility of BOKELA

Figure 1.   General schedule of the filter optimisation program.

filtration plant and exposes the "bottle-necks". A "Test Report" gives very concrete details about the optimisation potential, and first estimates of costs and potential profitability of improvements.

2.2   *Engineering step*

In the second phase, modifications to the filtration plant for re-engineering are worked out, presented in a "Modification report" and discussed with the customer. According to these suggestions, it can be decided which of the recommended measures shall be selected. BOKELA then develops the required specifications and drawings.

2.3   *Realisation step*

The third phase – i.e. the modification works and the commissioning of the filtration plant - starts when the specifications and drawings are checked. If the filtration plant consists of several filter units, only one filter unit will be modified initially. Most of the purchasing is organised by the customer, and the modifications are carried out in the customer's workshops as far as possible, while BOKELA supervises the modification work.

When the first filter modification is finished, the improved filter performance is determined in a test and compared to the calculated data from the tests of the first phase. All further filter units can be modified by the customer themselves in an analogous way.

3   TYPICAL RESULTS AND TYPICAL COSTS OF FILTER REVAMPING

Revamping of running filter plants can be realised faster and with significantly lower cost (ranging from 20% to 40%), compared to the cost of a new filter. Typical modification measures and typical costs of a filter revamping are presented in Figure 2 and Table 1 using the example of a vacuum disc filter. Figure 2 shows filter components subject to modification and gives values of throughput and moisture improvement in disc filters for seed filtration in an alumina refinery reference. Typically, filter revamping can be performed with a small or large package of modification measures depending on the individual situation and client's objectives. Table 1 gives typical costs for these modifications in percent of the cost of a new filter.

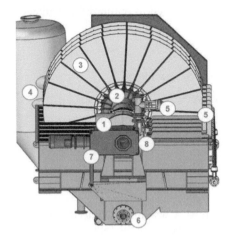

Large Package		
	1 Control head	
	2 Filtrate pipes / Shaft	
	3 Segments	
	4 Filtrate receiver	
Measures	5 Cake discharge	
	6 Feeding / Agitation	
	7 Drive	
	8 Level in Trough	
Capacity increase	60 – 150 %	
Costs compared to new invest	15 – 25 %	

Figure 2. Typical modification measures and improvements for a disc filter revamping (given values refer to Al-hydroxide seed filtration).

Table 1. Typical costs for a disc filter revamping.

	Modification measures (see Figure 2)	Throughput increase Δms	Costs in percent of costs for purchase of new filter	Costs in percent of costs for new filter incl. installation and commissioning
Small package	1, 4, 5	30 – 50 [%]	25 [%] *	approx. 10 [%] **
Large package	1, 2, 3, 4, 5, 6	60 – 100 [%]	75 [%] *	20 - 30 [%] **

* Only costs in percent of costs for new filter unit.
** Total costs of a new filter investment amount to 3 to 4 times the price of the new filter due to the additional costs for building, piping, auxiliary units and engineering.

This comparison is based on the total cost of a new filter finally amounting to 3 to 4 times the price of the new filter itself due to the additional necessary costs for building, piping, auxiliary units and engineering.

4 CASE STUDIES – EXEMPLARY REFERENCES

BOKELA is the No. 1 expert in filter revamping world-wide and has carried out a multitude of revamping projects in many industries on a variety of filter types of numerous OEMs. A main focus of BOKELA's revamping activities has been the alumina industry with many modifications of vacuum disc, pan and drum filters operated for seed filtration, product filtration or red mud filtration. Two exemplary case studies are briefly discussed below.

4.1 *Revamping of disc filters for seed filtration*

In an Alumina refinery, 5 disc filters of type A with 8 discs and 160 m² filtration area each, and 3 disc filters of type B with 3 discs and 110 m² filtration area each were operated for seed filtration. To improve the plant capacity BOKELA was requested to optimize these filters by a filter re-vamping.

Target of the client was:

– increase of the total plant capacity
– de-bottlenecking of the existing seed disc filters instead of new filters

Expectations with respect to the optimized disc filters were:

– solids throughput increases by the amount of two additional filters of type A
– moisture content may not increase
– secure and complete cake discharge
– improved filter availability
– more spare capacity
– reduced filter operating costs

To provide for these improvements one (1) filter of type A was modified according to the "large package" modifications (see Figure 2 and Table 1), comprising the following measures:

– new filter barrel (9 discs)
– new high performance segments (20 per disc)
– higher slurry level in the filter trough
– modifications to control head
– new control plate
– new filter drive
– new snap blow system for cake discharge
– filter operation without overflow

Four (4) filters of type A were modified according to the "small package" modifications (see Figure 2 and Table 1) comprising the following measures:

– higher trough level
– new control and wear plate
– new snap blow system for cake discharge

The results of these modifications in terms of operating and performance data before and after modifications are shown in Table 2. Figure 3 shows a modified type A disc filter in operation with highly effective cake discharge.

The data in Table 2 illustrate that a targeted capacity increase of nearly two times the additional capacity of one (1) type A disc filter could be achieved with the modified filters. The filter which received large package modifications has about 75 % more solids capacity after revamping while the four (4) filters which were modified with small package modifications show a solids capacity increase of 30% each. In total, the surplus capacity of the modified filters is equivalent to the capacity of two old type A disc filters. Cake moisture improved from 20.1 wt-% to 14.5 wt-% and 19.5 wt-%.

Table 2. Operating and performance data of seed disc filters of type A before and after modifications.

operating and performance data of type A seed disc filters		before	after modifications with large package (1 filter unit)	after modifications with small package (4 filter units)
filter speed	[rpm]	1	2.0	1.35
cake moisture content	[%]	20.5	14.5	19.5
specific slurry flow	[m³/m²h]	1.7	3.0	2.2
specific solids rate	[t/m²h]	3.0	5.4	4.0
cake discharge	[%]	80	95	95

Figure 3. Complete cake discharge of a type A disc filter after revamping with "large package" modifications.

The cost of modifying these 5 disc filters was less than 25% of the purchase and installation cost of two (2) additional type A disc filters.

4.2 *Revamping of disc filters for coal ultrafines filtration*

This coal disc filter optimisation project at the Saraji coal washery in the Bowen Basin, Australia was carried out together with the experts of Saraji. The original filters at the site were two Denver disc filters + (1 x EIMCO Belt Filter).

The disc filters were installed 30 years ago and had

– 14 discs each with 4.2 m disc diameter
– 12 sectors per disc (s/steel) – 40 kg per sector (Figure 4)
– total filter area: 280 m^2 per filter
– filter cloth: s/steel mesh
– air agitation (formerly mechanical agitation)
– filters operated with slurry overflow

Typically the filters rotated quite slowly at around 0.5 to 1 rpm and were fed with more fine coal slurry than they could process with the overflow reporting back to the feed system. The filters were also quite maintenance intensive because the original equipment designers did not recognise the flaws in their design. Routine access to the filters was cumbersome due to the need to build temporary access. And on top of the temporary access awkward maintenance

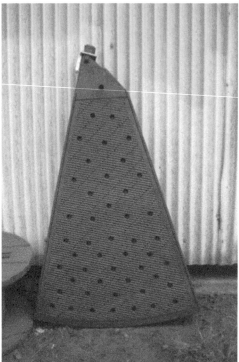

Figure 4.   Old and heavy (40 kg) filter segments (12 segments per disc).

activities such as replacement of heavy segments weighing at least 40 kg (90lb) was expected. This of course leads to maintenance being delayed until the filtering process is severely affected.

4.2.1 *Project Brief from the Client*
The client expected the following filter improvements from the revamping project:

– increase filtration capacity by approx 100% (as per testwork assessment)
– maintain or improve total cake moisture
– (target for coal moisture + free moisture = low 20's)
– improve operation (automated)
– reduce maintenance requirements

4.2.2 *Implementation of Modifications*
The following disc filter modifications were implemented by BOKELA to meet the design brief:

– 2 independent filters (each 7 discs) to replace each Denver filter, achieved by dividing trough with a new wall
– new filter shafts/barrels with removable filtrate pipes (Figure 6)
– new low wear sectors; 20 per disc, approx 19 kg each (Figure 5)
– new bayonet segment fixing for fast change out
– new low wear control heads
– new drive system (motor/slip-on gear box)
– new snap blow system
– increase of slurry level
– automatic level control so there is no overflow recirculation required

Figure 5.   New filter segments (19 kg per segment, 20 segments per disc).

Figure 6.   Lifting of new filter barrel.

- relocation of feed pipes
- no trough agitator required
- new permanent walkways
- modification of filtrate receiver to reduce carryover.

4.2.3 *Design and Performance of the Coal Filters after the Revamping*
The revamped filters run with up to 3 rpm and have 100 % more filtration capacity as can be seen from Table 3. Figure 7 shows one of the upgraded filters.

Figure 7.   Revamped filter with improved access along the filter by permanent walkways.

Table 3.   Filter data and filter performance before and after revamping by BOKELA.

	Before	After
disc diameter	4.2 m	4.2 m
number of discs	14	2 x 7
segments per disc	12	20
filter area per disc	20 m²	22 m²
segment fixing	bolted (tie rod)	bayonet fixing
filter speed	up to 0.75 rpm	up to **3 rpm**
total area	280 m²	2 x 154 m²
trough design	Air agitated trough	common trough with feed and disc agitation
volumetric flow	220 m³/h	**2 x 230 m³/h**
tonnage	76 t/h	**2 x 80 t/h**
specific throughput	270 kg/m²/h	**520 kg/m²/h**

5 CONCLUSION

Filter optimization with sensible modifications is a viable alternative to increase filter capacity and/or improve cake moistures. It can replace the requirement to purchase new filters plus the infrastructure for the new filter (building, piping, electrics, auxiliaries etc = 3 x filter cost). The optimization of filters requires a comprehensive understanding of all aspects of filter design and technology and a well founded knowledge of dewatering practice and hydrodynamics. Improving performance of filters can not be done by just tackling single 'soft spots and bottle-necks' of the filter itself. It rather requires a 'holistic' expert approach which takes into consideration the complex nature of design and performance parameters. Increasing the filter capacity is not the only benefit of filter re-vamping, an optimised filter design also improves the product quality and reduces operating and maintenance costs. Furthermore, an optimised filter design increases the filter flexibility and filter availability and leads to a significant improvement in filter operational control. Last but not least, the operators can work with familiar equipment.

ACKNOWLEDGEMENTS

Thank you to our clients and their staff for their great cooperation and input into the revamping projects featured in this report.

BIBLIOGRAPHY

Hahn et al. 2017 – Hahn, J., Egger, A. (2017) 'Report on Tailings Dewatering with High Performance Disc Filters' 12th International Symposium on Mining with Backfill, 19–22 February, Denver, USA

Hahn et al. 2016 – Hahn, J., Viet, Ch. (2016) 'Three Steps to Improved Filtration Performance and Reduced Cost in Times of Limited Capital' ICSOBA 2016, 34th Conference & Exhibition, 3–6 October, Quebec, Canada

Hahn et al. 2015 – Hahn, J., Bott, R., Langeloh., (2015) 'Size Matters, Weight Too - Disc Filter Design For the Future' proceedings of 10th AQW 2015, Perth, Australia.

Hahn et al. 2014 Hahn, J., Bott, R., Langeloh. T., (2014) 'Economical dewatering of tailings for mine backfill with high performance disc filters' proceedings of the 11th International Symposium on Mining with Backfill, 20–22 May 2014, Perth Australia.Conference papers

Minefill 2020-2021 – Hassani et al (eds)
© *2021 Taylor & Francis Group, London, ISBN 978-1-032-07203-6*

Theoretical method to improve the placement of backfill behind barricades

Chris Lee
Golder Associates Ltd., Canada

SUMMARY: Mine backfill is typically placed in stopes with drawpoints or access drifts at the bottom of the stope. To contain the fluid backfill, barricades are required to prevent the flow of backfill out into the access drift. Typically barricades are constructed of muck or reinforced shotcrete; however, there are some disadvantages with typical barricade design and this paper examines some theoretical possibilities for a new barricade design that may improve safety, productivity, cycle time and the mines economics.

Keywords: Backfill, pastefill, barricades, bulkheads

1   INTRODUCTION

Paste backfilL is a fluid material that is placed in stopes after the extraction of ore. Typically, the ore is extracted from an access drift at the bottom of the stope and therefore, when placing paste backfill it is required to barricade the access drift in order to prevent backfill from flowing out into the drift. Backfill barricades are typically constructed out of reinforced shotcrete or simply waste rock and must retain the hydraulic head produced by the placement of the backfill.

A typical sequence of backfilling will include the installation of the barricade and filling of the stope up until a point just above the barricade. This initial backfill placement acts as a plug and serves as a monolithic barricade for future pours on top of the plug pour. The constructed barricade is therefore only required to support the initial plug pour and then the plug pour supports subsequent pours.

This barricading method has been used for many years with very little change in methodology or materials of construction. It serves its purpose well; however, there are a number of drawbacks associated with this typical method of barricade construction under certain situations as discussed below:

– For underhand stopes (i.e. stopes mined underneath a previously mined and backfill stope), it is frequently required to develop back through the previously placed backfill in order to establish a drilling horizon for the drilling of the next stope below:
– This results in additional waste production since the mined backfill will need to be hauled and disposed of,
– It also results in increased cycle time due to the requirement to develop through the backfill using drill/blast/ground support,
– Before a drift can be developed through the backfill the barricade will need to be removed which typically means blasting of the shotcrete barricade which takes some time and requires removal of the reinforced shotcrete waste materials,
– Drifting through backfill results in increased backfilling volume if the drill drift will later be filled with backfill;

DOI: 10.1201/9781003205906-19

Table 1. Barricade related task durations – Note bolded items are related to areas where a new barricade method could reduce the backfill cycle time.

Barricade construction	3	Days
Plug pouring	2	Days
Plug curing	**7**	**Days**
Main body pouring	5	Days
Main body curing	28	Days
Removal of barricade	**1**	**Days**
Drifting through plug	**5**	**Days**
Additional pour time to fill drift void	**0.5**	**Days**
Total	51.5	Days

– A muck or shotcrete barricade requires a period of construction for the barricade which also delays cycle time. Typically, this construction period is 1-3 days depending on the type of barricade used;
– Conventional barricades require a delay period to cure the plug pour before the main body pour can commence. Typically, this curing period is 7 days or more; however, if there is a demand for a quick turnaround on cycle time, then it is possible to reduce the curing time by adding more cement which incurs additional binder costs.

The total, barricade related, task durations are listed below in Table 1. It should be noted that none of these tasks are typically performed concurrently and therefore any reduction in any one of these task durations will result in a reduction of overall cycle time. Although there is an option to perform some of these tasks concurrently there is additional risk associated with this and generally it is avoided.

Although there are parts of this backfilling cycle that are not improved by a change in barricading method, there are some portions of the cycle that can be improved such as the barricade construction time, plug curing time, removal of barricade, drifting through plug and additional time to fill drift void. The total time in the backfill cycle that is potentially available for improvement is significant (16.5 days of potential savings out of the 51.5 days total backfilling cycle time).

In addition to cycle time improvements, there are significant costs associated with these identified tasks. Particularly with the task of removing the shotcrete barricade and developing through the plug. These tasks are where the opportunity for improvement in the backfilling economics is most prevalent.

2 ALTERNATIVE BARRICADING METHOD CONCEPT DEVELOPMENT

In developing the alternative barricading concepts, the first step was to identify what properties the new barricade would have and relate those properties to the opportunities for improvement identified above. This relationship between barricade properties and opportunities is summarized below:

– Shorten the barricade construction period – reduce cycle time;
– Robust enough to support full weight of main body fill to eliminate the 7 days waiting period for plug curing – reduce cycle time;
– Provides a void at the access drift level so that no development in fill is required – reduce cycle time and reduce development and refilling costs;
– Reduce exposure of personnel to work near the mouth of the stope – Automate barricade construction to minimize personnel involved and keep their location far from the stope;
– Reduce the potential for barricade failure – Change the barricade design concept so that the mechanisms for failure are reduced or that additional redundancy is provided.

The list of opportunities and desirable outcomes above was used to direct the concept development of what a preferred barricade design would look like. Since the main drivers for improvement in cycle time and costs revolves around the ability to construct the barricade more quickly, eliminate plug curing time and eliminate the development through the backfill, the ideal barricade method will incorporate the ability to create a void in the stope that prevents a certain portion of the stope from being filled. Essentially the barricade should be a type of tunnel form that prevents the backfill from flowing into the area of the plug pour that will be required to be opened up in the future.

Options were developed and evaluated and are discussed below.

2.1 *Waste filled tunnel form*

This concept can be seen in Figures 1 through 12 below. The concept includes a type of remotely installed formwork that allows waste to be placed in the desired geometry to prevent the inflow of paste into the desired void. Essentially the formwork retains the waste rock in the desired geometry and the waste rock supports the load of the paste once the stope is filled.

The concept includes a set of sprung steel arches that provide containment for muck placed inside the form. Between the arches is a heavy duty reinforced rubber fabric (similar to flexible vent ducting). The steel arches provide the principal support to take the majority of the load and the rubber provides the secondary support and containment to prevent rocks from exiting the tunnel form.

With the tunnel form installed and waste rock placed in the tunnel form, the stope can be filled with cemented paste backfill or any other fill type without stopping to wait for the plug to cure since the waste rock will support the full load of the main pour. Once the backfill has cured, the waste can be extracted much more easily than drilling and blasting through fill.

Although simple in concept there are a number of challenges with the waste filled tunnel form concept as follows:

– Extending the tunnel form to the end of the stope will require mechanised equipment to move under unsupported ground. It is proposed that the best way to move the tunnel form assembly is by:

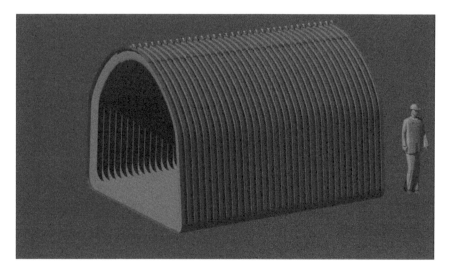

Figure 1.   Tunnel form linkage and pin arrangement.

Figure 2. Tunnel form in its packed position. The rod through the tabs keeps the sprung steel frame from expanding to its full diameter and the plastic liner is folded into the collapsed frame.

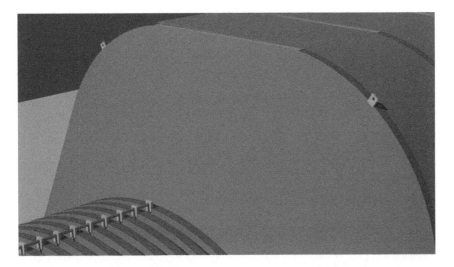

Figure 3. As the tunnel forms are pulled into the stope the tabs are pulled off the rod and the sprung steel snaps out to its full diameter, restrained by a cable that holds the ends of the steel frame together.

– The tunnel form assembly will be connected to a monolithic end weight that will be installed at the far end of the stope (away from the drawpoint). That monolith will be lifted by a remote scoop and the scoop will deposit it at the far end of the stope. Once the scoop retreats to the mouth of the drawpoint it will connect its bucket to the tunnel form pulley and then pull the tunnel form to the monolith by driving backwards out of the drawpoint. Alternatively, the tunnel form pulley assembly could have a mobile tugger that is used to pull the tunnel forms to the monolith;
– The tunnel form profile will ideally be the same profile as the access drift. Of course, this will mean that it is difficult to move the tunnel forms around the access drift or into the mine itself if they are too large. While it is possible that the tunnel forms could be made to

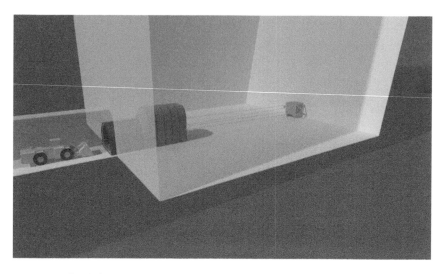

Figure 4.   A monolith is installed at the far end of the stope by the scoop in remote mode. The tunnel form is pulled to the monolith by pulley cables. As the tunnel form is pulled into the stope the sprung steel arches snap out to a larger dimension as the rod tabs on each arch are pulled off the rod and the arch snaps out to a dimension that is limited only by the cable tie between the ends of the arch.

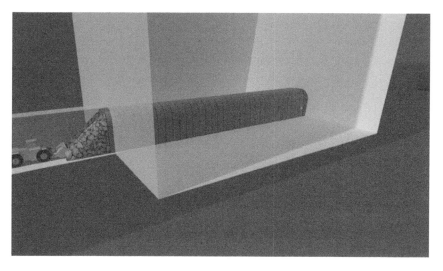

Figure 5.   Waste rock is placed in the tunnel form by remote scoop. The scoop will likely not fill the rock perfectly tight to the top of the form but the idea is to fill the majority of the tunnel form with rock.

be slightly smaller than the mine access drifts, it is likely that this will reduce the effectiveness of the drill drift and is undesirable. To address this the sprung steel tunnel form, members will be packed into a smaller cross section with a linkage and rod arrangement so that as the pins are extracted, the sprung steel members will snap outwards to the limits of the linkage (shown in Figure 1). The tunnel form assembly will be mounted on a skid with the rubber folded to the inside and as the tunnel form is extended to a point beyond the mouth

Figure 6. Paste poured to top of stope – With waste rock filling the tunnel form it can withstand the full pressure of the filled stope. The key is to seal the drawpoint area to ensure that fill can't short circuit from the stope through the mouth of the tunnel form.

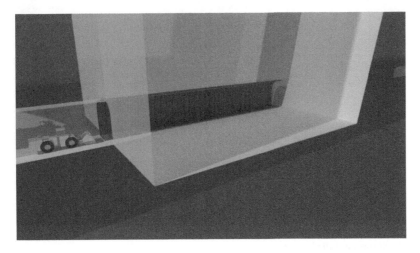

Figure 7. Waste rock and monolith removed from tunnel form by remote scoop.

of the access drift, the rod will be pulled out of their holes and the sprung steel will snap out to the full dimensions of the form.
- Filling the tunnel form will be accomplished by using a remote scoop which will bring waste rock or ore to the tunnel form and place it within the extent of the form (from the monolith to a point several meters outside of the stope). Although it is considered likely that the waste rock and tunnel form rubber alone will seal the stope it may be required to shotcrete the interface between the tunnel form and the drawpoint back in order to ensure a tight seal.
- Once the tunnel form is full the stope can be filled with backfill and allowed to cure.
- After curing, the waste rock can be mucked out by the scoop and ground support put in place.

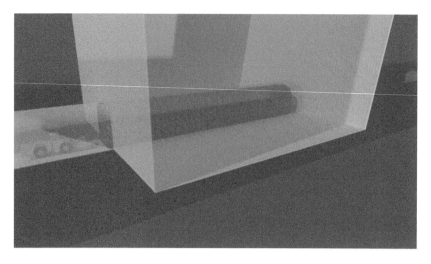

Figure 8. Ground support installed if required. Note that it is possible that ground support could be incorporated into the tunnel form design so that it is embedded in the paste during pouring.

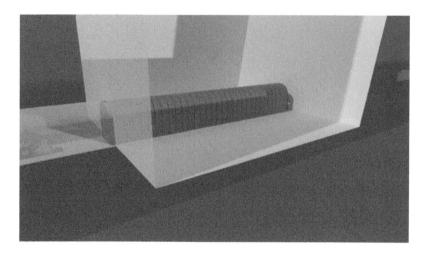

Figure 9. Void form installed.

2.2 *Empty tunnel form (void form)*

As an alternative to the waste rock tunnel form, it is possible that the form be used without waste rock and use the structural strength of the sprung steel members to prevent fill from entering the form. The installation method for placing the tunnel form would be exactly the same as with the waste rock filled tunnel form; however, there are several key differences in the properties of this type of barricade:

– Since the form would not be filled with waste rock, it would have to be substantially stronger than the waste rock tunnel form in order to support the collapse pressure (rather than hoop stress pressure) due to the weight of the main body fill above. It is likely that this method would be more practical if the pour height was limited to a plug pour height (typically 7 m);

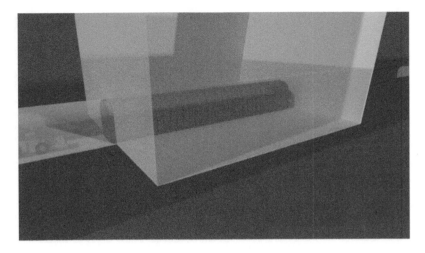

Figure 10.   Paste plug poured to 8m above floor (2m above tunnel form).

Figure 11.   Main body pour.

however, this compromise would limit the effectiveness of the tunnel form since one of the key advantages of the tunnel form is the potential to eliminate the 7 days curing period for the plug pour.

– With waste rock in the form it does not really matter if the tunnel form has rips and tears in it or whether there are leakage paths into the form from the exterior of the stope. All those leakage paths can fill the voids that remain inside of the waste filled tunnel form without compromising the overall concept. With the void tunnel form any major leakage path will prevent the tunnel form from working and filling would need to stop while the leaking area was repaired. Repairs to a leaking form when no personnel access is allowed would be difficult and in the end, the recovery plan may be the placement of waste rock and the use of the void form in the exact same way as the waste rock filled form.

– Because the void form is empty and the surrounding fill will have an SG of approximately 1.7 to 2.0, there is the potential that unless anchored to the floor of the stope, the void form

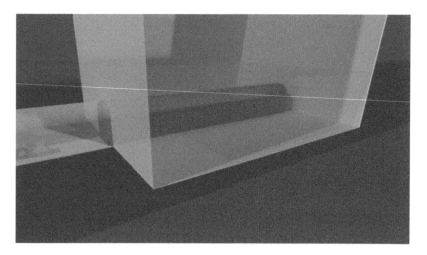

Figure 12. Ground support.

will float on the surface of the placed fill and may have an irregular floor or even a major heave in the floor in between the fixed points at the monolith and the stope drawpoint. Although this can be rectified by the placement of weights on the tunnel forms by some remote method, it is another step that contributes to the complication of this option.
- The major advantage of the void form is that there is no requirement to fill the form with waste rock or remove that waste rock. Although this is not a major cycle time component in comparison with waiting for the plug to cure or developing through fill, it would likely be at least 1 or 2 days of scoop time in order to fill with waste rock and then remove the waste rock.

2.3 *Other considerations*

Building from the basic concepts described above, there are other opportunities that may be achievable and beneficial for the mine to consider:

- Elimination of ground support step(s) – it is possible that with the design of a tunnel form that integrates elements of ground support into the form itself, that some, or all, of the post filling ground support could be eliminated. For example, it is possible that the sprung steel main form members could have both mesh and anchor bolts attached as part of the form structure. As the forms are pulled into the stope the mesh and bolts could swing into place (i.e. anchor bolts pop up to be perpendicular to the form face) so that when the backfill is placed it becomes integrated with the backfill and essentially provides reinforcement for the backfill. The downside is that this would make the tunnel form more complicated which will result in more potential for operational difficulties. The elimination of ground support steps could be complete (i.e. no bolts, shotcrete or screen since those would be replaced by the tunnel form structure) or it could be partial (maybe only bolts will be part of the tunnel form structure and shotcrete and screen will still be applied). It is also possible that ground support could be improved with the tunnel forms since the bolts and screen are embedded in the backfill. With embedded bolts rather than drilled and anchored bolts there are design options such as J hooks for bolts that could be considered to provide greater pullout resistance or reduce the bolt length.
- Re-usability of the barricade – It is possible that instead of the tunnel form being a consumable that it could be made to be re-usable. The potential to use a more robust, all steel tunnel form that could be put into place and then retracted could reduce the cost of tunnel form consumables. The method of extracting the form is the major hurdle to this

concept since it would likely be a very heavy steel structure that would be difficult to move around with limited access, cranage and adhesion to the backfill. In addition, it is likely that the steel forms would be damaged by the placement of waste rock and the time to repair the forms would likely be significant. It does, however remain a possibility.

3 CONCLUSION

The potential for using tunnel forms to improve barricading practice for underhand mining methods appears to be quite possible. Based on an initial review of the tunnel forming options above, it is likely that the use of waste rock filled tunnel forms will be a more viable alternative than void forms.

To advance this concept further to the piloting stage it would require a relatively small investment to perform the design and fabricate prototype tunnel forms. Due to the low cost of trialling this concept and the potential high payback for cycle time constrained underhand mining operations, this is a concept that mine operators should consider investigating in the near future.

Laboratory testing on static and dynamic behaviours of backfill

Minefill 2020-2021 – Hassani et al (eds)
© *2021 Taylor & Francis Group, London, ISBN 978-1-032-07203-6*

A novel approach to assessing the early age strength of fibrecrete, using shear wave velocity

Stephen Mcgrath & Matthew Helinski
Outotec Pty Ltd

SUMMARY: Prior to backfilling, fibrecrete barricades (which for the purpose of this paper refers to both barricades and bulkheads utilised for Minefill purposes) are typically cured to achieve a target strength. However, due to sensitivity of early age fibrecrete strength caused by variations in mix properties, curing environments and spraying techniques, the rate of strength development can be quite varied. Challenges associated with conventional curing, coring transportation, and destructive strength testing methods and mine location generally prevent early age quality control testing of fibrecrete barricades and the consequence is often conservative cure periods leading to extended fill cycle times.

This paper further develops a novel technique where laboratory testing is used to develop a unique relationship between strength and the shear wave velocity (or small strain stiffness) of the fibrecrete. The investigation considers a range of different variables encountered during the manufacture of fibrecrete to illustrate the unique relationship.

The presented non-destructive test method provides a practical method to define the relevant barricade strength and shear wave velocity relationship. Implementation of improved quality control techniques such as that proposed can reduce fill cycle times without compromising safety.

Keywords: Fibrecrete, Shear Wave Velocity, Non-destructive testing, Early Age Fibrecrete Strength

1 INTRODUCTION

Advancement in technology has resulted in the use of fibrecrete (fibre reinforced shotcrete), often labelled as shotcrete, for ground support, to increase the rate of development in mines. The introduction of fibres to shotcrete has led to an increase in toughness and impact resistance (Morgan & Bernard, 2017). Fibrecrete has also formed an integral part of the backfilling process, whereby the fibrecrete is used for barricade construction.

To date, the industry standard is to ensure that the fibrecrete reaches a target strength by 28 days hydration. However, there is usually limited ongoing early age quality control testing of the fibrecrete and a 'rule of thumb' approach, which includes a significant factor of safety, is typically used to extrapolate longer term quality control testing. Ultimately, it is the ground control engineer who determines the compressive strength required to provide safe re-entry (Rispin et al., 2017) and adequate bulkhead capacity. This duration can have significant impact on mine development and stope backfilling rates.

DOI: 10.1201/9781003205906-20

Numerous non-destructive test (NDT) methods are available for measuring the early age compressive strength of fibrecrete. These include the rebound hammer and ultrasonic pulse velocity (ISO 1920-7, 2004), penetration resistance (ASTM C403, 2016), needle penetrometer, Beam-End Tester and the Hilti gun-test method (AuSS, 2010). However, these all come with intrinsic issues and most do not measure accurately over the range required, which can significantly affect the relevance of results unless properly correlated (Moczko and Mockzo, 2018). The Australian Shotcrete Society (AuSS, 2010) suggest the only effective means of assessing direct compressive strength is using the Beam-End Tester. However, this only measures up to 8 MPa which is the highest of all . Due to the limiting range on these NDT methods. This strength is generally not suitable for the backfill process, which typically requires strength in the order of 10-30 MPa.

There is an abundance of data available on the early strength of fibrecrete and it is highly variable, as illustrated by Saw, Villaescusa & Windsor (2015). These variations are most likely a consequence of test method, variable mix designs, curing environment, application techniques, quality control measurements and fibrecrete design constituents. Therefore, it is imperative for any NDT method, that thorough and relevant correlations are made.

Stope backfilling typically involves placing a cemented slurry into an open stope, which is often contained using a sprayed fibrecrete bulkhead. The time at which backfilling can commence, post barricade construction, is based around the fibrecrete achieving a target strength, typically between 10-30 MPa, which is normally correlated to a curing period. This often leads to a conservative timeframe before backfilling can occur behind a fibrecrete barricade which can result in significant impacts to the mining schedule.

As illustrated by Helinski et al. (2011), for a given bulkhead geometry the aspect that most significantly influences the capacity of fibrecrete bulkheads is the fibrecrete strength. However, in order to accelerate stope cycle times and achieve economical benefits, it is often useful to begin filling behind minefill bulkheads as quickly as possible. The problem with loading fibrecrete bulkheads at early hydration periods is that even subtle variations in mix inputs or curing conditions, can have a significant impact on the rate of strength development. While this variability is widely recognised, due to the logistics of spraying and transporting quality control (QC) samples, it is difficult to regularly gather QC data at relevant hydration periods (1-3 days), which makes managing this risk challenging.

In addition to variations in fibrecrete mix properties and curing conditions, delays in heading availability, substantial transport distances from batch plant and changes to mine plans, can have a significant impact on the duration between batching and spraying of underground mine fibrecrete. This often results in the fibrecrete being discarded, or more water and/or chemicals being added to delay the cement hydration process. This can have a significant impact on the early age fibrecrete strength development.

This paper presents further development of a novel approach to assessing the fibrecrete strength using shear wave velocity across a range of different mix constituents to form a basis for future field research and implementation.

2 FIBRECRETE DESIGN AND SENSITIVITY STUDY

Fibrecrete is a designed material that comprises of cementitious binder, fine and coarse aggregates, fibres, water and often additional chemical and cementitious additives. It is usually mixed at an on-site batch plant and transported in an agitator bowl, where it continues to mix during transportation to the desired location. It is then pumped and sprayed to form a solid concrete layer on headings and barricades. The physical,

Batching Data and Calculations	Batching Data and Calculations
Raw Constituents	Raw Constituents

Cement 470 - For 1m^3		Cement 370 - For 1m^3	
Up front Water (Litres)	170.0	Up front Water (Litres)	160.0
Radmix 65 (kg)	6.0	Radmix 65 (kg)	6.0
Pozz 370c (Litres)	2.5	Pozz 322Ni (Litres)	1.1
Delvocrete Stabliser (Litres)	2.0	Delvocrete (Litres)	2.0
10/7mm Striling Nor (kg)	500.0	WKAL107 10/7mm	430.0
Crusher Dust (kg)	440.0	WKALCD Qsand (kg)	700.0
Fine Sand (kg)	680.0	WLENFS Leahys Sand (kg)	682.0
GP Cement (kg)	470.0	GP Cement (kg)	370.0
-	-	Simcoa Silica Fume (kg)	30.0
Rheobuild 1000 (Litres)	5.2	Super Plasticiser (Litres)	3.7
Accelerator (Litres)	30.0	Accelerator (Litres)	20.0
Maximum Total Water (litres)	190.0	Maximum Total Water (litres)	200.0

Figure 1.   Fibrecrete mix constituents for two different mine sites.

chemical and mechanical properties of fibrecrete effect the gelling, setting and hardening of the fibrecrete.

For this study, two different recipes from two different mine sites were investigated; Cement 470 and Cement 370, referring to 470 kg and 370 kg of GP cement per cubic meter of fibrecrete, respectively. The mix constituents are presented in Figure 1.

The mix constituents presented in Figure 1 are the "standard design" mix adopted for the laboratory testing. However, during the batching and transportation process, these constituents can vary. To investigate the impact of mix variations and curing conditions on strength, a sensitivity study was undertaken. This sensitivity study considered the following permutations:

1. Standard Fibrecrete Mix (Base Mix)
2. Base Mix, No Accelerator
3. Base Mix, –20% Cement
4. Base Mix, +20% Cement
5. Base Mix, +50% Water
6. Base Mix, +100% Water
7. Base Mix, Cured at ambient laboratory temperature
8. Base Mix, +20% or +50% Accelerator
9. Base Mix, +50% Water and –20% Cement
10. Base Mix, +50% Rheobuild

Each of these mixes were cast into 100 mm diameter × 200 mm long cylindrical moulds and a single sample was tested for strength (UCS) and seismic velocity (Vs) after hydration periods of 1, 2, 3, 7, 14 and 28 days. With the exception of Mix 7, all specimens were cured in a curing chamber that was set to achieve a temperature of 35°C ± 2°C and a relative humidity exceeding 95%. This is expected to be representative of underground conditions. Mix 7 was cured in ambient conditions where the temperature varied from 16-26°C.

3   SHEAR WAVE VELOCITY

Prior to testing each specimen in unconfined compression, the specimen's shear wave velocity was measured. The small strain shear stiffness was measured using piezo ceramic shear plates (Dyvik and Olsen, 1989, Baig et al. 1997, Fernandez and Santamarina, 2001). The shear plates are attached to opposite ends of the sample and for good results, the transmitter and receiver

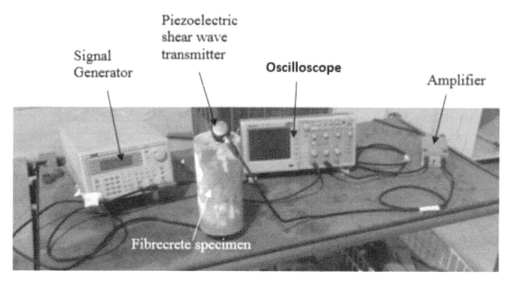

Figure 2. Photograph showing seismic testing apparatus of fibrecrete sample.

are properly coupled to the sample through a layer of viscous conductive gel (petroleum jelly). The shear plate at the top of the sample is used to generate a shear wave, which travels through the sample at a given frequency and are detected by the receiver element at the opposite end of the sample. A photograph of the experimental setup used in this testwork is presented in Figure 2.

The time taken for the shear wave to travel through the specimen is measured as time (Δt). Measuring the length of the specimen and using the time taken to travel from the sent transmitter to the received transmitter, the shear wave velocity, V_s, can be calculated. An example of a sent wave and received wave is presented in Figure 3, where the time taken to travel through the specimen is measured from peak to peak as illustrated.

Santamarina et al. (2001) provided an equation and reasoning for relating the shear wave velocity (V_s) to the small strain shear stiffness (G_0) and the mass density (ρ) through Equation (1):

$$V_s = \sqrt{\frac{G_0}{\rho}} \tag{1}$$

From the theory of continuum mechanics, the small strain shear stiffness is related to Young's modulus (E) and Poisson's ratio (v) in Equation 2:

$$G_0 = \frac{E}{2(1+v)} \tag{2}$$

Ciancio and Helinski (2010) showed there a clear relationship between the development of strength and Young's modulus with time. It was therefore expected that a similar relationship between compressive strength and shear modulus could be achieved and hence shear wave velocity and fibrecrete strength could be developed.

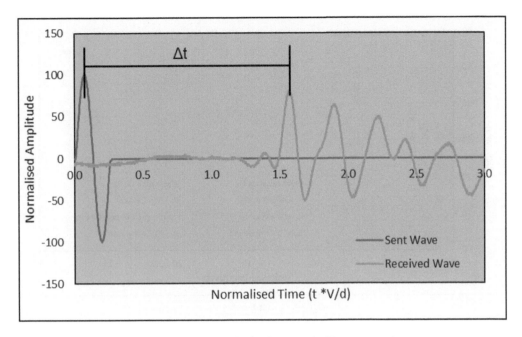

Figure 3.   Sent seismic wave and the measured seismic wave of a fibrecrete specimen.

4   EXPERIMENTAL RESULTS

The measured unconfined compressive strength (UCS) and the shear wave velocity (V_s) for the range of different fibrecrete mixes is presented against hydration time for 'Cement 470' recipe in Figure 4a and b and for the 'Cement 370' recipe in Figure 5a and b. The results for both recipes show that the mix constituents and curing conditions can have a dramatic impact on the strength of early age fibrecrete. This strength variation is most significant in the first 1-3 days, which is the relevant cure period for ground support and bulkhead applications. As an example, this is the impact of curing conditions, where the results show a 40% variation in 3 day fibrecrete strength (and therefore bulkhead capacity) due to curing condition. Notably, these figures also show a similar trend exists between shear wave velocity and hydration time.

To illustrate the correlation between strength and shear wave velocity the fibrecrete UCS is plotted against shear wave velocity in Figure 6 and Figure 7, for the 'Cement 470' and 'Cement 370' mixes, respectively.

The results presented in Figure 6 and Figure 7 show a relatively unique exponential relationship between the fibrecrete strength and the shear wave velocity.

5   DISCUSSIONS

The test results presented in this paper show a clear correlation between shear wave velocity and the strength of fibrecrete for a given mix recipe.

Whilst the results do appear somewhat scattered, which is expected given the significant changes to the base mix design, within the range relevant to backfill bulkheads (10-30 MPa) a lower bound curve could be used for NDT quality control testing. An

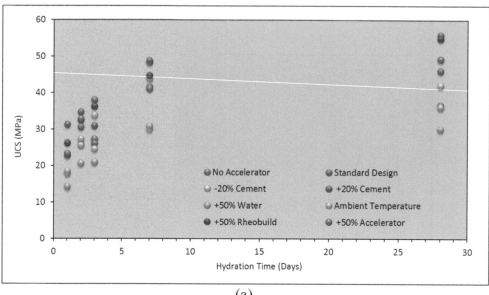

(a)

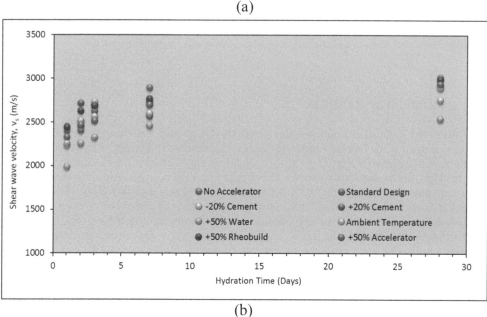

(b)

Figure 4. 'Cement 470' fibrecrete recipe for (a) UCS and (b) shear wave velocity versus hydration.

example of this is presented in Figure 8, for the 'Cement 470' recipe, whereby, the 'lower bound' trend can be used as the minimum required shear wave to confirm exceedance of a fibrecrete strength target. The proposed method would allow a specimen to be continuously monitored until the time that the target wave velocity is exceeded. For the example in Figure 8, if the design strength was 20 MPa, a shear wave velocity of 2,710 m/s would form the target.

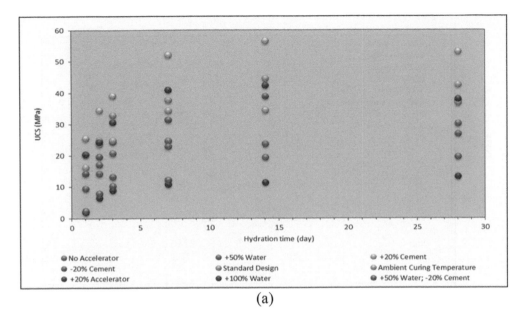

(a)

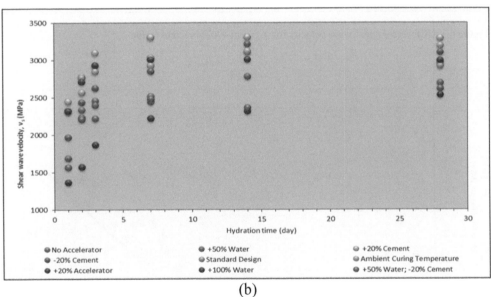

(b)

Figure 5. 'Cement 370' fibrecrete recipe for (a) UCS and (b) shear wave velocity versus hydration.

6 CONCLUSION

The results presented in this paper show a relatively unique relationship between shear wave velocity and the strength of fibrecrete for two different fibrecrete mixes. Given a suitable relationship, the shear plate method for measuring the shear wave velocity of fibrecrete could be used as a NDT to confirm fibrecrete strength. The approach provides a practical method to confirm suitability of early age fibrecrete strength.

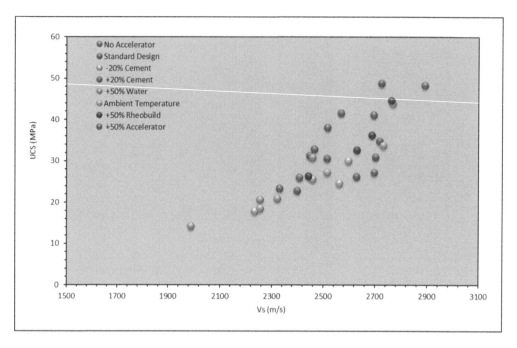

Figure 6. UCS versus shear wave velocity for 'Cement 470' fibrecrete recipe.

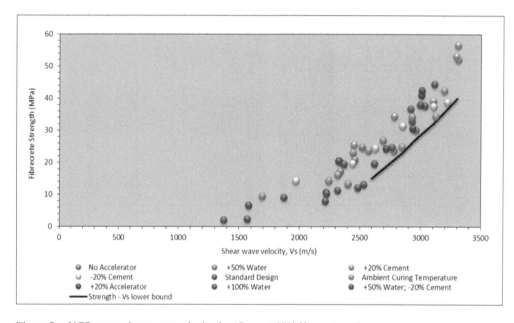

Figure 7. UCS versus shear wave velocity for 'Cement 370' fibrecrete recipe.

Implementation of improved quality control techniques such as that proposed can reduce development and fill cycle times where appropriate and also identify lower strength material, leading to increased mining efficiencies and reduced risk.

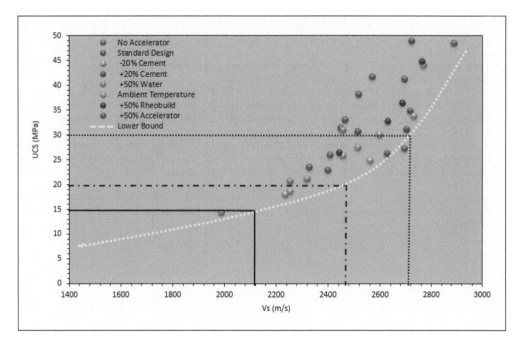

Figure 8.   UCS versus shear wave velocity where the shear wave velocity is used to measure the strength of the fibrecrete for the Cement 470 recipe.

This theory and application would benefit from a detailed field investigation, where the method is applied in an operating mine environment to define appropriate casting techniques and methods to develop a field testing procedure.

BIBLIOGRAPHY

ASTM C403/C403M. 2016. *Standard Test Method for Time of Setting of Concrete Mixtures by Penetration Resistance*, ASTM International, West Conshohocken, PA, 2016, www.astm.org

AuSS 2010. *Recommended practice: Shotcreting in Australia: Second Edition*. Concrete Institute of Australia and Australian Shotcrete Society, pp. 72–81.

Baig, S., Picornell, M. and Nazarian, S. 1997. Low strain shear moduli of cemented sands. *Journal of Geotechnical and Geoenvironmental Engineering*, ASCE. Vol. 123, No. 6, pp. 540–545.

Ciancio, D. and Helinski, M. 2010. The use of shear wave velocity for assessing strength development in Fibre Reinforced Shotcrete. [In:] Bernard, E.S. *Shotcrete: Elements of a system: Proceedings of the 3rd International Conference Engineering developments ins shotcrete*. Queenstown, New Zealand, 15–17 March 2010. London: Taylor & Francis, pp. 65–70.

Dyvik, R. and Olsen, T.S. 1989. G_{max} measured in oedometer and DSS tests using bender elements. *Proceedings of the 12th International Conference on SMFE*, Rio de Janeiro, Balkema, Rotterdam, Vol. 1, pp. 39–42.

Fernandez, A. and Santamarina, J.C. 2001. Effect of cementation on the small strain parameters of sand. *Canadian Geotechnical Journal*, 38, pp. 191–199.

Helinski, M, Wines, D, Revell, M & Sainsbury, D 2011, 'Critical factors influencing the capacity of arched fibrecrete bulkheads and waste rock barricades', in HJ Ilgner (ed.), Proceedings of the 10th International Symposium on Mining with Backfill, The Southern African Institute of Mining and Metallurgy, Johannesburg, pp. 293–304.

International Organization for Standardization. 2004. *Testing of Concrete —Part 7: Non-destructive tests on hardened concrete* (ISO 1920-7:2004(en). https://www.iso.org/home.html

Moczko, A. And Moczko, M. 2018. In-situ examintation of the concrete quality European standard approach. MATEC Web of Conferences 196, 02045 (2018). https://doi.org/10.1051/matecconf/201819602045

Morgan, D.R and Bernard, E.S. A brief history of shotcrete in the underground industry. *Shotcrete: 20th Anniversary Addition 19(4)*, pp. 24–29.

Rispin, M., Kleven, OB., Dimmock, R. and Myrdal, R. 2017. Shotcrete: early strength and re-entry revisited – practices and technology. [In:] M Hudyma & Y Potvin (eds), *Proceedings of the First International Conference on Underground Mining Technology*, Australian Centre for Geomechanics, Perth, pp. 55–70.

Saw, H.A., Villaescusa, E. And Windsor, C. 2015. Safe re-entry time for in-cycle shotcrete in underground mine excavations. [In:] Lu. M et.al. eds. *International Conference on Shotcrete for Underground Support XII*. Singapore, 11–13 October 2015. Engineering Conferences International: New York, pp. 173–192.

Santamarina, J.C., Klein, K.A., and Fam, M.A. (2001). Soils and waves: Particulate materials behavior, characterization and process monitoring. *Journal of Soils and Sediments*. 1. 130–130. 10.1007/BF02987719.

Minefill 2020-2021 – Hassani et al (eds)
© *2021 Taylor & Francis Group, London, ISBN 978-1-032-07203-6*

A design procedure for evaluation and prediction of in-situ cemented backfill performance

Xiaoming Wei & Lijie Guo
National Centre for International Research on Green Metal Mining, BGRIMM Technology Group, Beijing, China

ABSTRACT: To dispose of the smelting slag and reduce backfill cost in the mine, the aggregate mixed with smelting slag was selected as the filling aggregate, and then the filling industrial test of smelting slag was carried out in the stopes. Through the geological coring and strength test, the in-situ cemented backfill mass had been evaluated by the comparative analysis of RQD、P-wave velocity and UCS, which proved that the gradation of filling aggregate was optimized, and the quality of in-situ cemented backfill was improved by adding smelting slag. Based on internal links of physicomechanical parameters of cemented backfill, a relation between UCS and P-wave modulus was established from the perspective of dimensional balance. According to the test data of dry density, P-wave velocity and UCS of in-situ cemented backfill, the UCS prediction formula was obtained by the linear fitting method, which provided a strong research basis for the comprehensive quality evaluation of subsequent backfill.

Keywords: mixed aggregates, smelting slag, in-situ coring, quality of cemented backfill, prediction formula

1 INTRODUCTION

The Kalatongke Copper-Nickel mine was located in Fuyun County, Xinjiang. It was a non-ferrous metal company integrating mining, mineral processing and smelting (Bing and Jun, 2003; Xiaosu, 2009). At present, Gobi aggregate were used as filling aggregate in the mine (Chen et al., 2018). With the continuous improvement of national environmental protection requirements in recent years, procurement costs had risen sharply, and a large amount of smelting slag in the mine was also facing high disposal costs. In order to reduce the filling cost, the filling aggregate optimization test was carried out . Through in-situ coring and strength testing of cemented backfill in the underground industrial stope, the quality of in-situ backfill was systematically evaluated to guarantee the safety of stope (Ghirian and Fall, 2013; Di and Sijing, 2015).

2 TEST MATERIALS

The gobi aggregate and smelting slag were used as filling aggregates in the test stope, which the content of smelting slag was 12% of the total mass of aggregates. The cementitious material was cement, and the slurry concentration was 84%. The physical and mechanical parameters of Gobi aggregate and smelting slag were shown in Table 1. The particle size distribution were shown in Figure 1.

DOI: 10.1201/9781003205906-21

Table 1. The physical and mechanical parameters of Gobi aggregate and smelting slag.

Aggregate	Density / (g·cm^{-3})	Bulk density / (g·cm^{-3})	Porosity
Gobi aggregate	2.50	1.36	0.46
Smelting slag	3.57	1.99	0.44

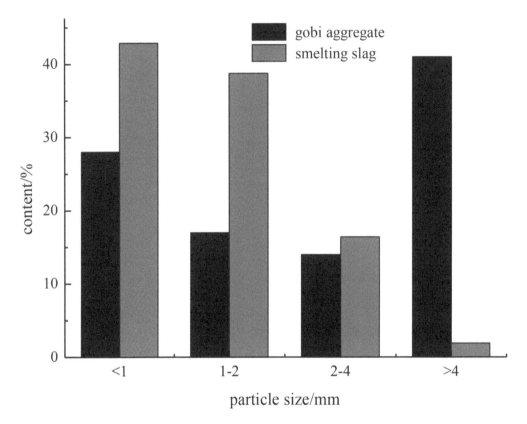

Figure 1. The particle size distribution.

3 FILLING INDUSTRIAL TEST

3.1 *Selection of test stope*

At present, the downward layered cemented filling method was adopted by the mine. The strength requirement of the bottom layer of backfill was greater than 3 MPa. Therefore, the E3 route was selected as the smelting slag test stope, S4 route was used as comparative test stope without smelting slag. The 1:5 of cement-sand ratio was designed in filling bottom layer, and the filling height is 2m.

3.2 *Filling process*

According to the existing conditions of the 1# filling station, the gravel silo was used as the smelting slag silo, and the discharge port was modified to realize the stable transportation of the smelting slag. Through the flow calibration test, a regulating valve was set at the discharge

Figure 2.   The regulating valve.

Figure 3.   The coring arrangement.

port, as shown in Figure 2. The smelting slag and Gobi aggregate were sent to the mixing tank. After being stirred, the filling slurry was transported to the stope.

3.3   *Sampling of test stopes*

In the filling process, the casting tests were performed in each stope. When the test blocks was placed in a underground stope for curing 60 days, then the uniaxial compressive strength was determined. After the cemented backfill of two test stopes were cured for 60 days, in the 926 m level a geological drill was used for coring detection, and the coring arrangement was shown in Figure 3. Meanwhile the return data and the integrity of the coring samples were recorded. Based on the RQD, P-wave velocity and uniaxial compressive strength (UCS), the quality of in-situ cemented backfill was evaluated .

4   QUALITY EVALUATION OF IN-SITU CEMENTED BACKFILL

4.1   *RQD*

The coring length of bottom layer at S4 route was 20 m, there were 3 sets of samples, as shown in Figure 4. A total of 67 standard samples were processed. From the perspective of sensory quality, the 0 ~ 11.5 m of coring samples were basically complete, even thickness, local fragmentations. After 11.5 m, the coring samples was relatively broken. The RQD of S4 route was 60%. The coring length of bottom layer at E3 route was 26.5 m, there were 5 sets of samples, as shown in Figure 4. A total of 120 standard samples were processed. From the perspective of sensory quality, the coring samples were complete, smooth surface, even thickness. The RQD of E3 route was 85%.

Through the comparative analysis of the RQD of the two test stopes, the RQD of E3 route was higher than S4 route, which proved that the integrity of the coring cemented backfill of the E3 route was better than S4 route.

4.2   *P-wave velocity*

The P-wave velocity distribution of S4 and E3 routes were shown in Figure 5. The P-wave velocity distribution of the S4 coring samples was discrete, and the average wave velocity was 2515 m/s. The P-wave velocity distribution of the E3 coring samples was concentrated, and the average wave velocity was 2649 m/s. Through the comparative analysis of the P-wave velocity of the two test stopes, the average wave velocity of E3 route was higher than S4 route, which proved that the compacting and homogeneity of the coring cemented backfill of the E3 route was better than S4 route.

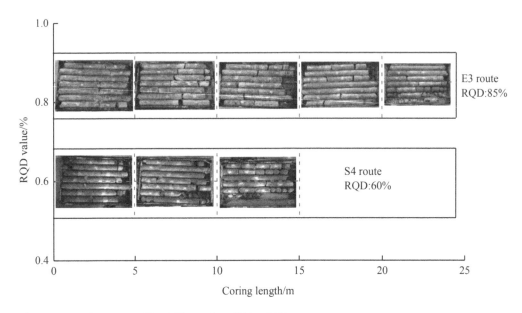

Figure 4.   In-situ cemented backfill samples of S4 and E3 routes.

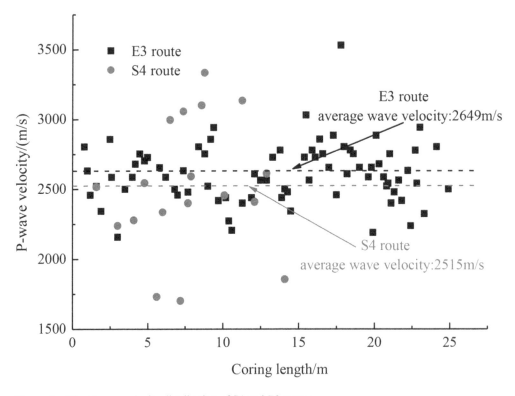

Figure 5.   The P-wave velocity distribution of S4 and E3 routes.

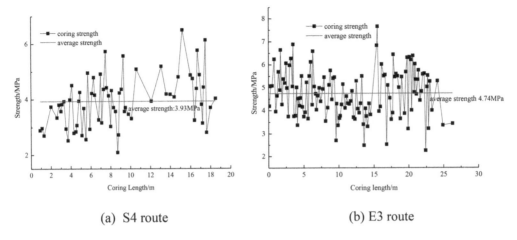

(a) S4 route   (b) E3 route

Figure 6.   The UCS distribution of S4 and E3 routes.

4.3 *UCS*

The UCS distribution of S4 and E3 routes were shown in Figure 6. The average strength of the S4 and E3 coring samples were 3.93MPa and 4.74MPa respectively. The average strength of the curing blocks were 4.1MPa and 5.2MPa respectively. Through the comparative analysis of the UCS of the two test stopes, the average strength of E3 route was higher than S4 route, which proved that the consolidation of the coring cemented backfill of the E3 route was better than S4 route.

5   UCS PREDICTION FORMULA OF IN-SITU CEMENTED BACKFILL

5.1 *Theoretical derivation*

In the elastic theory (Weiguo and Yulong, 2007; Moose et al., 2001), the three-dimensional wave equation in uniform, isotropic, and ideal elastic medium was

$$\left.\begin{array}{l}(\lambda + \mu)\frac{\partial \theta}{\partial x} + \mu\nabla^2 u - \rho\frac{\partial^2 u}{\partial t^2} = 0 \\ (\lambda + \mu)\frac{\partial \theta}{\partial y} + \mu\nabla^2 v - \rho\frac{\partial^2 v}{\partial t^2} = 0 \\ (\lambda + \mu)\frac{\partial \theta}{\partial z} + \mu\nabla^2 w - \rho\frac{\partial^2 w}{\partial t^2} = 0 \\ \theta = \frac{\partial u}{\partial x} + \frac{\partial v}{\partial y} + \frac{\partial w}{\partial z} \\ \nabla^2 = \frac{\partial^2}{\partial x^2} + \frac{\partial^2}{\partial y^2} + \frac{\partial^2}{\partial z^2}\end{array}\right\} \quad (1)$$

Where: u, v, and w are the displacements in the x, y, and z directions respectively; μ is the Lame constant; ρ is the medium density; θ is the volumetric strain; ∇^2 is the Laplacian operator.

If the medium deformation was caused by wave, only the change in volume without rotation, the equation was

$$\left.\begin{array}{l}(\lambda + 2\mu)\nabla^2 u - \rho\frac{\partial^2 u}{\partial t^2} = 0 \\ (\lambda + 2\mu)\nabla^2 v - \rho\frac{\partial^2 v}{\partial t^2} = 0 \\ (\lambda + 2\mu)\nabla^2 w - \rho\frac{\partial^2 w}{\partial t^2} = 0\end{array}\right\} \quad (2)$$

The wave by this equation was called P-wave (Qikun, 2009), and the wave equation of the P-wave could be written in the following simple form:

$$\frac{\partial^2 \theta}{\partial t^2} = v_p^2 \nabla^2 \theta; \ v_p = \sqrt{\frac{\lambda + 2\mu}{\rho}} \tag{3}$$

According to the literature, $(\lambda + 2\mu)$ was called P-wave modulus.

$$P = \lambda + 2\mu = \rho v_p^2 \tag{4}$$

In order to obtain the P-wave modulus of the cemented backfill, the density and the P-wave velocity were known. Aiming at the dimension consistency of P-wave modulus and UCS, the UCS prediction formula was established:

$$\sigma_c = aP + b \tag{5}$$

Where: a was a constant; b was the initial strength

5.2 UCS prediction formula

The P-wave modulus of S4 and E3 routes were shown in Figure 7. Origin numerical analysis software was adopted to fit linear of P-wave modulus. Based on R^2 coefficient, the fitting coefficients of S4 and E3 routes were 71% and 75% respectively, and the fitting effect was better. Therefore, it was proved that the UCS prediction formula was feasible.

S4 route:

$$y = 0.4 + 0.00015x \tag{6}$$

E3 route:

$$y = 1.2 + 0.00021x \tag{7}$$

Where: y was UCS, MPa; x was the P-wave modulus, MPa.

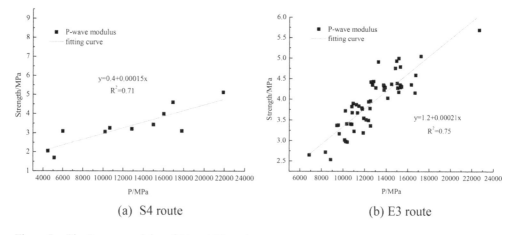

(a) S4 route   (b) E3 route

Figure 7. The P-wave modulus of S4 and E3 routes.

6 CONCLUSION

(1) The in-situ cemented backfill mass had been evaluated by the comparative analysis of RQD, P-wave velocity and UCS, the filling quality of E3 route was better than the S4 route (without smelting slag), which proved that the gradation of filling aggregate was optimized, and the quality of in-situ cemented backfill was improved by adding smelting slag.

(2) Based on internal links of physicomechanical parameters of in-situ cemented backfill, a relation between UCS and P-wave modulus was established from the perspective of dimensional balance. According to the test data of dry density, P-wave velocity and UCS of cemented backfill, the UCS prediction formula of cemented backfill based on P-wave modulus was obtained by the linear fitting method.

ACKNOWLEDGEMENTS

This work was supported by the National Key Research and Development Program of China (2017YFE0107000) and the Youth Innovation Fund of BGRIMM (04-2027).

BIBLIOGRAPHY

Bing, W. and Jun, L. 2003. Analysis of filling process and filling cost of Kalatongke Copper-Nickel Mine. *Mining Technology* (1), pp. 22–24.

Xiaosu, F. 2009. The application of Gobi aggregate cemented backfill of the underground drift in the Kalatongke mine. *Xinjiang Nonferrous Metals* (6), pp. 38–39.

Chen et al. 2018 – Chen, Y., Lijie, G., Yaping, Y. 2018. Rheological properties of coarse aggregate paste slurry and calculation of resistance in pipeline transportation. *China Mining Magazine* 27(12), pp. 178–182.

Chen et al. 2018 – Chen, Y., Lijie, G., Yaping, Y. 2018. Experimental optimization of filling aggregates in Karatungk Copper-Nickel mine. *Nonferrous Metals(Mine Section)* 70(6), pp. 38–41.

Ghirian, A. and Fall, M. 2013. Coupled thermo-hydro-mechanical-chemical behaviour of cemented paste backfill in column experiments. Part I: Physical, hydraulic and thermal processes and characteristics. *Engineering Geology* (164), pp. 195–207.

Di, W. and Sijing C. 2015. Coupled effect of cement hydration and temperature on hydraulic behavior of cemented tailings backfill. *Journal of Central South University* (5), PP. 1956–1964.

Weiguo G. and Yulong L. 2007. *Brief tutorial of stress wave foundation*. Xi'an: Northwestern Polytechnical University Press, 66pp.

Moose et al. 2001 – Moose, D., Zoback M., Bailey L. 2012. Feasibility study of the stability of open hole multilaterals, Cook Inlet, Alaska. *SPE Drilling and Completion* 16(3), pp.140–145.

Qikun, L. 2009. See world through dimension. *Mathematics Communication* 33(3), pp. 13–27.

Minefill 2020-2021 – Hassani et al (eds)
© 2021 Taylor & Francis Group, London, ISBN 978-1-032-07203-6

The uniaxial compressive strength of fiber-reinforced frozen backfill

Huiya Niu
State Key Laboratory of Geomechanics & Deep Underground Engineering, China University of Mining & Technology, Beijing, China

Ferri P. Hassani & Mehrdad F. Kermani
Mining Engineering Department, McGill University

Manchao He
State Key Laboratory of Geomechanics & Deep Underground Engineering, China University of Mining & Technology, Beijing, China

ABSTRACT: The possible application of using fiber-reinforced frozen (–15°C) mine backfill was investigated using a series of laboratory experiments. Sisal fiber, Basalt fiber, and Bio-filament were added to the tailings and water that constituted primary backfill, which served as the reference material. Samples differed in terms of fiber type, fiber content, and curing temperature. The mechanical properties of samples were investigated using the uniaxial compression strength (UCS) test. The UCS of fiber-reinforced frozen backfill exceeded that of primary backfill and reached as high as 5.95 MPa. The UCS stress-strain behavior of mine backfill was markedly affected by freezing and fiber addition. Frozen backfill exhibited more ductile behavior and a higher elasticity modulus than unfrozen backfill, which showed brittle behavior. Fiber-reinforced backfill could absorb more energy before total failure.

Keywords: Sisal fiber, Basalt fiber, Bio-filament, fiber-reinforced backfill, frozen backfill, uniaxial compressive strength

1 INTRODUCTION

The practice of filling the void created by underground mining activities with waste materials is termed mine backfilling and is generally an integral part of the underground mining process. Mine backfill is mainly used to increase ore extraction and underground mine stabilization and to reduce the volume of deposited waste materials. Mine backfill mostly comprises tailings (waste), water, and binders—usually Portland cement (PC) (Weaver & Luka 1970; Benzaazoua et al. 2004b; Wang & Qiao 2019). PC accounts for up to 75% of the cost of mine backfill (Hassani & Bois 1989; Potvin et al. 2005; Li et al. 2019), and PC production contributes a large amount of carbon dioxide to the atmosphere. Furthermore, increased demand for minerals and depletion of more accessible minerals forces mining companies to expand operations to more remote areas (e.g., higher latitudes in Canada). These challenges require innovative methods to meet the requirements of a mining operation in an environmentally friendly manner. Thus, exploring alternative binders to fully or partially replace PC to reduce the cost and carbon footprint (as well as improve the mechanical strength) of mine backfill has been

DOI: 10.1201/9781003205906-22

the focus of several research projects (Benzaazoua et al. 2004a; Kesimal et al. 2005; Mohamed Abdel-Mohsen et al. 2007; Singh et al. 2019).

Examples of alternative binders are pozzolanic products (Ercikdi et al. 2010), fly ash, and slag (Bouzoubaa & Foo 2004). Samples containing fly ash exhibited higher uniaxial compressive strength (UCS) than samples containing PC (Hassani et al. 2007). Cemented paste backfill (CPB) containing a mixture of fly ash (40%) and PC (60%) showed improved sulphate resistance (Benzaazoua et al. 1999). Among paste backfill samples prepared with PC, fly ash, and slag, slag-based samples had higher triaxial compressive strength (Belem et al. 2000). Other important factors affecting the physical properties of CPB are tailings properties (Kesimal et al. 2004; Fall et al. 2005; Kesimal et al. 2005), binder types and dosage (Kesimal et al. 2004; Ercikdi et al. 2009; Ercikdi et al. 2014), curing time (Yilmaz et al. 2015), and other additives (Mitchell & Stone 1987; Kermani et al. 2015; Koohestani et al. 2016; Mangane et al. 2018).

In cold and remote environments, application and transportation of cementitious materials are costly and logistically challenging. For instance, the hydration of cementitious material can dramatically decelerate or even cease in cold ambient temperatures (Jiang et al. 2020). Limited research has explored the feasibility of using ice as the main cementitious agent in mine backfill in cold climates. Although this method appeared to be feasible in terms of meeting strength and stiffness requirements, the cost of frozen backfill was low compared to conventional backfill (Daniel & Kazakidis 2012). Han (2011) compared the performance of frozen tailings backfill, with mechanical properties strongly influenced by water content, tailings fineness, and compaction pressure, to that of frozen cemented paste backfill with mechanical properties significantly affected by cement content, curing age, and binder type. The frozen tailings backfill displayed strain softening behaviors and a relatively brittle failure mode, whereas the frozen cemented paste backfill displayed strain hardening behaviors and was ductile.

Given the strength limitations of frozen backfill noted in previous studies, the aim of this work is to investigate the effect on the UCS and deformation behavior of reinforcing frozen backfill with Sisal fiber, Basalt fiber, and Bio-filament. An orthogonal experiment design was used to study the effect of fiber type, fiber content, and curing temperature. This study is expected to provide a reference to evaluate the performance of fiber-reinforced frozen backfill and help reduce the usage of PC in mine backfill.

2 MATERIALS

2.1 Tailings

Tailings were obtained from a mine in northern Quebec, Canada. The specific gravity (G_s), particle size (D_i where i=10, 30, 50, 60, and 90 mm), and coefficients of uniformity (C_u) and curvature (C_c) are summarized in Table 1. Figure 1 shows the particle size distribution of tailings, which were tested in according to ASTM C136/C136M-14 (2014). Table 2 lists the chemical composition analyzed by X-ray fluorescence.

2.2 Binder

Both primary and fiber-reinforced samples were made with 0% and 2% (dry mass of solid materials) General Use Type 10 Portland cement as binder.

Table 1. Physical properties of Westwood tailings.

G_s (g/cm³)	D_{10} (µm)	D_{30} (µm)	D_{50} (µm)	D_{60} (µm)	D_{90} (µm)	C_u	C_c
2.93	3.95	8.6	12.1	14.0	23.0	3.5	1.3

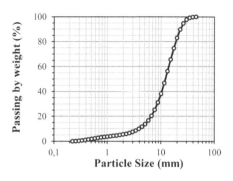

Figure 1.   Particle size distribution of tailings.

Table 2.   Chemical composition of Westwood tailings.

Oxides	SiO$_2$	Al$_2$O$_3$	Fe$_2$O$_3$	SO$_3$	CaO	MgO	K$_2$O	Na$_2$O
Content (%)	49.02	17.67	9.15	7.29	6.52	3.97	2.70	1.50
Metal element	P	Ni	Cu	Cl	Cr	Zr	Sr	Rb
Content (ppm)	677	648	406	319	115	108	90	44

2.3   Fiber

Sisal fiber (Figure 2) is extracted from the Sisal plant and has the advantages of high strength, easy processing, good dispersion, and low cost. The tensile strength and Young's modulus (Table 3) are mostly influenced by filament diameter and length (Deák & Czigány 2009; Jiang et al. 2014). Basalt fiber (Figure 2) is melted at 1450–1500°C and drawn through a platinum-rhodium alloy plate at a high speed, giving it high strength, electrical insulation, corrosion resistance, high

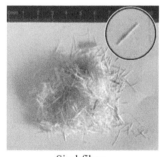

Sisal fiber

Basalt fiber

Bio-filament

Figure 2.   Photographs of Sisal fiber, Basalt fiber, and Bio-filament.

Table 3.   Physical properties of Sisal and Basalt fiber.

Fiber type	Young's modulus (GPa)	Tensile strength (MPa)	Filament length (mm)	Filament diameter (µm)
Sisal	9–26	5500–6500	10	300–500
Basalt	28–110	400–4800	10	7–15

Soure: Tolêdo Filho et al. (2000);Ku et al. (2011);Ratna Prasad and Mohana Rao (2011);Alves Fidelis et al. (2013);Ramesh et al. (2013);Sultan (2005);Deák and Czigány (2009);Lopresto et al. (2011);Colombo et al. (2012);Jiang et al. (2014).

temperature resistance, and degradability. The tensile strength is lower for short than continuous Basalt fibers (Table 3). The Bio-filament was supplied by Performance Bio-filaments Inc., who produce cellulose filaments from wood pulp. The process results in cellulose filaments of exceptional strength and purity, with an extraordinary high and unique aspect ratio. Bio-filament is a "wet fluff" (Figure 2) with 4% solid content and is dispersible in solvents like water.

3 SAMPLE PREPARATION AND EXPERIMENT SETUP

Fourteen batches of mine backfill samples were prepared with a solids concentration of 70% (Table 4). The fiber contents of the fiber-reinforced samples were 0.5 and 1.0 wt%. Reference samples contained no fiber.

To prepare the samples, dry materials were weighed and mixed for 2 min using a mixer with a stainless-steel wire whip blade. Water was added gradually and mixed for another 5 min. Cylindrical molds (100 mm deep and 50 mm diameter) were used to cast the specimens. Specimens containing 2 wt% of the binder PC (Group 1 in Table 4) were cured in a curing chamber where the relative humidity was kept constant at (90±2%) and the temperature was adjusted to (20±1°C) for 3 days and then moved to a freezer where the temperature was kept constant at –15°C. Samples without any binder (Group 2) were cured directly in a freezer at –15°C. Moreover, unfrozen samples in Group 2 were prepared and dried at ambient temperature (+20°C) for comparison to frozen samples.

The unfrozen samples in Group 2 was not consolidated until 50 days later. The UCS test was conducted after 56 days of curing to evaluate the mechanical properties of specimens according to ASTM C39/C39M-18 (2018). To minimize the influence of heat exchange between the specimen and the ambient environment on the UCS results of frozen backfill, the upper and lower plate of test device was also frozen at the same temperature as the samples.

4 RESULTS AND DISCUSSION

4.1 *Reinforced effect of fibers on unfrozen dry and frozen backfill samples*

Adding fiber to specimens dried and cured at +20°C did not affect the UCS (Figure 3). However, freezing all Group 2 specimens lacking PC binder (with or without fiber) enhanced the

Table 4. Experimental design.

Sample	Fiber Type	Content (wt%)	Binder (wt%)	Dry tailings (wt%)	Curing condition
1	None	0	2	98.0	Group 1: cured for 3 days at 20°C and then
2	Sisal	0.5	2	97.5	frozen at –15°C.
3	Sisal	1.0	2	97.0	
4	Basalt	0.5	2	97.5	
5	Basalt	1.0	2	97.0	
6	Bio	0.5	2	97.5	
7	Bio	1.0	2	97.0	
8	None	0	0	100	Group 2: frozen: cured at –15°C.
9	Sisal	0.5	0	99.5	Group 2: unfrozen:
10	Sisal	1.0	0	99.0	
11	Basalt	0.5	0	99.5	
12	Basalt	1.0	0	99.0	
13	Bio	0.5	0	99.5	Dried and cured at +20°C.
14	Bio	1.0	0	99.0	

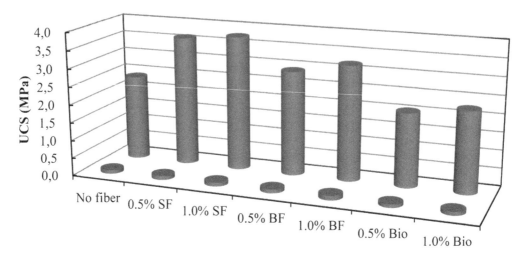

■ Natural dry ■ Frozen samples

Figure 3. UCS of unfrozen and frozen dried backfill samples without and with reinforcement with Basalt fiber (BF), Sisal fiber (SF), and Bio-filaments (Bio).

UCS: Sisal fiber was associated with the highest UCS (3.76 MPa), whereas Bio-filament add-ition resulted in lower UCS than the no fiber treatment. For a given fiber type, the UCS of frozen specimens was higher at 1% then 0.5% fiber concentration. This indicates that the bond between ice and fibers formed after the phase transition from water to the crystalline ice.

After 56 days of freezing, the UCS of fiber-reinforced frozen backfill specimens containing 2 wt% PC (Group 1 in Table 4) was 43% (Sisal fiber) to 143% (Biofilament) higher than the UCS of specimens lacking PC (Figure 4). In the presence of PC, the fiber content (0.5 vs. 1.0%) did not affect the UCS as much as it did when PC was absent. This phenomenon can be attributed to the hydration of cement, the phase transition of water, and enhanced bond inte-gration between tailings particles and fibers.

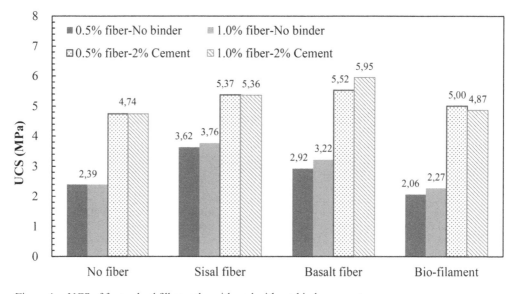

Figure 4. UCS of frozen backfill samples with and without binder cement.

The increase in UCS of backfill as result of inclusion of fiber reinforcement can be expressed as the UCS improvement factor, I_R, defined as:

$$I_R = \frac{\sigma_R - \sigma_U}{\sigma_U} \qquad (1)$$

Where σ_R is the UCS of fiber-reinforced backfill and σ_U is the UCS of unreinforced backfill. The I_R represents the relative gain in UCS of backfill due to the addition of fiber. Table 5 presents the I_R of frozen backfill without and with PC, using σ_U values of 2.39 and 4.74 MPa, respectively (Figure 4).

4.2 Deformation behavior of fiber-reinforced

The UCS stress-strain behavior of mine backfill was markedly affected by freezing (Figure 5). Frozen backfill not only obtained higher UCS values, but also exhibited more ductile behaviors and higher elasticity modulus, while the unfrozen backfill shows brittle behavior. The fiber-reinforced backfill is able to hold together for more deformation and therefore higher strength at failure. Therefore, the fiber-reinforced backfill can absorb more energy before total failure.

The brittle behavior of unfrozen tailings backfill (Figure 6) was transformed by freezing to a more ductile behavior (Figure 7). Unreinforced backfill samples failed at a relative low axial strain (approximately 2.4%), whereas backfill samples reinforced with Basalt fiber failed at approximately 3.3% axial strain and those reinforced with Sisal fiber failed at approximately

Table 5. Improvement factor, I_R, of frozen fiber-reinforced backfill.

No Cement		2% Cement	
Fiber	I_R (%)	Fiber	I_R (%)
0.5% Sisal	51.38	0.5% Sisal fiber	13.28
1.0% Sisal	57.13	1.0% Sisal fiber	13.03
0.5% Basalt	22.02	0.5% Basalt fiber	16.51
1.0% Basalt	34.36	1.0% Basalt fiber	25.43
0.5% Bio-filament	−13.83	0.5% Bio-filament	5.52
1.0% Bio-filament	−5.11	1.0% Bio-filament	2.63

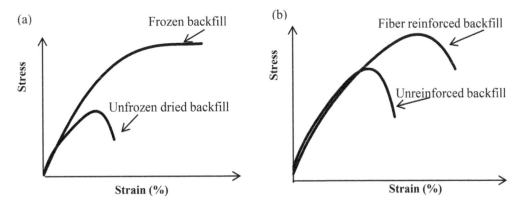

Figure 5. Typical unconfined compressive stress-strain curves for (a) unfrozen dried and frozen backfill; (b) unreinforced and fiber-reinforced backfill.

3.7% (Figure 6). The higher fiber content seems improve the UCS and elasticity modulus, but does not contribute to the ductility of unfrozen dried backfill.

The stress of frozen specimens rose faster than that of specimens dried without freezing during the elastic deformation stage (Figure 7). Fibers contributed to elevated UCS while the fiber content dominated the post-peak behavior. Among the three fibers, Sisal fiber had the strongest influence and Bio-filament had the weakest influence on the ductile behavior of backfill. The bond between tailings and fiber contributed to the enhanced ductility. All tests on reinforced specimens exhibited strain-hardening behavior with increasing strength as the strain increased.

The E_{50}, deformation energy, and post-peak strength of typical frozen samples were considered as indices to evaluate the performance of frozen backfill, in addition to UCS. E_{50} is the secant modulus at 50% UCS and is considered the elastic index. The energy absorbed by a unit volume of backfill before peak strength is calculated by:

$$M_0 = \int \sigma d\varepsilon \tag{2}$$

M_{0P} is the other index to evaluate the performance of frozen backfill:

$$M_{0P} = \int\limits_0^{\sigma_0} \sigma d\varepsilon \tag{3}$$

Where σ_0 is the peak strength of backfill.

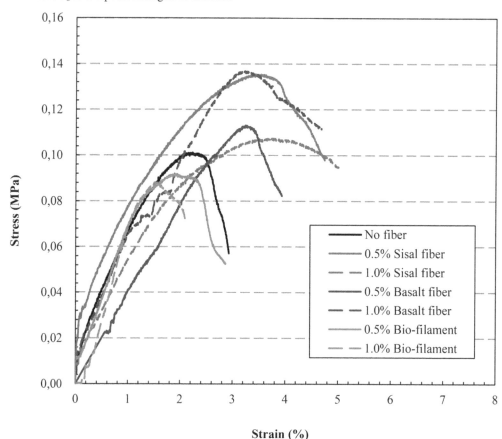

Figure 6.   Typical stress-strain curves for unfrozen dry backfill samples without binder cement.

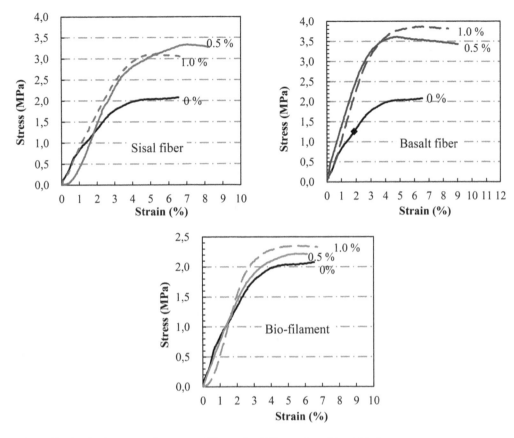

Figure 7. Typical stress-strain curves of frozen backfill samples without binder cement.

From Table 6, the fiber-reinforced frozen backfill absorbed more energy. The addition of Sisal and Basalt and fiber enhanced the UCS and residual strength, and improved the initial stiffness. The contribution of Bio-filament to frozen backfill performance was not significant.

Table 6. Strength performance of frozen fiber-reinforced backfill.

Frozen backfill type	E_{50} (MPa)	Energy absorbed before failure, M_{OP} (kJ)	Post-peak strength (MPa)
No fiber	0.74	71.4	2
0.5% Basalt fiber	1.24	117.84	3.42
1.0% Basalt fiber	1.14	184.81	3.81
0.5% Sisal fiber	1.02	147.61	3.27
1.0% Sisal fiber	0.87	96.16	3.05
0.5% Bio-filament	0.73	105.53	2.19
1.0% Bio-filament	1.02	86.97	2.33

5 CONCLUSION

A series of experiments explored potential application of frozen backfill with or without fiber (Sisal, Basalt, and Bio-filament) reinforcement as an alternative to CPB. The highest UCS (5.95 MPa) was achieved with frozen backfill containing 2% PC and 1% Basalt fiber. Freezing transformed the deformation behavior of backfill from brittle to more ductile. The fiber-reinforced frozen backfill absorbed more energy. Sisal and Basalt fiber addition not only enhanced the UCS and residual strength but also improved the initial stiffness. The contribution of Bio-filament to frozen backfill strength was not significant.

ACKNOWLEDGEMENTS

This work was supported by Geomechanics Laboratory in McGill University and Iamgold Inc. . Comments by Janice Burke improved an earlier version of the paper.

BIBLIOGRAPHY

Alves Fidelis et al. 2013. The effect of fiber morphology on the tensile strength of natural fibers. *Journal of Materials Research and Technology*, **2(2)**, pp. 149–157.

Belem et al. 2000. Mechanical behaviour of cemented paste backfill. *Proceedings of 53th*.

Benzaazoua et al. 1999. Cementitious backfill with high sulfur content physical, chemical, and mineralogical characterization. *Cement and Concrete Research*, **29(5)**, pp. 719–725.

Benzaazoua et al. 2004a. A contribution to understanding the hardening process of cemented pastefill. *Minerals Engineering - MINER ENG*, 17.

Benzaazoua et al. 2004b. The use of pastefill as a solidification and stabilization process for the control of acid mine drainage. *Minerals Engineering*, **17(2)**, pp. 233–243.

Bouzoubaa and Foo. 2004. Use of fly ash and slag in concrete: A best practice guide. *Materials and Technology Laboratory, MTL*, 16.

ASTM C39/C39M-18. 2018. *Standard test method for compressive strength of cylindrical concrete specimens*. West Conshohocken, PA: ASTM International.

ASTM C136/C136M-14. 2014. *Standard test method for sieve analysis of fine and coarse aggregates*. West Conshohocken, PA: ASTM International.

Colombo et al. 2012. Static and fatigue characterisation of new basalt fibre reinforced composites. *Composite Structures*, **94(3)**, pp. 1165–1174.

Deák and Czigány. 2009. Chemical composition and mechanical properties of basalt and glass fibers: A comparison. *Textile Research Journal*, **79(7)**, pp. 645–651.

Ercikdi et al. 2009. Cemented paste backfill of sulphide-rich tailings: Importance of binder type and dosage. *Cement and Concrete Composites*, **31(4)**, pp. 268–274.

Ercikdi et al. 2010. Effect of natural pozzolans as mineral admixture on the performance of cemented-paste backfill of sulphide-rich tailings. *Waste Management & Research*, **28(5)**, pp. 430–435.

Ercikdi et al. 2014. Strength and ultrasonic properties of cemented paste backfill. *Ultrasonics*, **54(1)**, pp. 195–204.

Fall et al. 2005. Experimental characterization of the influence of tailings fineness and density on the quality of cemented paste backfill. *Minerals engineering*, **18(1)**, pp. 41–44.

Han 2011. Geotechnical behaviour of frozen mine backfills. Master degre, University of Ottawa.

Hassani and Bois. 1989. Economic and technical feasibility for backfill design in quebec underground mines. Final report 1/2, canada-quebec mineral development agreement. *Research and Development in Quebec Mines. Contract No. EADM*, **1992**.

Hassani et al. 2007. Physical and mechanical behavior of various combinations of mine fills materials. *CIM Bulletin*, **100(3)**, pp. 1–6.

Jiang et al. 2014. Experimental study on the mechanical properties and microstructure of chopped basalt fibre reinforced concrete. *Materials & Design*, **58**.pp 187–193.

Jiang et al. 2020. Ultrasonic evaluation of strength properties of cemented paste backfill: Effects of mineral admixture and curing temperature. *Ultrasonics*, **100**. pp. 105983.

Kermani et al. 2015. Evaluation of the effect of sodium silicate addition to mine backfill, gelfill – part 1. *Journal of Rock Mechanics and Geotechnical Engineering*, **7(3)**, pp. 266–272.

Kesimal et al. 2004. Evaluation of paste backfill mixtures consisting of sulphide-rich mill tailings and varying cement contents. *Cement and concrete research*, **34(10)**, pp. 1817–1822.

Kesimal et al. 2005. Effect of properties of tailings and binder on the short-and long-term strength and stability of cemented paste backfill. *Materials Letters*, **59(28)**, pp. 3703–3709.

Koohestani et al. 2016. Experimental investigation into the compressive strength development of cemented paste backfill containing nano-silica. *Cement and Concrete Composites*, **72**. pp. 180–189.

Ku et al. 2011. A review on the tensile properties of natural fiber reinforced polymer composites. *Composites Part B: Engineering*, **42(4)**, pp. 856–873.

Li et al. 2019. Evaluation of short-term strength development of cemented backfill with varying sulphide contents and the use of additives. *Journal of Environmental Management*, **239**. pp. 279–286.

Lopresto et al. 2011. Mechanical characterisation of basalt fibre reinforced plastic. *Composites Part B: Engineering*, **42(4)**, pp. 717–723.

Mangane et al. 2018. Influence of superplasticizers on mechanical properties and workability of cemented paste backfill. *Minerals Engineering*, **116**.pp. 3–14.

Mitchell and Stone. 1987. Stability of reinforced cemented backfills. *canadian geotechnical journal*, **24(2)**, pp. 189–197.

Mohamed Abdel-Mohsen et al. 2007. Evaluation of newly developed aluminum, lime, and fly ash technology for solidifaction/stabilization of mine tailings. *Journal of Materials in Civil Engineering*, **19(1)**, pp. 105–111.

Potvin et al. 2005. *Handbook on mine fill.* Crawley, WA, Australia: Australian Centre for Geomechanics.

Ramesh et al. 2013. Mechanical property evaluation of sisal–jute–glass fiber reinforced polyester composites. *Composites Part B: Engineering*, **48**. pp. 1–9.

Ratna Prasad and Mohana Rao. 2011. Mechanical properties of natural fibre reinforced polyester composites: Jowar, sisal and bamboo. *Materials & Design*, **32(8)**, pp. 4658–4663.

Singh et al. 2019. Optimisation of binder alternative for cemented paste fill in underground metal mines. *Arabian Journal of Geosciences*, **12(150** pp. 462.

Sultan. 2005. The effect of fibre content on the mechanical properties of hemp and basalt fibre reinforced phenol formaldehyde composites. *Journal of Materials Science*, **40(17)**, pp. 4585–4592.

Tolêdo Filho et al. 2000. Durability of alkali-sensitive sisal and coconut fibres in cement mortar composites. *Cement and Concrete Composites*, **22(2)**, pp. 127–143.

Wang and Qiao. 2019. Coupled effect of cement-to-tailings ratio and solid content on the early age strength of cemented coarse tailings backfill. *Geotechnical and Geological Engineering*, **37(4)**, pp. 2425–2435.

Weaver and Luka. 1970. Laboratory studies of cement-stabilized mine tailings. *Canadian Mining and Metallurgical Bulletin*, **63(701)**, pp. 988.

Yilmaz et al. 2015. Curing time effect on consolidation behaviour of cemented paste backfill containing different cement types and contents. *Construction and Building Materials*, **75**. pp. 99–111.

Minefill 2020-2021 – Hassani et al (eds)
© 2021 Taylor & Francis Group, London, ISBN 978-1-032-07203-6

Incorporating the monolithic nature of paste backfill into self heating assessments

Chris Lee, Bret Timmis, David Brown & Valerie Bertrand
Golder Associates Ltd. (Canada)

Matthew Stewart
Vale Newfoundland and Labrador Inc

SUMMARY: Tailings have exhibited self heating behaviour due to exothermic sulphide oxidation. Self-heating in paste backfill has caused dangerous temperatures and SO_2 concentrations. Self-heating potential is commonly evaluated using a laboratory test in which heated air is forced through crushed material. However, due to the monolithic nature of backfill, including a very low permeability to oxygen and moisture, laboratory tests using a pulverized sample may be conservative and conducting tests with monolithic samples may be more representative. Field trials were conducted at Voisey's Bay Mine, Canada.

Keywords: Self-heating, tailings, sulphide, oxidation, geochemistry

1   INTRODUCTION

Mine tailings, ore, waste rock, and cemented paste backfill (CPB) have exhibited self-heating behaviour largely due to the exothermic oxidation of sulphide minerals, most notably pyrrhotite, with oxygen and water. The by-products of this reaction include heat and SO_2 gas. The potential for self-heating in paste backfill has also been documented in several mines (Bernier and Li, 2003; Good, 1977; Fong et al., 2009; Nantel and Lecuyer, 1983; Patton, 1952) and has been seen to result in dangerous temperatures and SO_2 gas concentrations which have affected production. Therefore, characterization of self-heating potential is an important component of mine planning with respect to health & safety as well as identification of risks to continuity of mine operations.

Presently, the potential for self-heating is typically evaluated using a laboratory scale test initially developed by Noranda (now Glencore), and then further refined at McGill University by Dr. Jan Nesset. Within this paper, this laboratory test procedure is referred to as the Noranda Test. The Noranda test uses a modified calorimeter apparatus and relies on the use of heated air which is forced through tailings that are at the worst-case moisture content for self-heating potential. The heat produced by the reaction of the tailings is measured and a comparison is made against a substantial database of other materials to assess the potential for self-heating. However, due to the monolithic nature of the paste backfill which includes a very low permeability to both oxygen and moisture it is intuitively expected that the Noranda test which uses a pulverized sample may be conservative and that conducting tests with monolithic samples may be more representative of the actual field conditions. Before investing billions of dollars of capital, mining companies require a high level of certainty regarding the potential for self-heating and its impact on mine production and safety and until such a test is

DOI: 10.1201/9781003205906-23

available, the Noranda test is the best available method for determining self-heating potential. However, application of the Noranda test alone may result in overly conservative assumptions by mine developers with negative implications for mine economics.

Golder Associates Ltd. (Golder) was retained by Vale Newfoundland and Labrador Inc. (Vale) to evaluate the self-heating potential of paste backfill proposed for use in underground mine development, beginning with laboratory scale tests, which advanced to field scale tests upon identification of strong potential for self-heating in laboratory scale testing. This paper presents an evaluation of a new field-scale methodology for evaluation of self-heating potential of paste backfill alone or in conjunction with ore in order to provide a greater degree of certainty than the Noranda laboratory test alone. Field-scale testing was conducted at Voisey's Bay Mine, Newfoundland and Labrador, Canada.

2   SELF-HEATING OF SULPHIDE BEARING MINE WASTES

Sulphide mineral oxidation is an exothermic reaction. Pyrrhotite ($Fe_{1-x}S$) is of particular concern for self-heating due to its fast oxidation reaction rate upon exposure to air (oxygen) and moisture (20-100 times faster than pyrite) and the ability of pyrrhotite to self-heat on its own whereas other self-heating cases require a mix of sulphide minerals (i.e. pyrite and chalcopyrite, Wang, 2007). When pyrrhotite content is elevated or when pyrrhotite occurs with other sulphide minerals where galvanic interaction increases the oxidation rates, the heat generated by oxidation may not dissipate quickly enough and can result in self-heating (Wang 2007, Payant et al, 2012). Heat generation at a rate insufficient to overcome cooling by ambient air is not classified as self-heating. Pyrrhotite oxidation can generate elemental sulphur ($S°$) that will oxidize into sulphur dioxide gas (SO_2) through direct oxidation reaction involving proton acidity (H^+):

$$Fe_{1-x}S + (1-x)/2O_2 + 2(1-x) H^+ = (1-x) Fe^{2+} + S^O + (1-x) H_2O \text{ (with x = 0.1)} \tag{1}$$

$$S^O + O_2 = SO_2 \tag{2}$$

Alternatively, hydrogen sulfide gas (H_2S) can be generated through an oxidation reaction involving H^+:

$$Fe_{1-x}S + H^+ = (1-3x) Fe^{2+} + 2xFe^{3+} + H_2S \text{ (with x = 0.1)} \tag{3}$$

Several papers have reported self-heating events leading to SO_2 emissions in underground mines. Some of them identified backfill as the source (Bernier and Li, 2003; Nantel and Lecuyer, 1983; Patton, 1952) and some discussed ore as the source (Good, 1977; Fong et al., 2009). The purpose in reporting these papers is to highlight that self-heating events have occurred in the past with backfill containing pyrrhotite, sometime cemented and sometimes not. Bernier and Li (2003) presented the results of three case histories where highly oxidized CPB samples were recovered underground at two mines in northern Quebec and one mine in New Brunswick. The less oxidized, green-greyish paste reacted at temperatures less than 100 °C and did not emit SO_2. The more oxidized, orange to red paste reached the temperature range of 250-400 °C and emitted SO_2 during its oxidation process.

Patton (1952) reported heating of underground backfill made of 75% of crushed reverberatory furnace slag (0 to 4 inches) and 25% pyrrhotite at Noranda mine. Top stope backfill reached 87°C while centre and bottom reached respectively 204°C and 68°C; occasional odour of SO_2 in underground mine was also reported. Fong et al. (2009) reported in an internal memorandum a CPB self-heating event at Brunswick mine occurred in 2008. CPB used at this time at Brunswick contained 57.1% pyrite and 1.0% pyrrhotite and achieved 1MPa at 28 days with 5%wt. of T50 cement. For that case, the CPB in and/or around the muck fire stope was deemed unlikely to be the source of self-heating initiation. However, the paste fill present in this stope contributed to SO_2 emission once the self-heating process was taken beyond the ignition temperature of about 350 – 400 °C for pyrite.

CPB is generally considered to have reduced self-heating potential as compared to the source ore, tailings, or waste rock. The addition of cement and binder supports the formation of a low-permeability monolith with added neutralisation capacity to prevent mobilisation of acidity generated through sulphide oxidation.

3 SITE CHARACTERISTICS

The Voisey's Bay Underground Mine project is expected to contain a very high level of pyrrhotite content in the ore (66 to 72% in high-grade ore, [Golder 2018a, Golder 2018b]). As part of development of underground operations, Vale proposes to use milled tailings from the processing of underground ore to generate CPB to place in mined out stopes for underground structural support and improved ore recovery. This approach will also serve to minimise the surface disposal of the potentially acid-generating tailings.

Milled tailings contain high pyrrhotite contents (32-75%, [Golder 2018a, Golder 2018b]) remnant from processed ore. Cement and slag binder (10% normal Portland cement and 90% ground granulated blast furnace slag, respectively) is to be added to CPB to provide strength and to help control self-heating reactions. The addition of cement adds neutralisation capacity to the reactive tailings while the hydration products of the binder that partially coat the tailings can reduce the effective permeability and increase the water retention capacity of the CPB and thus limit the ability of oxygen to penetrate into the CPB mass and to perpetuate the oxidation reactions. These factors are expected to help control the oxidation of pyrrhotite, limit self-heating and the subsequent generation of acid rock drainage (ARD).

During engineering design studies, laboratory tests were completed to assess the self-heating capacity of CPB with 2% to 4% binder content, as discussed in Golder (2017). The samples were prepared by crushing, drying, and re-crushing and grinding the cured paste cylinders; determination of self-heating capacity was conducted as described in Nessetech (2018). These test results indicated that CPB with 2% binder had a propensity for self-heating and therefore were not recommended for backfill, while CPB samples with 4% binder were not anticipated to self-heat. However, an increase in binder content from 2% to 4% would have implications on mine economics. Golder (2017) identified that the Noranda test may be overly conservative because it does not account for the properties of the CPB as planned to be placed in underground development at Voisey's Bay, which will consist of a water saturated monolith where oxidation reactions will be limited to surfaces exposed to air. Further studies were recommended to evaluate the self-heating potential of CPB with 2% binder at field-scale.

4 METHODOLOGY

4.1 *Design overview*

The test cells were designed to simulate a monolithic mass of CPB that would more closely resemble a backfilled underground stope but in a controlled, above ground setting. CPB and ore were placed in shipping containers (hereafter referred to as "test cells"). The test cell was selected to allow testing of a suitably large monolithic specimen of CPB and allow the installation of required instrumentation and equipment to control the temperature and moisture conditions. To evaluate sensitivity of the test materials and align with evaluation techniques employed in the Noranda test, the test concept was to heat the test cells to create conditions favourable to the development of self-heating. This method is meant to accelerate the self-heating reaction pathway and permits evaluation of self-heating potential over a practical period of time.

246

Each test cell was instrumented with sensors to monitor oxygen, sulphur dioxide, temperature, and humidity. Humidity was controlled by a mister system which sprayed water (as mist) into the cell at multiple locations. The mister system was controlled by an automatic timer to activate it at pre-set intervals, with user intervention as needed to increase or decrease activation frequency based on monitoring data. Temperature was controlled by a heating and ventilation system with a user-adjustable thermostat. The thermostat was adjusted as necessary to align with planned heating cycles as described in Section 4.3. Sulphur dioxide and oxygen were monitored to identify geochemical reactions potentially indicative of self-heating. Sensor data from all test cells were recorded and stored in data loggers in an adjacent shipping container used as a field office for the test. The SO_2 sensors used have a sensitivity of 0.6 ppm. Bernier and Li (2003) reported an odour threshold of 3-5 ppm for SO_2. Newfoundland & Labrador prescribes short-term occupational exposure limits (OEL-STEL) of 0.25 ppm. O_2 sensors used have a sensitivity of 0.5%.

The test design intended to replicate as accurately as possible the anticipated conditions in an underground stope. However, the following characteristics could not be fully represented:

1) Underground mining at Voisey's Bay has not commenced and as such the volumes of tailings and ore required for the test were not available from underground sources. However, geochemical and mineralogical evaluation of ore performed by Voisey's Bay staff from the presently mined Ovoid Pit have demonstrated these characteristics to be similar to those of rock from exploration core from the underground development zones. As a result, ore and tailings derived from present operations in the open pit were used in the self-heating tests.
2) Underground temperature and humidity are expected to be variable inside the mine. According to current knowledge on self-heating, the test aimed at creating an environment that would promote self-heating (i.e. higher temperatures than expected inside the mine) and measure the reaction.
3) An underground stope is larger than the test cells. However, the spatial dimensions of the test CPB is considered to be adequate to represent surface interactions.
4) Variability in cement content of the CPB. Testing was performed with 2% cement (Table 1) representative of the current CPB recipe. If cement content is decreased, the behaviour may be different than that described in this paper.
5) Variability of the tailings and ore sulphide content. Behaviour observed is strongly related to materials used and cannot account for variability over the time. Especially for tailings, particle size is an important parameter of oxygen diffusion behaviour and a change would involve a different performance for the backfill than that described in this paper.
6) As observed in the memorandum of Fong et al. (2009), explosives can be an ignition source of self-heating. This aspect was not part of the testing program.

4.2 Test cell preparation

The current study was conducted prior to construction of an on-site paste plant. As an alternative, paste was prepared in the mill by pumping fresh tailings into a concrete ready-mix truck. Some variability in paste composition is a potential result of the time between collection of tailings for each test cell; although not known at the time of paste preparation, notable differences in paste composition (specifically sulphide content) were identified through laboratory testing and are discussed in Section 5.2. The mixing truck drum was filled with tailings

Table 1. Masses applicable to binder addition.

Cell #	Content	CPB curing time	Tailings dry mass (kg)	Binder mass (kg)	Binder (%wt.)
Cell 1	CPB only	31 days (28 planned)	8,970	183	2%
Cell 2	CPB and ore	28 days	15,548	317	2%
Cell 3	CPB and ore	19 (14 planned)	16,768	342	2%

and allowed to settle. After settling, water above the tailings was removed, and the space previously occupied by water was filled with fresh tailings. This process was repeated until the drum was more than 90% full of tailings and paste technical parameters (moisture content and slump) were achieved. The required binder mass was calculated and added to the drum (Table 1). The drum was rotated for a minimum of one hour to fully mix binder and tailings to produce paste. The paste was then poured into lined formwork in the test cell, the cell was closed, and the curing period initiated. Paste in each test cell was permitted to cure in a closed environment for the period of time defined in Table 1; curing time was varied between Cells 2 and 3 to evaluate if curing time affected the development of self-heating.

Following completion of the curing period, the test cells were opened, formwork was removed, and ore was placed beside and in direct contact with the paste (Cells 2 and 3 only). Cell 1 is intended for the purposes of this evaluation to be a "control" cell, to evaluate the potential for self-heating conditions in the paste alone without the presence of ore. Ore was sourced from the Ovoid Pit stockpile, not screened but coarse particles were rejected when sampled with the result of having a wide particle size distribution from zero to approximately 200 mm. Subsequently, ore was stored under a tarp near the test cells for 4 weeks, until the time to be placed in contact with the CPB by a backhoe. The volume of ore added to Cell 2 and Cell 3 was similar (+/- 15%) to the volume of paste; however, the mass was not recorded. At the time of placing the ore in the test cells, visible ore oxidation was minimal to absent. The test cells following completion of construction and paste placement are shown in Figure 1. Only the three left cells were used for testing; the fourth cell was compromised and not included in the test.

4.3 Test operation

The test was conducted as per the dates presented in Table 2. Sensor data was logged continuously and evaluated for indications of self-heating throughout each day.

Over the evaluation period, each cell was submitted to nine heating periods of 24 hours. During heating cycles, the heater unit was activated to reach the target temperature, monitored via internal sensors (Figure 2 & Figure 3). Once the target temperature was reached the temperature was maintained for 24 hours after which the heater unit was turned off for 24 hours. Following the 24-hour no-heat period, the next heating cycle began. For the first three

Figure 1. External photograph of test cells.

248

Table 2. Testing phases.

Event	Cell 1	Cell 2	Cell 3
CPB poured	June 11, 2018	June 14, 2018	June 22, 2018
Ore placement	N/A	July 12, 2018	July 12, 2018
Start of test (no heat)		July 12, 2018	
Initiation of heating cycles		July 18, 2018	
End of heating cycles		August 8, 2018	
End of test		August 28, 2018	

heating cycles, the target temperature was 40 °C. The remaining 6 heating cycles were operated to reach the maximum temperature possible via throughput of the heater (45 to 55 °C inside air temperature), with this temperature limit influenced by external climate.

The heating cycle from July 25th through 29th was extended to determine if running the heater for an extended period would permit reaching higher air temperatures (the finding was that the extended operation did not yield higher air temperatures than normal cycle durations, and the delayed peak during this period is due to low outside air temperatures). The objective of these temperature increase time periods was to generate favourable conditions for self-heating (high temperature, high humidity, and presence of ore in contact with the paste). It should be noted that the maximum temperatures tested are beyond maximum safe temperatures for an underground mine work environment; however, the higher temperatures were used to promote reactions which may lead to self-heating. Relative humidity inside the test cells was targeted to be in a range 80-90% over the testing period using automatic sprayers to represent the underground mine environment.

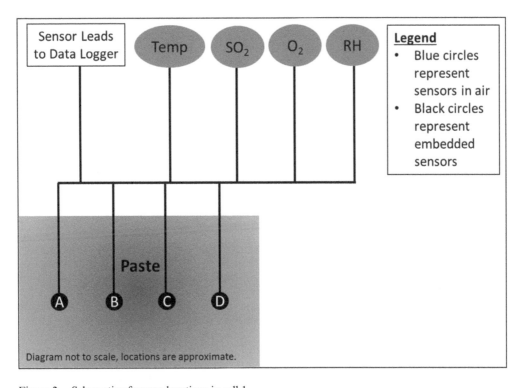

Figure 2. Schematic of sensor locations in cell 1.

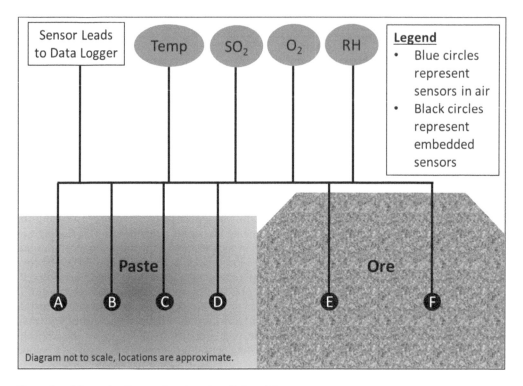

Figure 3.   Schematic of sensor locations in cells 2 and 3.

5   RESULTS

5.1   *Test cell instrumentation results*

Instrumentation measurements were recorded from the start of test (July 12) to end of test (August 28, 2018), monitoring temperature, oxygen, sulphur dioxide, and relative humidity. Sulphur dioxide measurements were within the standard error of the sensor at all times during the test. Accordingly, these values are not discussed further and are inferred to indicate that no appreciable generation of sulphur dioxide occurred. At the initiation of the test period following removal of formwork (all cells) and placement of ore (Cell 2 and Cell 3), no notable changes in temperature, oxygen, or relative humidity were observed; temperature changes noted are attributed to external (natural) variation in air temperature. Therefore, discussion of instrumentation results is focussed upon the period from initiation of heating (July 18th, 2018) to the end of the test (August 28th, 2018).

5.1.1   *Temperature*
Temperature results are presented graphically in Figure 4 and Figure 5, with statistics in Table 3. Figure 4 permits evaluation of relative changes in air temperatures inside each cell and outside air temperatures. Outside air temperatures influence internal (inside cell) air temperatures, particularly when heating is not activated, and also as a control on the upper limit of inside air temperatures during heating. Figure 5 presents internal paste and ore temperature results for each temperature sensor, showing differences with respect to sensor location and differences between each cell.

Based on the results presented in Figure 4 and Figure 5, activation of the heater is the primary control upon internal air, paste, and ore temperatures. The secondary control is external air temperatures in the natural environment. At the end of the heating cycles through to the

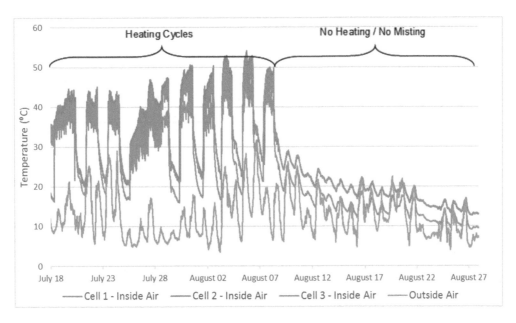

Figure 4.   Temperature of air inside test cells and outside air temperature.

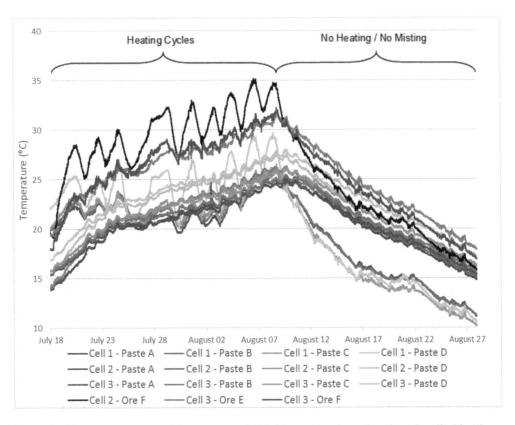

Figure 5.   Temperatures reported by sensors embedded in paste and ore (locations described in Figure 3). A to F refer to sensor location.

Table 3. Range of measured air temperatures over the test period (°C) after initiation of heating.

	Outside Air	Cell #1	Cell #2	Cell #3
Minimum	3.4	9.1	12.7	12.6
Average	12.5	23.9	28.0	27.6
Maximum	31.6	47.8	54.1	52.9

end of the test, a steady decrease in temperatures is observed, with minor increases in temperature aligned directly with instances of higher external air temperatures (Figure 4). During periods when the heater is not active, changes in internal cell temperatures are directly relatable to external air temperatures and sunlight on test cells. Temperatures in Cell 1 generally exhibited greater variability and lower values than in the other cells; this is interpreted to be primarily due to the lack of insulating effect from ore. It is possible that heat generated by oxidation of ore contributed to higher temperature values but if so, the effect was insufficient to overcome passive cooling during the no-heat periods. Because the sidewall of the paste in Cell 1 is in direct contact with air, the paste temperature at that location changes more rapidly.

No evidence exists for sustained increases in temperature other than as a direct result of heating applied intentionally in the test procedure. If any heat was generated from sulphide oxidation, it occurred at an insufficient rate to overcome cooling towards ambient conditions during no-heat periods.

5.1.2 Oxygen

Relative humidity and oxygen inside the test cells were influenced by operation of the heating and ventilation unit which introduced outside air to each cell when active. Evaluation of results from the three test cells (Figure 6) indicates that relative humidity decreases when internal temperature increases (ventilation blows out humidity) while O_2 increases (ventilation brings fresh air). Analysis of recorded oxygen concentrations (Figure 6) do not show evidence of sustained consumption of oxygen from oxidation of sulphides (and potential self-heating) during any phase of the test. The test cells are not airtight and ingress of oxygen over time is expected even while the cells are sealed. The cause of increased variability in Cell 2 results is unclear and potentially related to inconsistency in the sensor given the lack of correlating changes in any other parameters. Cell 3 results are interpreted to be lower due to a calibration error but appear to be consistent with Cell 1 over time.

5.1.3 Relative humidity

From the complete dataset recorded, relative humidity data ranged between 19.2% and 99.9% (Figure 7). The relative humidity inside the test cells was targeted to be in the range of 80-90% over the testing period using automatic sprayers to represent the underground mine environment; however, when the heating and ventilation unit was active, added humidity was quickly lost and the high humidity was not maintained, while excess humidity was present when the heating and ventilation was not active. Actual moisture content of paste was assumed to be at a higher level than heated air due to ingress of water during high humidity periods which would not be withdrawn from paste during the 24-hour heating cycles. This assumption was confirmed during the post-test visual inspection which determined that the first 1 cm of paste was unsaturated, while all paste beyond 1 cm depth was water saturated.

Cell 1 generally was observed to have the highest humidity values, which may be due to quicker settling of the mist in the space next to the paste, whereas in the other cells, airborne water could only circulate in the limited headspace above both paste and ore resulting in increased loss through ventilation. Cell 2 generally has higher humidity values than Cell 3; paste and ore placement is the same in both cells and there is no clear driver for this variation.

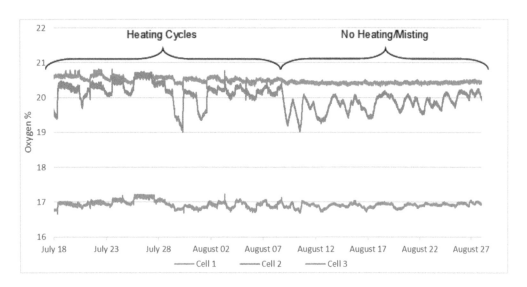

Figure 6.   Oxygen content over evaluation period.

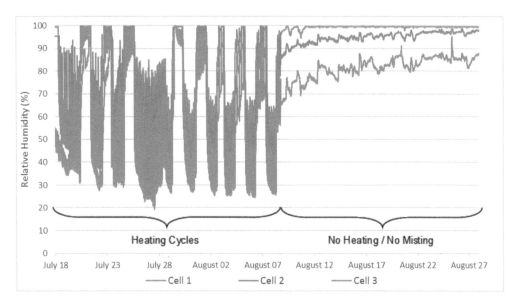

Figure 7.   Relative humidity over evaluation period.

5.2 *Mineralogical and geochemical results*

Prior to the initiation of testing, samples of ore and wet paste were collected for geochemical and mineralogical analysis. Samples collected following completion of the field test were submitted for the same analyses. Pre-test results are used to confirm the test materials are consistent with expected characteristics for paste backfill underground at Voisey's Bay, and as a baseline reference for post-test samples. Post-test paste samples were collected at surface (<1 cm depth) to 75 cm depth from the closest edge of the paste block to evaluate changes in geochemical and mineralogical composition resulting from the field test. Results for selected parameters of interest are summarized in Table 4.

Table 4. Comparison of average paste properties prior to and after test (mineralogy as reported by XRD).

After Test			1						2						3				Test 2 Cell II	
Sample ID	2A	2B	1-1	1-2	1-3	1-4	1-5	1-6	2-1	2-2	2-3	2-4	2-5	2-6	3-1	3-2	3-3	3-4	3-5	3-6
Sulphide %	20	22	14	14	15	15	15	15	23	24	25	25	24	24	25	27	25	27	25	26
Sulphate %	10	7	11	11	10	11	11	10	19	19	17	15	19	18	9	11	9	9	11	10
Neutralizing Potential (t CaCO$_3$/1000 t)	25	26	58	48	51	49	49	50	36	46	29	31	30	30	35	24	26	22	23	24
Acid Potential (t CaCO$_3$/1000 t)	613	672	428	444	456	459	453	456	731	763	769	766	747	759	794	828	788	841	794	813
Pyrrhotite/Troilite %	77	73	35	34	39	Not analysed	Not analysed	Not analysed	67	63	72	Not analysed	Not analysed	Not analysed	74	76	74	Not analysed	Not analysed	Not analysed
Other Sulphides %	0.6	0.6	1.5	1.3	-	Not analysed	Not analysed	Not analysed	-	-	-	Not analysed	Not analysed	Not analysed			1.2	Not analysed	Not analysed	Not analysed
Sulphates %	-	-	1.2	0.5	-	Not analysed	Not analysed	Not analysed	-	-	-	Not analysed	Not analysed	Not analysed			0.5	Not analysed	Not analysed	Not analysed
Silicates %	16	21	53	58	56	Not analysed	Not analysed	Not analysed	27	27	21	Not analysed	Not analysed	Not analysed	15	16	17	Not analysed	Not analysed	Not analysed
Oxides %	6	6	6	7	4	Not analysed	Not analysed	Not analysed	5	7	6	Not analysed	Not analysed	Not analysed	9	9	8	Not analysed	Not analysed	Not analysed
Carbonates %	-	-	2.6	-	1.3	Not analysed	Not analysed	Not analysed	1.4	3.0	0.9	Not analysed	Not analysed	Not analysed	2.7	-	-	Not analysed	Not analysed	Not analysed

Table 4 indicates some differences in geochemical and mineralogical characteristics for post-test samples collected from Cell 1 as compared to samples from Cell 2 and Cell 3. No pre-test samples were analysed from paste used in Cell 1 for comparison. Cell 1 samples have lower abundance of sulphide minerals and greater abundance of silicate minerals than Cell 2 and Cell 3. Because this change in mineral composition cannot occur simply as a transformation during the test, heterogeneity in the tailings feed from the mill at the time of preparation is the probable cause. Surface shavings samples (VB2-2, VB3-1) which were anticipated to potentially demonstrate stronger effects from testing due to directly exposed surface area, did not yield any notable differences compared to samples collected at greater depth. It is important to consider that some extent of sample heterogeneity is expected and that small variations in results (i.e. relative percent difference less than 20%) may not be indicative of any trends.

5.3 *Visual inspection of test cells after termination*

Following the completion of the field test, a visual inspection of the test materials was undertaken to identify any evidence of test related effects (Figure 8). In both Cell 2 and Cell 3, ore was observed to have taken a reddish tarnish, indicating that surficial oxidation of the sulphides had occurred. In contrast, paste showed no evidence of evolution over the testing period and no oxidation products were evident. Two samples of isolated white precipitate were selected for a targeted QEMSCAN analysis. Results of this analysis found that in one case this material was comprised of calcite, while in the other case the precipitate was no longer visible, but a cementing material comprised of calcite and gypsum was observed. Calcite is likely derived from lime and/or cement binder addition, while gypsum is a common precipitate following sulphide oxidation. The constituents required for gypsum precipitation were likely present in the paste prior to pouring and may not represent additional sulphide oxidation during testing. Regardless, the volume of precipitate observed was extremely small and not representative of bulk characteristics of the paste. Paste was water saturated at shallow depth (less than 1 cm from surfaces), and no defined crust was present. The paste appearance was uniform across horizontal and vertical position.

6   DISCUSSION

Key findings of the test cell monitoring results can be summarized as follows:

– Paste and ore temperatures rose during heating periods, and fell during all periods when heating was not active;
– Sulphur dioxide values were not detected above minimum sensor sensitivity at any time;
– Oxygen was not consumed at a rate to indicate rapid oxidation reactions and self-heating; and
– If heat was generated from sulphide oxidation, it occurred at an insufficient rate to overcome cooling towards ambient conditions during no-heat periods.

The test results indicate that sulphide oxidation over the test period was primarily limited to ore, with no evidence of oxidation in CPB. Self-heating conditions (generation of heat at a rate sufficient to overcome cooling with ambient air during no-heat periods) were not observed through any of the measurements made during and after the test. Visual inspection of the CPB determined that it was water saturated, which is an important characteristic for the prevention of sulphide oxidation and therefore self-heating.

The test results indicate that CPB with 2% binder did not develop significant self-heating conditions in the field test cells, given that paste and ore temperatures fell during all periods where the external heat source was no longer applied. All instances of increasing paste and ore temperatures can be directly related to elevated ambient air temperatures, either due to activation of the heating unit or due to external (natural) temperature increase. The tailings and the

Figure 8.   Cell 2 paste with partially excavated ore.

addition of the binder are favourable for maintaining high water saturation in paste pore space that in turn limit oxygen diffusion. The test was designed as an analogue to underground paste backfill; however, the test cells are not a perfect representation of the underground environment due to factors including, but not limited to the following:

– Duration of testing (shorter than underground exposure);
– Differences in ventilation and ambient humidity and temperature;
– Presence of ore likely to occur on multiple sides of the paste in underground development rather than only one side in the test cells;
– Blasting (heat generated by blasting) conducted adjacent to paste in underground development;
– Minor differences in mineralogy and geochemistry of paste derived from underground ore compared to ore derived from the open pit used in the test cell; and
– Differences in paste preparation procedure (batches produced in a mix truck for test cells vs. continual production in a paste plant for underground use).

7   CONCLUSIONS

Field study results indicate that the potential for self-heating of the CPB under evaluation in this study was low under the study test conditions. This finding is in conflict with previous laboratory scale testing completed (i.e. The Noranda Test), which indicated high potential for the development of self-heating conditions in the materials tested. Therefore, evidence from the present study supports the hypothesis that laboratory scale testing of pulverised material may be overly conservative relative to actual conditions. Long-term monitoring of full-scale paste backfill deposition is required to validate this hypothesis.

The unique composition of minerals present at any given site may result in different outcomes than those described herein. Regardless of further refinement and validation of the procedures described in the current study, laboratory scale testing methods are anticipated to remain as important preliminary screening tools to determine whether additional study is required. However, the current findings suggest that laboratory scale test results of self-heating should not be relied upon in isolation to determine that an increased binder content is required to prevent the development of self-heating. Rather, the proponent should review the potential for field-scale testing to identify a lower binder content relative to long-term costs of increased binder content. If field-scale testing is deemed a desirable approach to evaluate whether binder content can be reduced, the proponent should engage with qualified experts in mine backfill technology and sulphide geochemistry.

ACKNOWLEDGEMENTS

Golder Associates Ltd. appreciates the support of Vale Newfoundland and Labrador in sharing the results of this innovative field trial with the scientific community and advancing the body of knowledge in the field of self-heating.

BIBLIOGRAPHY

Bernier, L. R., and Li, M. 2003. High temperature oxidation (heating) of sulfide paste backfill: a mineralogical and chemical perspective. [In:] *Proceedings of Sudbury Mining and the Environment Conference*.

Fong et. al. 2009 - Fong, G., Isagon, I., Lebonté, G., and Wheeler, R. 2009. *Self-Heating Issues due to Oxidation of High-Sulphide Content in Ore, Waste and Paste Backfill, Notes from the Site Visit Conducted at Brunswick Mine and Smelting, Bathurst, NB, on Dec. 17–18, 2008*.

Golder, 2017. *Voisey's Bay Underground Mine – Self-Heating of Cemented Paste Backfill*. Report prepared for Vale Newfoundland and Labrador. Project No. 1665718.

Golder, 2018a. *Voisey's Bay Mine – Evaluation of Tailings Self-Heating Properties*. Report prepared for Vale Newfoundland and Labrador. Project No. 1665718.

Golder, 2018b. *Voisey's Bay Self-Heating Evaluation – Field Scale Trials*. Report prepared for Vale Newfoundland and Labrador. Project No. 1665718.

Good, B.H. 1977. Sulphide Fires - The "Hot Muck" Problem, The Oxidation of Sulphide Minerals in the Sullivan Mine, CIM Bulletin, June 1977.

Nantel, J. and N. Lecuyer. 1983. Assessment of slag backfill properties for the Noranda Chadbourne Project Noranda, Quebec, *CIM Bulletin*, V. 76, No. 849.

Nessetech, 2018. *Self-Heating Tests – Vale Voisey's Bay Samples*. Report prepared for Golder Associates Ltd.

Patton, F. E. 1952. Backfilling at Noranda. *Canadian Institute of Mining & Metallurgy Transactions*, V. 55, pp. 137–143.

Payant, R., F. Rosenblum, J.E. Nesset, J.A. Finch. 2012. The self-heating of sulphides: galvanic effects. In Minerals Engineering. Volume 12, p 57–63.

Wang, X. August 2007. Exploring Conditions Leading to Self-Heating of Pyrrhotite-Rich Materials. A thesis submitted to McGill University in partial fulfilment of the requirements of the degree of Master of Engineering.

Minefill geomechanics, numerical modeling, mitigation of subsidence

Minefill 2020-2021 – Hassani et al (eds)
© *2021 Taylor & Francis Group, London, ISBN 978-1-032-07203-6*

Dem numerical modeling of longwall extraction of coal in "Mysłowice" colliery

Grzegorz Smolnik
Silesian University of Technology, Gliwice, Poland

SUMMARY: The paper presents the results of the numerical DEM analysis of the extraction of coal by longwall system both with roof caving and with hydraulic backfilling. The exploitation of the most productive in the basin coal seam 510 was undertaken by the Mysłowice colliery in a very rockburst prone strata at the depth of app. 600 m. The successful extraction with backfilling of the last longwall panel in the section of the mine was a key to any other future mining activity in the area as the stress concentration and resulting rockbursts would prevent it otherwise. The deformation of boreholes measurements as well as geophysical monitoring was widely applied by the mine. The UDEC numerical model of the extensive section of the mine was built and all the extraction of coal activities ranging the period of nearly 45 years were simulated. The changes of the state of stress and strain as well as movements of the blocks in the vicinity of the longwall workings were analyzed. The strain changes in the numerical model were determined at the positions corresponding to the location of borehole measurements in-situ. The rate of advance and productivity of the longwall face turned out to be directly related to the changing geomechanical conditions just as they were determined by the numerical simulation.

Keywords: longwall mining, hydraulic backfill, numerical modelling, Distinct Element Method, UDEC

1 INTRODUCTION

The most productive in the whole Upper Silesian Coal Basin seam has been extracted in Myslowice colliery for many years. The average seam thickness in that mine was about 9.5 m and so it was exploited in stages (slices) by longwall mining methods both with roof caving and hydraulic backfilling of the goaf area. The stoping was undertaken simultaneously both in the eastern and in the western part of the mine. The longwall extraction commenced at the borders of mine lease area and gradually, with each next panel, it was moved towards the central portion of the seam. Finally, the strip of a width of 200 – 320 m was left surrounded by the goaf from each side. The successful extraction with backfilling of the last longwall panel in the section of the mine was a key to any other future mining activity in the area as the stress concentration and resulting rockbursts would prevent it otherwise. Not only the exploitation of the above lying layers of seam 510 would have been impossible but the plans for stoping in seam 501 of the average thickness of 8 m laying above and separated from the 510 seam by the 2.5 m layer of shale might have also been abandoned.

Challenging, mainly due to the rock burst hazard extraction of the mining panel 411b resulting in the distressing of the coal seams and rocks in its vicinity was of a huge importance for the mine.

DOI: 10.1201/9781003205906-24

Numerical modelling allows to investigate the response of the rock mass subjected to extensive stoping in a way far beyond the other analytical methods. Nevertheless both the determination of the rock properties in the laboratory and the in-situ measurements of the rock mass properties are very welcomed sources of data which make the modelling process useful for the mining company. Without those the boundary and initial condition determination as well as the material properties input is much less reliable. As the extensive program of measurements of the deformations of the boreholes was undertaken in the mine and seismic methods including seismic tomography, micro-gravimetry, seismic acoustic techniques were deployed for years the challenge to create a model of a significant portion of the mine and the simulation of the exploitation of coal carried out for decades in the area was undertaken.

The numerical modelling of mine strata and the computer simulation of longwall mining with hydraulic backfilling and roof caving was done using the two-dimensional distinct element method code *UDEC*. *UDEC* is based on a Lagrangian calculation scheme that is well-suited to model the large movements and deformations of blocky systems. This code is extremely useful in the simulation and analysis of the effect of rock joints, faults, bedding planes, *etc.* on the behaviour of rock masses and the stability of excavations and structures built in rocks or on rocks. It has been used successfully at the Rock Mechanics Laboratory of the Silesian University of Technology for years to study the behaviour of mine strata affected by the influence of longwall mining (see for example Kwaśniewski, 2008).

2 NUMERICAL MODELLING OF MINE STRATA

2.1 Structural model of mine strata

In an attempt to study the behaviour of the mine strata in the vicinity of the longwall panel in the high stress conditions, a numerical model of a large part of the rock mass in the vicinity of longwall panel 411b in coal seam 510 in the Mysłowice Colliery was built. This was a two-dimensional rectangular model 1500 m long and 750 m high. The upper boundary reached the land surface and the floor of the seam 510 was at the depth of 592 m (Figure 1).

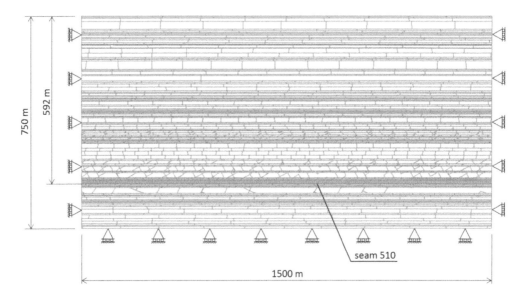

Figure 1. Structural model of mine strata in the vicinity of coal seam 510 in longwall panel 411b in the Mysłowice Colliery: 112 rock layers (including 9 coal seams and a soil (clay) layer at the upper boundary).

Above and beneath the seam 510 one hundred and nine rock layers (including eight coal seams) of a thickness from 1.9 to 29.0 m were modeled. These were the layers built of different varieties of sandstones, mudstones, sandy shales, clayey shales, coaly shales and coal. On the land surface of the mining area (upper boundary of the model) there was also one soil layer (clay) of the thickness 2.5m and it was likewise modeled.

Individual rock layers, separated by parallel contacts (joint system I, $\alpha_1 = 0°$), were divided into blocks by joint systems oriented at angles $\alpha_2 \approx 30°$ and $\alpha_3 \approx 120°$ (joint system II) and $\alpha_4 \approx 60°$ and $\alpha_5 \approx 160°$ (joint system III)relative to joint system I. It was only the to-be-mined parts of coal seams that were divided into rectangular blocks by a system of joints oriented to the bedding planes at an angle (α_6) of 90° (joint system IV). All these discontinuities were the so-called fictitious joints, *i.e.* the joints whose mechanical properties were the same as the mechanical properties of the rock material. Such joints may become active only if stresses reach either the tensile or shear ultimate strength at a given location. The decision to divide the model into blocks with fictitious joints was taken deliberately in order to be on the conservative side and not to facilitate the fracturing of the modeled mine strata.

Four systems of joints of different orientations divided the model of the mine strata into 4101 fictitious blocks of an average size of 274.3 m^2, which contacted each other at 41,292 contacts. The blocks were discretised into finite-difference triangular elements. The number of deformable zones and the number of gridpoints were equal to 32,079 and 37,837, respectively (Table 1).

2.2 *Constitutive models and material parameters*

The strain-softening (elastic-brittle-plastic) constitutive model with ultimate and residual strength defined by the modified Mohr-Coulomb strength criterion was assigned to the rock blocks. The following material parameters occur in this model: bulk modulus (K), shear modulus (G), angle of internal friction (φ), angle of residual internal friction (φ_r), cohesion (c), residual cohesion (c_r), tensile strength (σ_T), angle of dilatancy (ψ) and angle of residual dilatancy (ψ_r). Values of these parameters for different rocks that build the mine strata in the vicinity of seam 510 in longwall panel 411b in the Myslowice Colliery are listed in Table 2. Values of the remaining parameters were calculated using the following relationships: $c_{Mr} = 0.1c$, $\varphi_r = 2/3\varphi$, $\psi = 0.5\varphi$ and $\psi_r = 2/3\psi$.

The model adopted makes it possible to describe any arbitrary non-linear behaviour of rock material in the post-failure domain. In contrast to the Coulomb model of plasticity, where cohesion, friction and dilation remain constant, in the brittle-plastic model these quantities are certain functions of plastic shear strain (e^{ps}). When modelling the softening behaviour of rocks in the post-shear-failure domain, it was assumed in the present study that: (i) these functions have a piecewise-linear form and (ii) the cohesion, angle of internal friction and angle of dilatancy drop to their residual values at $e^{ps} = 0.001$. A mechanical model capable of: (i) normal and shear elastic deformation, (ii) dilation, (iii) irreversible slip with displacement softening and (iv) tensile fracturing, was assigned to the joints between blocks.

It was assumed that the joints behave in a linear manner in the pre-failure domain and slip is governed by the Coulomb criterion. The adopted values of the mechanical parameters of

Table 1. Basic structural parameters of the model of mine strata.

Parameter	Value
Model dimensions, L [m] × H [m]	1500.0 × 750.0
Number of rock layers, including coal seams	112
Number of fictitious blocks	4101
Block size [m^2]: min./max./av.	0.23/7868/274.3
Number of fictitious contacts	41,292
Number of finite-difference zones	32,079
Number of gridpoints	37,837

Table 2. List of rocks and soil distinguished in the mine strata in the Mysłowice Colliery and values of their mechanical parameters.

Rock type	ρ kg m^{-3}	K_M MPa	G_M MPa	c_M MPa	φ deg	σ_{TM} MPa
Coarse-grained sandstone	2475	5460	3926	1.56	38	0.11
Medium-grained sandstone	2500	5556	4167	1.79	39	0.13
Fine-grained sandstone	2525	5469	4449	2.03	40	0.15
Very fine-grained sandstone	2550	5392	4741	2.20	42	0.17
Sandy mudstone	2570	5655	3893	2.14	38	0.21
Mudstone	2590	5455	3673	2.17	36	0.22
Sandy shale	2610	5247	3455	1.41	34	0.14
Sandy-clayey shale	2600	4938	3252	1.06	34	0.11
Hard clayey shale	2625	4487	2823	1.17	33	0.13
Clayey shale	2650	4000	2400	0.89	32	0.10
Soft clayey shale	2400	2083	1190	0.47	29	0.08
Coaly shale	2200	1852	980	0.24	26	0.04
Coal	1400	1667	5563	0.64	30	0.22
Clay	1950	15	7	0.008	20	0.0008

Explanations: ρ – bulk density, K_M – rock mass bulk modulus, G_M – rock mass shear modulus, c_M – rock mass cohesion, φ – angle of internal friction, σ_{TM} – rock mass tensile strength.

this Coulomb slip (area contact) model of joints, which include: coefficients of normal and shear stiffness (k_n and k_s, respectively), friction angle (φ_j), cohesion (c_j), tensile strength (σ_{Tj}) and dilation angle (φ_j), were obviously different for different rocks; in Table 3 only the maximum and the minimum values of these parameters are given. As was mentioned above, the mine strata were divided into blocks by the so-called fictitious joints; therefore, the assumption was made that $c_j = c_M, \varphi_j = \varphi$ and $\sigma_{Tj} = \sigma_{TM}$.

2.3 Physical model of hydraulic backfill

To model the hydraulic backfill the approach proposed by Clark, 1991 was followed. The double-yield constitutive model was chosen in which permanent volume changes caused by the application of isotropic pressure are taken into account by including, in addition to the shear and tensile failure envelopes in the strain softening/hardening model, a volumetric yield surface (or "cap"). For simplicity, the cap surface, defined by the "cap pressure" $p_c > 0$, is independent of shear stress; it consists of a vertical line on a plot of shear stress versus mean stress (ITASCA, 1996). The hardening behaviour of the cap pressure is activated by volumetric plastic strain, and follows a piecewise-linear law prescribed in a user-supplied table. According to results obtained by Clark the cap pressure position is described by the formula:

$$c_p = W \left[\frac{\varepsilon^{pv}}{H - \varepsilon^{pv}} \right]^\alpha + c_p^0$$

Table 3. Values of the mechanical parameters of discontinuities in the model of the mine strata (Fictitious joints between rock and soil blocks).

k_n MPa m^{-1}	k_s MPa m^{-1}	c_j MPa	φ_j deg	σ_{Tj} MPa	ψ_j deg
14.4 – 8000	5.5 – 3296	0.0008 – 2.2	20 – 42	0.004 – 0.22	10 – 21

Explanations: k_n – normal stiffness coefficient, k_s – shear stiffness coefficient, c_j – cohesion, φ_j – friction angle, σ_{Tj} – tensile strength, ψ_j – dilation angle.

Table 4. Values of the mechanical parameters of the double-yield model of the backfill.

	Volumetric yield						
Strain ε^{pv}	0.00	0.02	0.04	0.06	0.08	0.10	0.12
Pressure c_p	0.01	0.2	0.8	1.3	3.5	5.5	8.5
Strain ε^{pv}	0.14	0.16	0.18	0.20	0.22	0.24	0.28
Pressure c_p	11.5	19.0	34.0	50.0	80.0	170.0	540.0

	Shear yield	
Strain ε^{pv}	0.0	0.1
Cohesion c, MPa	0.004	0.005
Internal friction angle ϕ, deg	28	32
Dilation angle Ψ, deg	7	8

where $C_p{}^\circ$ denotes initial cap pressure and W, H and α are constants for a given material. The relationship $c_p = f(e^{pv})$ is given in a Table 4, and cohesion, internal friction angle and dilation angle changes as a function of plastic shear strain are also listed there.

The distinct element method code *UDEC* is able to simulate the excavation of the seam with backfilling in a quite realistic manner. It is performed by changing the seam blocks to a null material, and backfilled by changing the null material to a double-yield material.

2.4 *Model of the longwall support*

The support of the 2.6-m-high longwall face in seam 510 was modelled using a 10-element support member. The maximum load-bearing capacity of the model of the longwall support ($P_{max} = 1.03$ MN) and axial stiffness coefficient (30 MN/m) were equal to the nominal load-bearing capacity and stiffness of the two-leg mechanised shield support FAZOS-21/31Op that was used in longwall 411b in the Myslowice Colliery. The model of the support was positioned at a distance of 0.45 - 0.75 m from the coal face and supported the roof over a length of 3.5 m.

2.5 *Boundary and initial conditions*

The horizontal displacement of the model was fixed to zero at the left and right vertical boundaries and no vertical displacement boundary condition was applied to the bottom boundary of the model. The top of the model was left unstressed, as it was assumed to represent the land surface of the mining terrain.

When initializing the primary state of stress in the model it was assumed that the vertical stress in rock masses results from body forces and is determined by the depth and bulk density of the overlying rocks and coal, while the horizontal stress – according to the classic solution based on the theory of elasticity – is a fraction of the vertical stress governed by Poisson's ratio. After stepping the model to initial equilibrium, which required 3000 calculation steps, the primary state of stress was established in the model with the maximum principal stress in the region of the future setup entry in seam 510 equal to 14.38 MPa and the minimum principal stress in the same area equal to 6.18 MPa.

Only then was the computer simulation of longwall mining and investigation of the behaviour of mine strata affected by mining begun.

3 COMPUTER SIMULATION OF THE EXTRACTION OF MULTIPLE SEAMS IN THE MYSŁOWICE COLLIERY

3.1 *Longwall mining in 9 seams extracted over the period of 45 years*

Before the detailed study of changes in the state of stress, strain and displacements of mine strata that accompany the extraction of longwall panel 411b commenced all the mining of coal in the area, in 9 coal seems, which took place over the period of 45 years was simulated. Progressive extraction of two seams(405 and 416) with roof caving and seven (418, 501/1, 501/2, 501/3, 510/1, 510/2, 510/3) with backfilling was simulated in the same order as it was done in the mine. In each mining simulation stage, the seam was extracted over a length of 25.0 m by removing five blocks 5.0 m long. Each time the constitutive model of coal was replaced by the so-called null model and then the blocks were either deleted (roof caving) or their constitutive model was changed to the double-yield (backfilling). When an extraction in one seam took place in the same quarter of the given year as the extraction in another longwall panel in a different coal seam the simulation of those exploitations was done simultaneously.

As a result of mining, very significant changes in the stress field in the vicinity of the mined seam occurred. A substantial redistribution of stresses, including both a change of the directions of the principal stresses and a change in stress values took place. The maximum principal stresses (σ_1) that were originally oriented vertically assumed a horizontal direction above and below the mined-out section of the seam, whereas the minimum principal stresses (σ_3) changed their direction from horizontal to vertical. Both these components of the stress tensor decreased significantly in this region; at the same time, the minimum principal stresses became tensile in many places.

Extensive exploitation of 9 coal seams resulted in the fracturing of nearly two thousand three hundred and thirteen contacts between the initially fictitious blocks; this number is about 5.6% of the total, 41,292, number of contacts. The number of fractured contacts, *i.e.* the contacts that do not convey normal and/or shear forces, increases as the mining progresses.

The size of the area where rocks underwent failure due to excessive shear or tensile stresses reached 206,900 m^2, it was 18.4% of the total, 1 125,000 m^2, area of the model being investigated (Figure 2). The area of failed rocks is particularly large in the strata lying above extracted seams, it reaches the upper boundary of the model in some places and, moreover, it extends beyond the edges of the extraction area. In many cases rocks located at a significant distances from the longwall panels underwent damage. Failure of rocks also occurred in layers directly beneath the longwall panels.

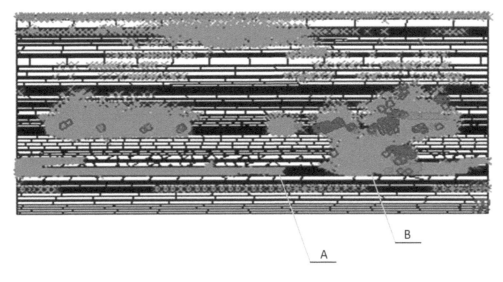

Figure 2. Failure of rocks due to excessive shear stresses (displayed in light red when the state of stress is at the yield surface and in green when yielded in the past) or tensile stresses (denoted by circles in magenta) after nine seams had been extracted over 45 years and before the longwall panel 411b exploitation took place. (A – position of the start of the simulation of the extraction of coal in longwall face 411b, B – recovery room of longwall face 411b).

The extent of the fractured zones in the rock mass is much more profound in the vicinity of the longwall panels extracted with roof caving, when compared to those extracted with backfilling.

3.2 *Extraction of longwall panel 411b with hydraulic backfilling in a very rock burst prone area*

The first months of extraction of longwall panel 411b resulted in an increased seismicity in the area and when a few very strong tremors occurred in the vicinity of the panel its operation was halted for several months. The mine suffered in the past form the strong rock bursts/coal bumps resulting also in some fatalities so the State Mining Authority demanded special extensive monitoring measures to be adopted before the permission to resume the mining of coal in longwall panel 411b could be issued. The deformation of boreholes measurements as well as geophysical monitoring was widely applied by the mine. It is this stage of exploitation when the detailed numerical analysis of the extraction of coal in longwall panel 411b with hydraulic backfilling was commenced.

Progressive extraction of seam 510 in longwall panel 411b was simulated over a length of 300 m starting from the position x=845 m and advancing towards the right boundary of the model (see Figure 2). At the first simulation stage one 5.0-m-long block was removed from the seam and the support members simulating the shields were placed. At the second simulation stage next 5.0-m-long block was removed from the seam and the roof support was moved five meters. At the third simulation stage the constitutive model of the next block was changed to null, roof support moved five metres and the hydraulic backfill was introduced by changing the constitutive model of a block where roof support was placed at the first stage, to double-yield.

At each of the subsequent 57 mining simulation stages, one 5.0-m-long block was removed from the seam, the model of the shield was moved and a hydraulic backfill was placedaccordingly. Next, 2000 calculation steps were executed. The last two stages simulated the operation in the recovery room of the longwall i.e. backfilling of the area directly behind the roof support units, removing the shields and complete backfilling of the recovery room.

4 DISCUSSION OF THE SIMULATION RESULTS

Simulation of over forty years of extensive extraction of coal revealed the mining-induced changes in the state of strain and stress in rock blocks and the location and size of the areas where rock material failed in shear or tension and underwent transition into the post-failure range in the direct vicinity of the longwall panel 411b.

Initially uniform state of stress has been altered significantly; the whole 300 m long portion of the seam 510 to-be-mined-out by longwall panel 411b can be divided into two parts (halves) of approximately the same length.

Major principal stress values in the first 130-150 m portion of the seam increased from the initial values of 14.5 MPa to 18-19 MPa. In the second half of the seam σ_1 values decreased quite rapidly to about 8 MPa and closer to recovery room slightly increased but to not more than 10 MPa.

Extraction with roof-caving of 3.2 m thick seam 416laying 60 m above, which took place a year and a half before was a major contributor to those changes. It was also a cause of the failure of considerable parts of the rock mass between the seams 416 and 510 and even the layer of shale directly beneath the seam 510. Noteworthily the coal of the seam 510 due to its high strength did not yield and remained in the pre-failure, elastic range and thus was able to accumulate the elastic strain energy.

When the simulation of the operation of longwall panel 411b began the state of stress, strain and displacement was monitored in its vicinity and the major principal stress changes in a coal block directly adjacent to the shearer, armoured face conveyor and shields (i. e. longwall face) are shown, as an example, in Figure 3. Horizontal axis represents the position of the longwall face; abscissa x=0 m gives the position of the face at the start of the

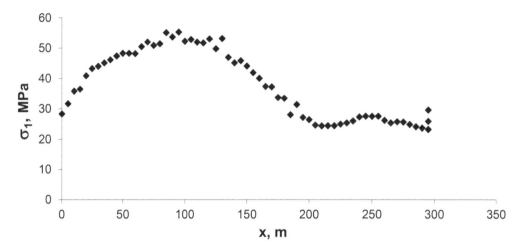

Figure 3. Distribution of the maximum values of the major principal stress in seam 5107 ahead of the longwall face corresponding to different lengths of the mined-out section of the seam.

simulation of the extraction of coal in longwall face 411b (point A in Figure 2) and x=295 m represents the final position of the face reaching the recovery room (point B in Figure 2).

After the commencement of the extraction of coal σ_1 values increase from 28 MPa to more than 55 MPa for the first 100 m of the seam and then the major principal stress decreases gradually. When the longwall face is at the position x≈200 m σ_1 values drop to about 25 MPa. The changes of major principal stress in the final 100 m are not significant, σ_1 values are in a range of 24.5 – 27.5 MPa.

The above mentioned changes of the major principal stress values are only one of the indicator of the geomechanical conditions in which extraction of the panel 411b took place.

High σ_1 values in the first half of the extracted portion of the seam 510 are accompanied by substantial movement of rock blocks, both located in the roof but also in the floor of the seam, towards the working area. There are also significant closures of the joints in the gob and compaction of the hydraulic backfill. Directly above the coal face there are considerable separations and shear displacement of blocks resulting in a significant loading of the longwall shields.

Conditions in the second half of the extracted portion of the seam 510 are appreciably different. Destressing due to the roof-caving longwall extraction of the above laying seam 416 although did not result in failure of coal in the seam 510 but did cause the adjacent layers of shales (manly clayey and sandy shales) to reach the post-failure regime of their behavior and very significant drop of σ_1 values. In a result there are much smaller movements of rock blocks towards the working area; the displacement of blocks in the roof decreases so substantially that the floor up-heave becomes more pronounced. Closures of the joints in the gob, compaction of the hydraulic backfill, shear and normal displacements (separations) in the roof above the longwall face are also much smaller. Not surprisingly, loading of the hydraulic powered roof support units (shields) is likewise significantly smaller.

The results of the numerical simulation were confirmed by the conditions experienced in the longwall face and as an example, monthly advance of the longwall face 411b is given in Figure 4.

During the first five months of the exploitation of the panel 411b the monthly advance rate was quite regular but in the final portion of the seam it increased twice. The beginning of the significant increase of the advance rate (December) was achieved when the longwall face reached the position x≈120-135 m corresponding to the location where major principal stress values as well as shear and normal displacements of blocks and joints and shields loading started do decrease.

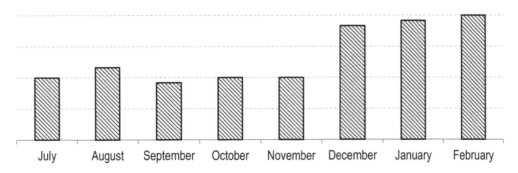

July   August   September   October   November   December   January   February

Figure 4. Average monthly advance rates of longwall panel 411b.

Displacements and strains were monitored at ten positions in the model which corresponded to the spots where extensometers were placed in drilled holes in the mine.

The results obtained were very similar in all the places of monitoring which were located close (both in vertical and horizontal direction) to the longwall face. Recorded displacements and strains at sites positioned far away from the active longwall were much smaller and the pattern of changes was not so regular and was different from that which is common to the points located in the vicinity of the longwall panel.

As an example of the recorded data Figure 5 displays the general pattern of the behavior of roof layers above the longwall face with hydraulic backfill as revealed by the numerical simulation. The actual measurements results are of the same nature although not so regular and

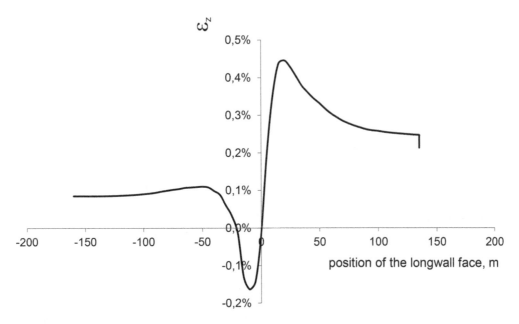

Figure 5. Vertical strain (ε_z) changes induced by the longwall face 411b passing beneath a D1 monitoring station recorded in a numerical model. D1 monitoring station position in the model, about 16 m above the extracted seam, corresponded to the D1 measurement site location in the Myslowice colliery rock mass.

269

not in every monitoring station. The reason for this is to be believed the difficulty to conduct measurements in a very challenging environment of the underground coal mine.

Horizontal axis gives the position of the longwall face relative to the position of the monitoring point. Negative values mean that the longwall face is located in front of the monitoring spot, positive values indicate that the longwall face passed beneath the recording point and moves away from it. Vertical axis displays vertical strain (ε_z), negative values mean shortening, positive – elongation.

Figure 5 display the strain values from the time when the longwall face was located 160 m ahead and till the time when it was 135 m behind the position of D1 monitoring station.

Initially, the strain increasers slightly, about 0.02% and when the longwall face is about 45 m away it suddenly decreases and becomes negative when the longwall face approaches D1 monitoring station at a horizontal distance of 20 m. Then the rapid increase of the shortening strains follows till the minimum value (of about -0.16%) is reached when the approaching longwall face is 10 m in front of the monitoring point. Strain during the movement of the longwall face from that point to the place directly beneath the monitoring station decreases and assumes 0.0% value when the longwall face passes under the monitoring point position. The subsequent movement of the longwall face, moving away from the monitoring station location induces positive strain which reaches its maximum (about 0.45%) when the horizontal distance from the longwall face to the monitoring point is 20 m.

After the maximum value is reached the strain decreases, at a declining rate to 0.24%. Vertical portion of the curve $\varepsilon_z = f(x)$ for x=135 m represents the reduction of strain during the halt of the longwall in a recovery room. The above mentioned pattern of strain changes obtained in the numerical simulation is very similar to the in-situ strain changes described in literature (e.g. see Knothe, 1984).

5 SUMMARY

Using the distinct element code *UDEC*, numerical model of a large part of a rock mass was built. This was two-dimensional model of the strata in the vicinity of coal seam 510 in longwall panel 411b, extracted with hydraulic backfilling, at the depth of app. 600 m in the Mysłowice Colliery in the Upper Silesian Coal Basin, Poland.

Before the detailed simulation of the extraction of longwall panel 411b in a very rock-burst prone conditions was started, all the mining of coal in the area, in 9 coal seems, which took place over the period of almost 45 years was simulated. Progressive extraction of two seams with roof caving and seven with backfilling was simulated in the same order as it was done in the mine.

When analyzing the computation results, discontinuous deformations of rock masses in the vicinity of the longwall faces were investigated with particular attention being paid to the formation of the caving zone above the seam being mined and loading of the backfill and longwall support, and to the fracturing, separation and heave of the floor strata. Normal and shear displacements along joints were monitored carefully, with the main focus on joint slip and separation. Mining-induced changes in the state of strain and stress in rock blocks and the location and size of the areas where rock material failed in shear or tension and underwent transition into the post-failure range were determined as well.

Displacements and strains were monitored at ten positions in the model which corresponded to the spots where extensometers were placed in drilled holes in the mine.

Strain changes obtained in the numerical simulation are very similar to the published results of the in-situ strain changes measurements in the roof layers located above longwall workings.

The results of the numerical simulation were confirmed by the conditions experienced in the 411b longwall face.

The rate of advance and productivity of the longwall turned out to be directly related to the changing geomechanical conditions just as they were determined by the numerical simulation.

Although the induced-seismicity level in the 411b longwall panel was high no strong (dangerous) tremor occurred and the extraction of the whole mining panel 411b was executed safely.

The successful extraction with backfilling of the last longwall panel in that section of the colliery in the near-floor slice of seam 510 did open a way for mining activities in the area in the overlaying layers of the seams 510 and 501 of the total thickness of 15 m or even more. *UDEC - DEM* numerical simulation proved to be a useful tool supporting the assessment of the rock mass behavior when extensive longwall mining with roof caving and hydraulic backfilling operations are undertaken.

ACKNOWLEDGMENT

The author also expresses gratitude to the National Agency for Academic Exchange of Poland (under the Academic International Partnerships program, grant agreement PPI/APM/2018/1/00004) for financial support of the internship at Montanuniversität in Leoben, Austria, which enabled the execution of a part of the work.

BIBLIOGRAPHY

Clark I.H., 1991, *The cap model for stress path analysis of mine backfill compaction processes*. Computer Methods and Advances in Geomechanics (Proceedings of the 7th International Conference, Cairns, Australia, May 1991), Rotterdam, A. A. Balkema, Vol. 2, pp. 1293–1298.

Itasca Consulting Group, Inc., 1996, UDEC 3.0 *User's Manual* (3 vols), Minneapolis.

Matwiejszyn, A. & M. Ptak: Measurements of borehole deformations for assessment of rockburst hazard. In *Seismogenic Process Monitoring* (H. Ogasawara, T. Yanagidani & M. Ando, eds), pp. 63–74. A. A. Balkema, Rotterdam 2002.

Lasek S., Matwiejszyn, A. & M. Ptak: Completing exploitation with hydraulic backfilling in conditions of coexisting natural hazards: A case study. *Proceedings of the 5th International Symposium on Rockburst and Seismicity in Mines (RaSiM5)- 2001, Johannesburg, South Africa, September 17-20, 2001* (Edited by G. Aswegen, A.J. Durrheim & W.D. Ortlepp), pp. 343–349. The South African Institute of Mining and Metallurgy, Johannesburg, 2001

Knothe S., 1984, *Prognozowanie wpływów eksploatacji górniczej*, Katowice, Wydawnictwo „Śląsk" (in Polish).

Kwaśniewski M.: Numerical analysis of strata behavior in the vicinity of a longwall panel in a coal seam mined with roof caving. In *Proceedings of the 1st International* FLAC/DEM *Symposium on Continuum and Distinct Element Numerical Modeling in Geo-Engineering - 2008, Minneapolis, Minnesota, August 25-27, 2008* (Edited by R. Hart, C. Detournay & P. Cundall), Paper No. 07-08. Itasca Consulting Group, Inc., Minneapolis 2008. (CD-ROM)

Salamon M. D. G., 1968, *Elastic moduli of a stratified rock mass*. Int. J. Rock Mech. Min. Sci., Vol. 5, pp. 519–527.

Wardle L. J. and Gerrard C. M., 1972, *The "equivalent" anisotropic properties of layered rock and soil masses*. Rock Mech., Vol. 4, pp.155–175.

Minefill 2020-2021 – Hassani et al (eds)
© *2021 Taylor & Francis Group, London, ISBN 978-1-032-07203-6*

Deformations of the mining area surface as a result of exploitation with sealing of caving gobs

Violetta Sokoła-Szewioła, Aleksandra Mierzejowska & Marian Poniewiera
Silesian University of Technology

SUMMARY: In Polish conditions of conducting underground hard coal mining in the past, in order to limit the negative impacts of conducted mining, systems with gobs filling with dry or hydraulic backfill were used. In recent years, this method of managing the floor has been abandoned, while as part of fire prevention and in order to enable waste placement in old gobs, caving mining with sealing gobs is used. It seems that this solution also results in a reduction in the amount of surface deformation. The authors presented in the article the results of own research on the subject matter. The research covered the area of one of the mines conducting exploitation in the area of the Upper Silesian Coal Basin. It was determined that the use of exploitation with sealing caving gobs reduces the subsidence coefficient, and the values of the subsidence observed in the analyzed area may be even 25% lower than when conducting exploitation without sealing and the strain extreme values over time can be up to 40% smaller.

Keywords: deformation, sealing, caving gobs, subsidence

1 INTRODUCTION

Underground mining of useful mineral deposits causes deformations of the rock mass and the surface area. The values of these quantities depend primarily on the geological structure of the rock mass, mechanical properties of the rocks forming the rock mass, as well as the exploitation systems used and methods of filling the post-mining void. The above can be described using geological, geomechanical and mining factors. Among geological factors, in the conditions of underground hard coal mining in Poland, the exploitation depth, the thickness of the overburden rocks above carbon, the dip of the seam and layers in the rock mass, the presence of tectonic dislocations. Geomechanical factors determine the rock mass's ability to transfer the influence of mining exploitation. The ability to measure this can be the rock mass parameter, depending on the type and mechanical properties of the rocks, such as compressive and tensile strength, deformation properties and the ratio of Carboniferous rock thickness to the overburden, and the extent of the rock mass violation by previous exploitation. The most important mining factors include the dimensions and shape of mining fields, mutual location of fields and mining fronts, thickness of seams extraction, mining systems and methods of gobs filling, velocity and conditions for moving fronts as well as the order of exploitation and cleanliness of mining of deposit and activation of old gobs (Szpetkowski, 1995). Appropriate mining prevention is used to limit the negative impacts of conducted mining. In this case, it is important, among others, to know the description of the effects of exploitation and to be able to determine optimal exploitation solutions on the basis of the predicted effects of designed exploitations. Limiting the value of rock mass and surface deformation can be obtained by

DOI: 10.1201/9781003205906-25

changing the thickness of exploitation and using an appropriate way of liquidation of gobs, which is characterized by the subsidence coefficient a, also called the exploitation coefficient. Its value determines the ratio of the largest subsidence observed to the thickness of the selected seam or layer. For the conditions of conducting mining in the region of the Upper Silesian Coal Basin (USCB), if the deposit is selected with a fall of roof, its value is from about 0.7 with a predominance of strong rocks in overburden to 0.9 with a weak overburden. An additional factor affecting the value of the coefficient a is the degree of rock mass violation through previous exploitations. Experiments have shown that in the case when the value of the maximum subsidence is less than the thickness of the mining, the parameter a takes the value <1, with the value of the subsidence equal to the thickness of exploitation a takes the value 1, in the case of the maximum subsidence value greater than the thickness of the exploitation a reaches the value> than 1. This variation depends on the type of rock in the rock mass, rock compactness and divisibility into blocks. In the case of exploitation of a deposit with backfill, the value of the coefficient is reduced. The process of deposit deformation is milder and the effects of exploitation on the surface are quantitatively smaller. Until now, in the conditions of exploitation of hard coal deposits in the Upper Silesian Coal Basin region, a hydraulic backfill was used, which was a mixture of sand and water, being a means of transport, dry backfill, which was rock material supplied from the surface or from mining excavations, a hardened backfill, which was a multi-component mixture, which hardens after some time, obtaining a certain compressive strength, and to the binding material was added e.g. gypsum or cement. The amount of surface subsidence depends in this case primarily on the compressibility of the backfill, determined by the change in layer height under load. It was found that regardless of the mining system, the observed surface deformations were even 4 times smaller than those caused by exploitation with fall of roof (Ostrowski, 2015). Table 1 presents the approximate value of the subsidence coefficient a for hard coal mining in the Upper Silesian Coal Basin region, depending on the method of gobs filling.

Currently, exploitation with backfill is not used in Polish hard coal mines, while as part of fire prevention and in order to allow waste to be deposited in old gobs, exploitation with sealing caving gobs is used. Studies indicating the impact of using this type of gobs filling on the value of surface deformation are few, so the authors in this paper presented the results of their own research on this subject. The research was conducted in the area of one of the hard coal mines conducting mining in the Upper Silesian Coal Basin area. To determine the value of parameter a and to develop computer simulations of the value of deformation indicators, the geometric-integral theory of forecasting the impacts of mining exploitation on the surface was used, data on mining exploitation performed in the analyzed region and the results of subsidence measurements made using geodetic methods. The research abandoned the method allowing the determination of the coefficient a based on the results of measurements of maximum subsidence, due to the fact that such observations were not available due to the location of the measuring points. The values of surface deformation indicators were determined for two computational areas. The research areas were selected so that in the analyzed period the impact of other exploitation on the values of subsidence found by geodetic measurements in these regions was not observed.

Table 1. Approximate value of the subsidence coefficient a in hard coal mining in the Upper Silesian Coal Basin area.

Methods of gobs filling	Subsidence coefficient a
Fall of roof	0,7-0,85
Dry full backfill	0,5-0,6
Dry pneumatic backfill	0,4-0,5
Hydraulic sand backfill	0,15-0,25
Hydraulic crushed stone backfill	0,3
Partial exploitation with belts (50% with hydraulic backfill)	0,02-0,03
Partial exploitation with belts (50% with fall of roof)	0,1

Source: Protection of building structures in mining areas. (Collective work edited by Kwiatek. 1998).

2 PREVIOUS STUDIES OF THE IMPACT OF SEALING GOBS ON VALUES OF DEFORMATION

The problem of sealing caving gobs as part of fire prevention in Poland was the subject of research conducted at the Silesian University of Technology. The degree of filling caving gobs and sealing caving gobs was determined using backfill mixture of dusts (Plewa et al, 2008). The degree of gobs filling was determined as the ratio of the volume of waste deposited in gobs to the volume of gobs defined as the volume of the exploited seam. The degree of sealing was determined as the ratio of the volume of deposited waste to the theoretical absorbency of caving gobs. The theoretical absorbency of gobs was determined by the product of the gobs absorbency coefficient and the volume of the exploited seam. In work (Piotrowski and Mazurkiewicz, 2006), the authors based on the analysis of 39 cases determined that the absorbency of gobs can range from 0.07 to 0.27. Research on the impact of sealing caving gobs on the value of the surface deformations presented in work (Zych et al, 1993) showed that the estimated value of the subsidence coefficient in the case of sealing gobs may be reduced by 0.1 compared to the exploitation without sealing gobs. In 2019, T. Rutkowski determined that in the area of exploitation carried out in the area of the Ruda-Ruch Pokój coal mine, the value of the coefficient a when exploiting with sealing caving gobs was less by 0.07 than the coefficient a determined for exploitation with fall of roof (Rutkowski, 2019). Among foreign research in the field of the above issue can be mentioned work (Yu Yang et al). In the field of sealing caving gobs, a number of studies were also carried out, the main purpose of which was to determine the impact of organic compounds on the parameters of mixtures injected into caving gobs. As a result of research carried out recently, relevant from the point of view of the subject discussed, among others, criteria were selected for the selection of waste materials in terms of their granulation, improvement of the ability to increase penetration in caving debris and the degree of filling and sealing caving gobs (Świnder, 2014).

3 THE METHOD OF DESCRIBING THE SURFACE DEFORMATION DUE TO MINING EXPLOITATION USED IN THE RESEARCH

Research on the effects of mining exploitation on the rock mass and the surface are conducted in several basic directions (Litwiniszyn, 1969). In the first, the results of the research are empirical formulas, on the basis of which the predicted values of selected deformation indicators are determined. When considering the impact of mining exploitation on the rock mass and the surface, it is also possible to use methods based on deductive schemes, which include methods assuming that the rock mass is a stochastic medium, methods based on the assumptions of the theory of elasticity and plasticity, or based on geometric assumptions regarding the distribution of impacts. In theories based on geometrical assumptions regarding the distribution of extraction impact, called geometric-integral theories, each point on the surface or point inside the rock mass has been assigned a function describing the impact of extracting deposit volume with a unit surface area on this point, depending on the location of the extracted deposit volume relative to the subject point. These theories differ in the form of functions. Examples of such theory include, among others, works (bals, 1931/1932, Beyer, 1945). The most important works of Polish researchers include the works of S. Knothe (Knothe, 1953), W Budryk (Budryk, 1953), and T. Kochmański (Kochmański, 1949), as well as further modifications and extensions of these theories developed by J. Białek (Białek 2003), B. Dżegniuk (Dżegniuk, 1979), K. Greń (Greń, 1981), J. Zych (Zych, 1987), A. Sroka (Sroka, 1999), R. Hejmanowski (Hejmanowski, 2004), A. Kowalski (Kowalski, 2007). Among the numerous works in this field, one should also mention those that contain the results of research that allow to improve the accuracy of the description of the mining area surface deformation in time contained in the works (Białek and Sokoła-Szewioła, 2012, Sokoła-Szewioła and Kowalska-Kwiatek, 2013).

In order to make calculations in the research, which is the subject of the article, the geometric-integral theory of influences of S. Knothe (Knothe, 1953) was used, which was extended by W. Budryk in 1953 (Budryk, 1953). This theory was based on the normal Gaussian distribution of the effect of mining an elementary volume. The choice of theory was conditioned by several reasons, primarily the commonness of its application for forecasting the impact of exploitation in Polish mining and the possibility of using extensive computer packages developed at the Silesian University of Technology by J. Białek. These programs allow to make prognostic calculations with the presentation of results for individual points in tabular form} and in the form of deformation maps, and also allow the analysis to be carried out consisting in determining the parameters of forecasting theory based on the results of measurements of subsidence of the mining area over time and the gobs geometry corresponding to these measurements. The most important feature of the programs is to take into account the development of exploitation over time. The calculations used extensions of S. Knothe's theory formulas proposed by J. Białek. In the classical form of the theory (Knothe, 1953), the subsidence of w_k at point P, i.e. the vertical component of the rock mass displacement vector caused by the exploitation of the surface of the S seam, is expressed by the formula:

$$w_k = \iint\limits_{S} \frac{ag}{r(z)^2} \exp \left(-\pi \frac{(\xi - x)^2 + (\eta - y)^2}{r(z)^2} \right) d\xi d\eta \tag{1}$$

where:
w_k – final subsidence of point P, calculated without taking into account the time delay,
x, y, z – coordinates of the calculation point,
a - subsidence factor, depended on the method of the roof management,
g - exploitation thickness of the seam or layer,
S – surface of the exploited seam, assuming that it is a function of time
$S=S(t)$,: then the subsidence is a function of time $w_k=w_k$ $(x,y,z,$ $S(t))$,
$\xi,,\eta$ - coordinates of the element of dS surface.
Impact dispersion radius:

$$r(z) = \frac{h}{tg\beta} \left(\frac{z - z_0}{h - z_o} \right)^n \tag{2}$$

$tg\beta$- parameter of Knothe's theory,
h – depth of the exploited element of the surface of dS seam
z – height of the calculation point above the element of dS surface,
z_o – parameter proposed by B. Drzęźla,
n – exponent in the formula (assumed for these calculations $z_o=5ag$, $n=0,665$).
The horizontal components of the displacement vector are calculated based on the Awierszyn hypothesis according to which the horizontal displacement at P is proportional to a certain constant B and the dip T:
- horizontal displacements in the direction of the x axis:

$$u_{xk} = -BT_x = -B\frac{\partial w_k}{\partial x} \tag{3}$$

- horizontal displacements in the direction of the y axis:

$$u_{yk} = -BT_y = -B\frac{\partial w_k}{\partial y} \tag{4}$$

w_k- subsidence calculated from formula (1),
B - horizontal strains factor.
The B value was calculated by the formula developed by Popiołek and Ostrowski (Popiołek and Ostrowski, 1981) in the form:

$$B = 0,32r \tag{5}$$

J. Białek's (Białek, 2003) formula adopted in the calculations takes the form:

$$W_k = (1 - a_w)w(r_1) + a_w w(r_2) - A_{obr} \frac{5.3333 \cdot w(r_1) \cdot [r_1 \gamma(r_1)]^2}{6.666 \cdot [0.5 \cdot w(r_1) + 0.5 \cdot w(r_2)]^2 + [r_1 \gamma(r_1)]^2} \tag{6}$$

where: A_{obr},- parameter including the asymmetry of the subsidence trough profile,
$w(r_1)$, $w(r_2)$ – subsidence calculated according to S. Knothe's formula (1) for the radii of impacts scattering $r = r_1$ and $r = r_2$;
$\gamma(r_1)$ deformation calculated by the formula:

$$\gamma(r_1)^2 = \left[0.25 \cdot r_1 \cdot \left(\frac{\partial^2 w(r_1)}{\partial x^2} + \frac{\partial^2 w(r_1)}{\partial y^2}\right)\right]^2 + \left(\frac{\partial w(r_1)}{\partial x}\right)^2 + \left(\frac{\partial w(r_1)}{\partial y}\right)^2 \tag{7}$$

$$a_w = 0,4 - 1,25 A_{obr}; \tag{8}$$

$$r_1 = \frac{h}{\mathrm{tg}\,\beta} \, F(A_{obr}); \quad r_2 = 2r_1 \tag{9}$$

Table 2 shows the value of the F function.

Table 2. Values of the F function.

A_{obr}	0	0,050	0,100	0,150	0,200	0,250	0,300
$F(A_{obr})$	0,800	0,844	0,916	1,003	1,099	1,200	1,303

Source: 4.Białek J. Algorithms and computer programs for the prediction of mining ground deformation Silesian University of Technology, Gliwice 2003 (Białek, 2003)

In order to assess the impact of exploitation, the basic and most accurately described deformation ratio are subsidence. In order to describe the deformation process more accurately, derivative values of displacements, i.e. deformation, dip and curvature, are determined. Research focused on the analyzes of the mining area subsidence. In addition, the analyzes were performed for horizontal deformations due to the fact that they are the basic indicator significant for the impact of mining exploitation effects on objects located on the surface of the mining area. Horizontal strains were calculated by the formula in the general form:

$$\varepsilon_{xk} = \frac{\partial}{x}\left(-B\frac{\partial wk}{\partial x}\right) \tag{10}$$

4 MATERIALS AND METHODS

The research was carried out in the area of hard coal mine conducting exploitation in the area of the Upper Silesian Coal Basin. As a result of the analysis of exploitation data, two regions and time periods were identified for which detailed analyzes were carried out. In area I (exploitation of seam C), in the selected period, exploitation was carried out successively with three longwalls with sealing of caving gobs, using a mixture of flotation waste, flying ashes and water. Data on the sealing process is given in Table 3. The height of the mixture in

Table 3. Data characterizing the sealing process.

Seam	Longwall	The volume of caving gobs V [m³]	The volume of the sealing mixture [m³]	The height of the mixture in the excavation M [m]
C- region I	C1	581824	24223	0,120735
	C2	813622	31594	0,112611
	C3	1083136	28620	0,076627

Source: own study

the excavation was determined assuming its even distribution. The volume of caving gobs was calculated based on mining data from the region.

In region II (exploitation of seam B) in the assumed period, two longwalls with fall of roof were exploited in succession. The research areas were selected so that the impact of exploitation carried out in other regions did not appear in the analyzed period. The availability of measurement results of observation p

Points subsidence was also taken into account (in the analyzed area, above seam C - observation lines a, b, c, in the analyzed area, above seam B - lines *d, e*). Mining and geological data of exploitation fields in the analyzed regions are presented in Table 4. The contours of the exploitation plots in the areas in question and the location of points of the observation lines are shown in Figures 1 and 2.

Further considerations were made using the geometric-integral theory of forecasting the impacts of mining exploitation on the surface, discussed in detail in Chapter 3. The calculations were carried out in the program of J. Białek in the following ranges:

– values of parameters of the theory *a, tgβ* and A_{obr} were determined separately for research regions I and II,
– computer simulation of subsidence and horizontal strains for regions I and II using the determined parameters was developed,
– a computer simulation of subsidence and horizontal strains for region I was developed assuming the values of parameters determined for exploitation with fall of roof (region II).

The calculation results were presented in the form of subsidence increments maps and strain extreme values maps over time. The analyzes that are the subject of the research are described in chapter 6.

Table 4. Mining and geological data of exploitation fields in the analyzed regions.

Seam	Longwall	Startup date	End date	Average depth [m]	Length of long-wall [m]	Excavation advance [m]	Exploitation thickness g[m]	Dip of the seam [g]
Exploitation with sealing of caving gobs								
C- region I	C1	1.09.2011	31.03.2012	825	245	799	2,90	2-7
	C2	20.03.2012	30.10.2012	840	240	1168	2,90	
	C3	1.10.2012	31.08.2013	862	250	1470	2,90	
Exploitation with fall of roof								
B- region II	B1	15.07.2005	28.02.2006	581	250	900	2,70	1-8
	B2	15.10.2006	31.07.2007	585	250	1082	2,80	

Source: own study

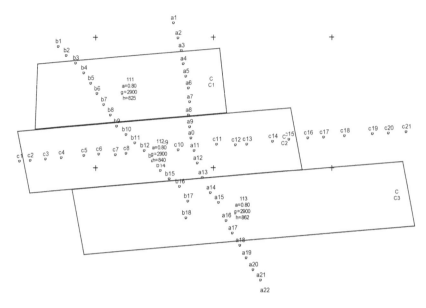

Figure 1.   Contours of exploitation carried out in the region I (seam C) in the period under research and location of points of the observation lines - lines *a, b, c*.

Source: own study

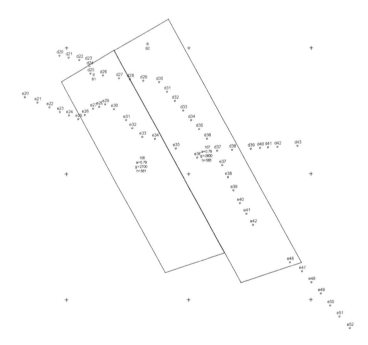

Figure 2.   Contours of exploitation carried out in region II (seam B) in the period under research and location of points of the observation lines *d* and *e*.

Source: own study

5   RESULTS

The parameters of the established calculation model: a, $tg\beta$, A_{obr} determined on the basis of geodetic measurements, geometry of exploitation fields, thickness and depth of exploitation are presented in Table 5. These parameters were determined based on the adjustment of the theoretical profile of the subsidence trough to the real trough using the least squares method. For the first region (exploitation of seam C), the parameters were determined on the basis of the measured subsidence for three calculation variants. In the first variant, the parameters were determined after exploitation of the parcel C1 in seam C. In the next variants, the parameters were determined respectively, after exploitation of the parcels C1 and C2, and after exploitation of all the parcels, i.e. C1, C2 and C3. In the region II (seam B), the parameters were determined for the situation after exploitation of parcels B1 and B2.

The distribution of real subsidence and subsidence calculated by the formula (6) on the points of the measuring line b for the state after completing the selection of parcels C1, C2, C3 is shown in Figure 3.

The surface distribution of increments of the area subsidence and maximum strains extreme over the period covered by the research in regions I and II are shown in Figures 4, 5, 7 and 8. The spatial distribution of subsidence increments in the period mentioned above is presented in Figure 6.

Table 5.   The results of matching theoretical subsidence to the subsidence determined by geodetic measurements on observation lines - determined parameter values.

Exploitation stage	$tg\beta$	a	A_{obr}	Standard deviation [mm]	Correlation coefficient	W_{max} -measured [m]
Fall of roof with sealing						
C1	1,48	0,473	0,131	13,6	0,9915	-0,338
C1 and C2	2,20	0,566	0,120	95,3	0,9591	-1,101
C-1, C-2 and C-3	2,3	0,691	0,177	114,2	0,9695	-1,534
Fall of roof						
B-1 and B-2	3,00	0,793	0,131	194,9	0,9540	-2,083

Source: own study

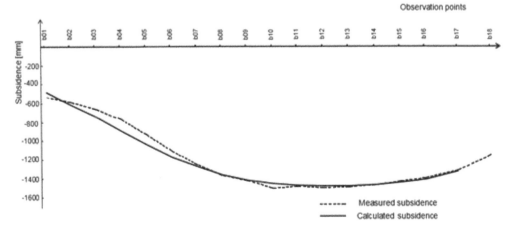

Figure 3.   The distribution of measured subsidence and subsidence calculated on the points of the measuring line b in the region I for the state after exploitations of parcels C1, C2, C3.

Source: own study

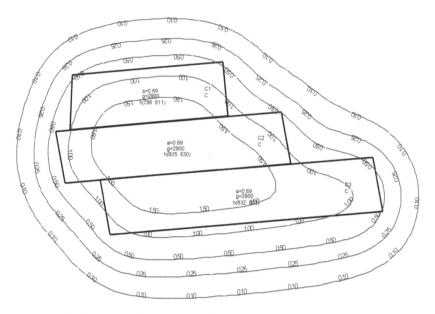

Figure 4.   Surface distribution of increments of the subsidence [m] in region I after exploitation of the parcels C1, C2, C3.

Source: own study

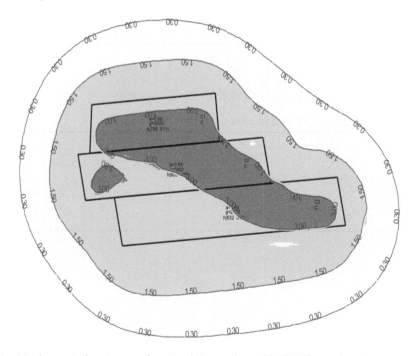

Figure 5.   Maximum strain extreme values [mm/m] over time - C1, C2, C3 parcels region.

Source: own study

The results of computer simulation of subsidence and strains extreme over time in the region I, assuming the value of forecasting parameters as for fall of roof, are shown in Figures 9, 10.

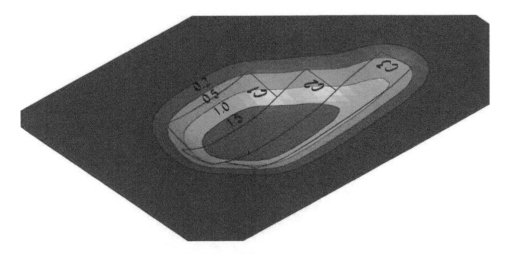

Figure 6.   Spatial distribution of subsidence increments - region I, parcels C1, C2, C3.
Source: own study

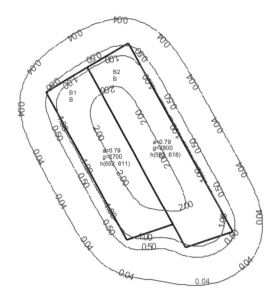

Figure 7.   Surface distribution of subsidence increments [m] in region II after exploitation of the parcels B1, B2.
Source: own study

6   ANALYZES

A comparative analysis of values of the subsidence coefficient *a* determined in regions I and II showed that in case of exploiting with sealing of caving gobs, this coefficient ranges from 0.473 to 0.691 and is not less lower than 0.1 than the subsidence coefficient for exploitation with fall of roof.

The value of the coefficient *a* after choosing the first longwall with sealing of caving gobs was 0.473, after selecting the first and second longwall with sealing of gobs 0.566, and after selecting three longwalls it was 0.691, which confirms the significance of the impact of the

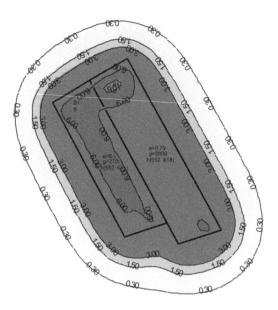

Figure 8. Maximum strain extreme values [mm/m] over time in the region II - parcels B1, B2.
Source: own study

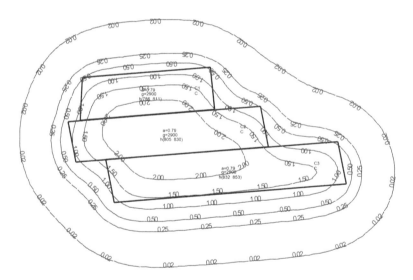

Figure 9. The surface distribution of subsidence increments [m] in the region I assuming that C1, C2, C3 parcels are exploited with a longwall system with fall of roof.
Source: own study

volume of extraction on its value. The dependence of the value of the coefficient *a* on the volume of the extraction can be described by a linear functional dependence in the form presented in Figure 11.

Computer simulations of subsidence for longwalls exploitation with sealing of caving gobs, assuming the values of the parameters of forecasting theory determined for exploitation with fall of roof, showed that using a solution with sealing of caving gobs, the values even up to 20% lower can be obtained when exploiting without sealing. The maximum value of the subsidence calculated for the exploitation with the fall of roof was 2.00 m. The surface

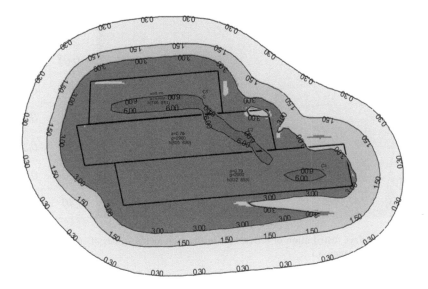

Figure 10. Maximum strain extreme values [mm/m] over time in the region I assuming the exploitation of parcels C1, C2, C3 with a longwall system with fall of roof.

Source: own study

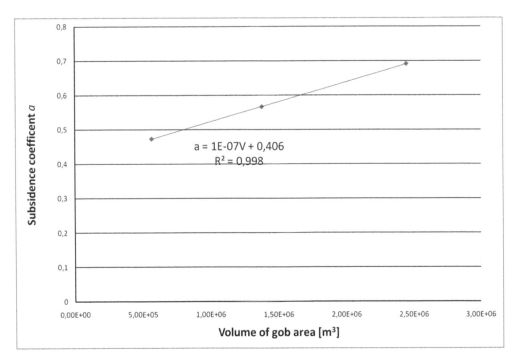

Figure 11. Dependence of the settlement coefficient value *a* on the selected volume V of gob area.

Source: own study

distribution of differences in the values of subsidence during the analysis period is shown in Figure 12. In the case of extreme strains with the use of sealing of caving gobs solution, the maximum strain extreme values were found to be lower of up to 40% than the values found

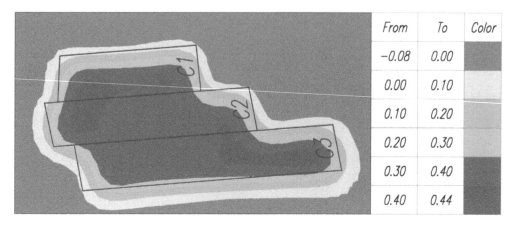

From	To	Color
−0.08	0.00	
0.00	0.10	
0.10	0.20	
0.20	0.30	
0.30	0.40	
0.40	0.44	

Figure 12.   The surface distribution of differences in the increments values of subsidence [m] (gobs sealing - fall of roof) in the region I.
Source: own study

assuming the use of non-sealing exploitation. The maximum value of strains extreme over time for fall of roof exploitation was 5 mm/m. The surface distribution of extreme strains differences over time is presented in Figure 13.

Further considerations were made in determining the dependence of increments changes of the determined subsidence coefficient a on the degree of sealing gobs defined as the ratio of the height of gobs filling M to the height of exploitation g. The relationship between these quantities is shown in Figure 14. The obtained results allow us to assume that increasing the degree of sealing results in a decrease in the subsidence coefficient. In the examined region, when filling about 12% of the selected space, the value of the coefficient was about 0.47, which is about 40% of the value of the coefficient of subsidence with fall of roof that is assumed in the region. However, when 10% of the selected space is filled, the subsidence coefficient value is about 12% lower than for exploitation with fall of roof.

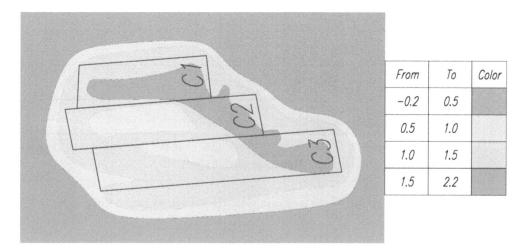

From	To	Color
−0.2	0.5	
0.5	1.0	
1.0	1.5	
1.5	2.2	

Figure 13.   The surface distribution of differences of the values of maximum strains extreme [mm/m] over time (gobs sealing - fall of roof) in region I.
Source: own study

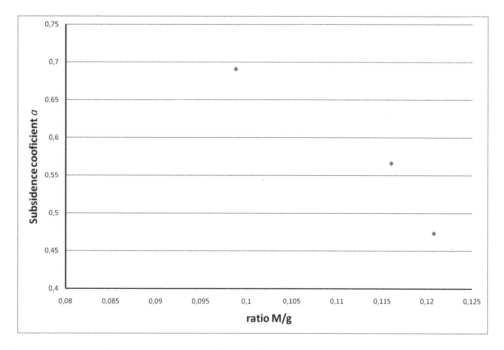

Figure 14.   The relationship between the coefficient of subsidence a and the degree of sealing caving gobs.
Source: own study

7   CONCLUSIONS

Based on the conducted research it can be stated:

− the use of exploitation with sealing of caving gobs significantly reduces the values of mining area surface deformation coefficients. In the example under consideration, the subsidence recorded in the conditions of exploitation conducted with sealing of caving gobs reached values up to a maximum of 20% lower than with exploitation without sealing, in the case of deformations, the values of maximum strains extreme over time have reached a value up to 40% lower than without sealing,
− the use of a solution with caving gobs sealing causes the reduce of the value of the subsidence coefficient by at least about 0.1. Considering the relationships of the calculated degree of sealing with changes in the coefficient a showed that when filled with sealing mixture of about 10% of the selected space, its value decreases by about 0.12. At 12% filling, the coefficient value was as much as 40% lower than in the case of fall of roof exploitation,
− the dependence of the coefficient a on the extraction volume V can be described by a linear functional dependence.

Therefore, it follows from the above that in the case of using extraction with gobs sealing, a reduction in the values of deformation rates can be obtained, which should be taken into account at the stage of forecasts of surface area deformations developed for the planned mining exploitation, which is currently not practiced.

The issue that is the subject of the article requires further research, in particular taking into account the parameters of the sealing mixture.

BIBLIOGRAPHY

Bals, R. 1931/32. *Beitrag zur Frage der Vorausberechnung bergbaulicher Senkungen.* Mitteilungen aus dem Markscheidewesen 42–43. *(in German)*.

Beyer, F. 1945. *Über die Vorausbestimmung der beim Abbau flachgelagerter Flöze autretenden Bodenverformungen.* Postdoctoral dissertation. Berlin. *(in German)*.

Białek, J. 2003. Algorithms and computer programs for the prediction of mining Ground deformation. (*Algorytmy i programy komputerowe do prognozowania deformacji terenu górniczego*). Gliwice: Silesian University of Technology, 199 pp. *(in Polish)*.

Białek, J. and Sokoła-Szewioła, V. 2012. Short-term approximation of observed subsidence of a mining ground point. (Krótkookresowa aproksymacja zaobserwowanych obniżeń punktu terenu górniczego) *Przegląd Górniczy, 68 (8)*,pp. 148–153. *(in Polish)*.

Budryk, W. 1953. Determining the size of horizontal land strains (Wyznaczanie wielkości poziomych odkształceń terenu). *Archiwum Górnictwa i Hutnictwa 1(1)*. *(in Polish)*.

Dżegniuk. B. 1975. Some non-linear effects in the process of subsidence over mining exploitation. (Niektóre efekty nieliniowe w procesie osiadania nad eksploatacja górniczą). Kraków: *The Bulletin of AGH University of Science and Technology, s. Geodezja* 34. *(in Polish)*.

Greń, K. 1981. An attempt to include the asymmetry of mining exploitation impacts at horizontal lying of seams (Próba ujęcia asymetrii wpływów eksploatacji górniczej przy poziomym zaleganiu pokładów). *Prace Komisji Górniczo-Geodezyjnej. Geodezja 29.* Kraków: PAN. *(in Polish)*.

Hejmanowski, R. 2004. *Dynamics of mining exploitation from the point of view of mining damage.* (*Czasoprzestrzenny opis deformacji górotworu wywołanych filarowo-komorową eksploatacją złoża pokładowego*). Kraków: AGH University of Science and Technology Publishing House, issue 131. *(in Polish)*.

Knothe, S. 1953. Equation of the profile of the finally formed subsidence trough. (Równanie profilu ostatecznie wykształconej niecki osiadania). *Archiwum Górnictwa i Hutnictwa, 1(1)*. *(in Polish)*.

Kochmański, T. 1949. Horizontal and vertical movement of a terrain due to mining. (Przesunięcia terenu w pionie i w poziomie pod wpływem odbudowy górniczej). *Hutnik*, 7–8. *(in Polish)*.

Kowalski, A. 2007. *Unspecified mining surface deformations in the aspect of forecast accuracy. (Nieustalone górnicze deformacje powierzchni w aspekcie dokładności prognoz*). Katowice: GIG, Prace GIG 871. *(in Polish)*.

Litwiniszyn, J. 1969. On directions of theoretical research on the impact of underground mining exploitation on rock mass movements (O kierunkach badań teoretycznych wpływu podziemnej eksploatacji górniczej na ruchy mas skalnych). *Proceedings of the Ist National Symposium on Ochrona powierzchni przed szkodami górniczymi*. SITG, WUG. *(in Polish)*.

Ostrowski, J. 2015. *Deformations of the mining area* (*Deformacje powierzchni terenu górniczego*). Kraków: AGH University of Science and Technology. *(in Polish)*.

Piotrowski, Z. and Mazurkiewicz, M. 2006. Absorption of caulking of caving gobs (Chłonność doszczelnianych zrobów zawałowych). *Górnictwo i Geoinżynieria (3*, .pp, 37–45. *(in Polish)*.

Plewa et al. – Plewa, F .Mysłek, Z. and Strozik, G. 2008. The use of energy waste to solidify rock debris (Zastosowanie odpadów energetycznych do zestalania rumowiska skalnego). *Polityka energetyczna 11(1)*, pp. 351–360. *(in Polish)*.

Popiołek, E. and Ostrowski, J. 1981. An attempt to determine the main reasons for the discrepancy of forecasted and observed post- mining deformation rates. (Próba ustalenia głównych przyczyn rozbieżności prognozowanych i obserwowanych poeksploatacyjnych wskaźników deformacji). *Ochrona Terenów Górniczych (58)*. *(in Polish)*.

Praca zbiorowa pod redakcją J. Kwiatka (Collective work edited by J. Kwiatek). 1998. *Protection of building structures in mining areas* (*Ochrona obiektów budowlanych na terenach górniczych*). Katowice. *(in Polish)*.

Rutkowski, T. 2019. Impact of caulking of caving gob on deformation of the mining area of the Ruda-Ruch - Pokój coal mine (Wpływ doszczelniania zrobów zawałowych na deformacji powierzchni terenu górniczego KWK Ruda – ruch Pokój). *Przegląd Górniczy 2*, pp. 13–20. *(in Polish)*.

Sokoła-Szewioła, V. and Kowalska-Kwiatek, J. 2013. A complex method of indication of parameters of the model describing the mining area depression in time (Kompleksowa metoda wyznaczania parametrów modelu opisującego obniżenia terenu górniczego w czasie). *Przegląd Górniczy* 69(3), pp. 142–148

Sroka, A. 1999. *Dynamics of mining exploitation from the point of view of mining damage (Dynamika eksploatacji górniczej z punktu widzenia szkód górniczych)*. The Mineral and Energy Economy Research Institute of the Polish Academy of Sciences, Studia, Rozprawy, Monografie 58. *(in Polish)*.

Świnder, H. 2014. Influence of selected organic compounds on the parameters of mixtures injected into goaf caving equipment (Wpływ wybranych związków organicznych na parametry mieszanin zatłaczanych do zrobów zawałowych). *Przegląd Górniczy 12*, pp.67-73. *(in Polish)*.

Szpetkowski, S.1995. *Forecasting impacts of seam deposits exploitation on the rock mass and surface (Prognozowanie wpływów eksploatacji złóż pokładowych na górotwór i powierzchnię terenu)*.Katowice: Śląskie Wydawnictwo Techniczne. *(in Polish)*.

Yu Yang, Xie-xing Miao, Wen-sheng Liu and Xing-hua Li. 2008. Deformation of mining caving zone grouting compound rock under overlying strata pressure. *Journal of Coal Science and Engineering (*China*)* 14, pp. 594–596.

Zych et al. 1993- Zych, J. Żyliński, R. and Strzałkowski, P. Impact of caving gobs caulking on the amount of surface deformation (Wpływ doszczelniania zrobów zawałowych na wielkość deformacji powierzchni). *Proceedings of the Conference on II Dni Miernictwa Górniczego i Ochrony Terenów Górniczych.* GIG, pp.307–311. *(in Polish).*

Zych, J. 1987. A method of forecasting the impacts of mining exploitation on the surface area, taking into account the asymmetrical course of the deformation process. (Metoda prognozowania wpływów eksploatacji górniczej na powierzchnię terenu uwzględniająca asymetryczny przebieg procesu deformacji). *The Bulletin of Silesian University of Technology, s. Górnictwo,164. (in Polish).*

Minefill 2020-2021 – Hassani et al (eds)
© 2021 Taylor & Francis Group, London, ISBN 978-1-032-07203-6

Geomechanical safety aspects in hard rocks mining based on room-and-pillar and longwall mining systems

Witold Pytel
KGHM CUPRUM R&D, Wrocław, Poland

Bogumila Palac-Walko
Faculty of Geoengineering, Mining and Geology, Wrocław University of Science and Technology, Poland

Piotr Mertuszka
KGHM CUPRUM R&D, Wrocław, Poland

SUMMARY: The article discusses geomechanical aspects and numerical simulations associated with the mining methods dedicated for flat or flatly dipping orebody, i.e. room-and-pillar and longwall mining systems with regard to development of hypothetical mining panel. It was assumed that the overburden strata consists of several homogeneous rock plates reflecting the typical lithology in the Lower Silesia area. The results of the computer 3D simulations permitted elastic-plastic rock mass stability analysing, identifying the areas being more susceptible to damage. Each model has been analysed from the point of view of different methods of roof control. The geomechanical risk assessment procedure utilized so called safety margins which were defined as a distance between the point characterized by the actual local strain/stress conditions and the instability (limit) surface(s) which location in the 3D stress/strain space could be determined using different strength theories. The obtained results have proved that in the considered geological conditions, room-and-pillar approach has a significant advantage from point of view geomechanical safety, over the mechanized longwall excavation system.

Keywords: numerical modelling, geomechanical hazard, room-and-pillar and longwall mining systems

1 INTRODUCTION

Currently, exploitation of hard rock deposits in deep underground mines is performed primarily by the use of blasting technology. However presently, new technologies based on a mechanized type of rock mass excavation technologies are also under development (Spisak and Zelko, 2015). An important factor affecting the mining method selection is the hardness of the rock and the adapted mining system. Apart from the mechanical excavating system, different varieties of room-and-pillar, sublevel, shrinkage and other mining methods are used. The implementation of mechanical excavation systems in world's hard rock mines haven't reached a level of success so far. Several attempts have indicated that due to the presence of hard rocks in the deposit, proposed road headers and continuous miners haven't proved themselves to be so effective as they were expected.

DOI: 10.1201/9781003205906-26

Therefore the main goal of the presented analyses is to evaluate geomechanical aspects associated with two mining methods, i.e. longwall mining and room-and-pillar mining system with regard to excavating of hypothetical mining panel. Both methods are well suited to extracting the relatively flat deposits. Room-and-pillar mining system is quite effective, relatively safe, and fits well under tight geomechanical conditions, which are typical for flat deposits. However, this is a longwall mining which is still more efficient from point of view of productivity, compared with room-and-pillar mining (Spisak and Zelko, 2010). This is because the longwall mining is in principle a continuous operation which require smaller crew and allow for a high rate of production.

Today, 3D computer simulations of extensive mining operations performed underground permit elastic-plastic rock mass stability analysing aimed for identifying the areas being more susceptible to damage, presently and in a far future. The geomechanical risk assessment procedure utilized so called safety margins which were defined as a distance between the point characterized by the actual local strain/stress conditions and the instability (limit) surface(s) which location in the 3D stress/strain space could be determined using the well-known strength theories (Palac-Walko and Pytel, 2015).

2 GENERAL DESCRIPTION OF CONSIDERED MINING SYSTEMS

Room-and-pillar mining system (Figure 1) is a dominant technology utilized in underground metal and chemical raw material mines (Hustrulid and Bullock, 2001). Mine workings' rational design requires the ability for pillar bearing capacity prediction as well as its load-deformation characteristics, since these factors permit control of pillar loads and surface and roof deflection. On the other hand, the load acting onto a given pillar depends on some effects created by surrounding geological and mining environment. Yield pillars offer a number of advantages that may result in improved stability in gate and tail entries, reduced seismic risk, and reduced subsidence curvature and slope values. This was proven in the deep coal mines in USA (e.g. Mark, 1990) as well as in Polish mines where so called 'roof deflection' technology, based on yield pillar mechanics, has been successfully implemented (Butra et al., 2001). However, since the empirical type of the technique has been used for yield pillar size selection (KGHM Guidelines, 1994), the parameters and design criteria suitable for those regions have found a limited application under different conditions of other fields.

Room-and-pillar mining system (Figure 1b)is based on one- or two-phase almost complete extraction approach with the first phase creating elastic pillars after driving rooms, while during the second phase these pillars are mined out on the retreat. The room-and-pillar mining method is a type of open stoping used in near horizontal deposits in reasonably competent rock, where the roof is supported primarily by pillars and, depending on the roof conditions, by rock bolts or other types of support. The ore is excavated as rectangular shaped rooms or entries in the ore body seam, leaving parts of the ore between the entries as remnant pillars to support the hanging wall or immediate roof strata.

Longwall mining (Figure 1a) is the most efficient mining system especially suitable for mining thin seams, usually for soft rock (coal, potash) but also for hard rock mining (metals) – Hustrulid, 1982. It enables mining of nearly the complete resource. In this method, the width of the panel varies from 150 to 300 m and length from 1,000 up to 3,500 metres. The average thickness of the deposit varies from 2 to 5 metres. In hard rock deposits, longwall mining aims to maintain near-continuous behaviour of the near-field rock mass, which requires a strong and competent hanging wall and footwall rock mass. The area located in front of the face needs to be supported by a series of hydraulic roof supports.

They temporarily hold up the roof strata and protect the working space for the continuous miner and chain conveyor. When each slice of rock is removed, the set of longwall, i.e. hydraulic roof supports, the conveyor and the longwall shearer machinery are moved forward. As a longwall equipment advances along a panel, the roof behind is allowed to fall down.

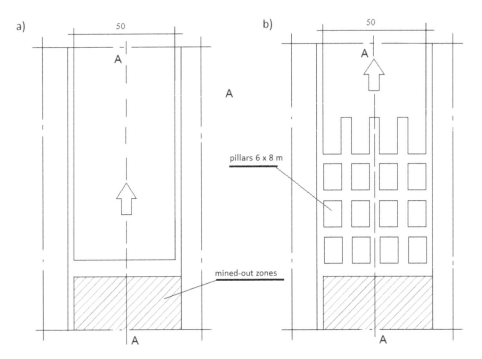

Figure 1.   General schemes of (a) longwall mining system and (b) one-phase room-and-pillar mining system – both geometries of mine workings used in numerical analysis.

3   INTRODUCTION INTO THE GEOMECHANICAL MODELING

As a basic physical model for the problem, the multi-plate overburden model has been accepted with the following simplifying assumptions (Figure 2):

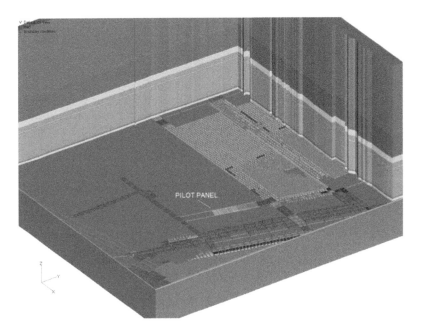

Figure 2.   General view of the numerical model.

- overburden strata consists of several homogeneous rock plates reflecting the real lithology in the area,
- technological and remnant pillars work effectively within post-critical phase (elastic-plastic with strain softening behaviour),
- the value of carried loads depends on pillar size and actual extraction ratio.

Geomechanical problem solution and results visualization were based on the NEi/NAS-TRAN computer program code utilizing FEM in three dimensions. It was understood that all but the pillars' materials reveal a linear-elastic characteristics, whereas copper ore rock mass behaviour is represented by elastic-plastic with strain softening kind of mechanical model. The entire numerical model general boundary conditions were described by displacement based relationships. More information on the applied solution method may be find elsewhere (Pytel, 2003).

Numerical experiments modeling the three-dimensional mine layout were performed, using two boundary condition sets defined by presence or absence of horizontal additional stress. The determined stress/deformation states were used afterwards for quantitative characterization of the effect of horizontal tectonic stress on system behavior and safety (Orlecka-Sikora and Pytel, 2013) using the indicators called safety margins related to well-known shear type failure criterion based on the Coulomb's theory (Jaeger et al., 2007):

$$\sigma_1 = \sigma_{cm} + A \cdot \sigma_3 \tag{1}$$

where: σ_{cm} – unconfined compression strength in rock mass scaled down acc. to the Hoek's (2007) approach, in [MPa], σ_1, σ_2 - major and minor principal stress respectively (compression – positive), in [MPa], A– positive material constants dependent on angle of internal friction $\emptyset$ in rock mass:

$$A = \frac{1 + \sin \emptyset}{1 - \sin \emptyset} \tag{2}$$

Thus, the respective safety margin has been formulated as follows:

$$M_c = \sigma_{cm} + A \cdot \sigma_3 - \sigma_1 \tag{3}$$

Rock mass instability potential is indicated by safety margin value smaller than zero. Instead of safety margins one may use so called safety factors based on the same failure criterion and formulated as:

$$F_c = \frac{\sigma_{cm} + A \cdot \sigma_3}{\sigma_1} \tag{4}$$

One may use another type of safety factor, based on Hoek-Brown theory (Pariseau, 2009).

The technological pillars, presented in this paper, located within this development areas are characterized by $\sigma - \varepsilon$ relationship shown in Figure 3, with the rectangular pillar bearing capacity value given by Pytel and Chugh (1992) in the following form:

$$\sigma_p = 0.294 \cdot \sigma_c \left(1 - \frac{B}{5L}\right) \cdot \sqrt{\frac{B}{H_p}} \tag{5}$$

where: σ_c – uniaxial compressive strength of rock specimen tested in the laboratory, in [MPa], B and L – pillar dimensions, in [m], H_p – pillar height, in [m].

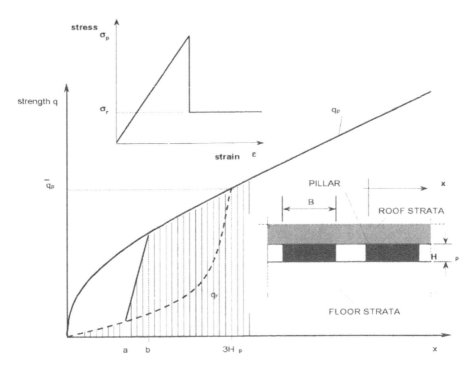

Figure 3.   Stress-strain relationship for yield pillars.

The uniaxial compressive strength should be calculated for rock sample having cube size or diameter $d = 5.08cm$. It is assumed that Eq. 5 deals with all kinds of rocks, therefore it may be utilized not only in coal but also in metal mines. Furthermore it is assumed that with load increasing, pillar behaviour characteristics transform from critical into a post-critical type with bearing capacity reduction due to pillar yielding (elastic-plastic model with softening). It is also assumed that residual pillar strength σ_r, being a portion of the pillar bearing capacity (pillar critical strength) σ_r, may be determined by integrating rock strength (peak or residual) over the horizontal cross-section of the pillar (Pytel, 2003). Geomechanical problem solutions were based on the planned geometry of the hypothetical mining parcel. The considered panel mine workings' geometries with its detailed spatial finite element approximation is presented in Figure 4. Two different mining systems were analyzed. The first one (variant A) is based on longwall system, while the second one (variant B) based on room-and-pillar mining system. Height of the assumed mining drifts, located at the depth of 751 m, in both cases was accepted to be 2 m.

Each model has been analysed from the point of view of different methods of roof control, i.e. roof deflection, wooden chock support, Tekpak support (Amick et al., 1993; Barczak et al., 2003; Minova, 2017)) and cemented backfilling with 10% of effectiveness (deflection factor = 0.1). To analyze geomechanical conditions within the hypothetical mining panel, typical lithology in the Lower Silesia area (Kijewski and Lis, 1994) was assumed (Table 1).

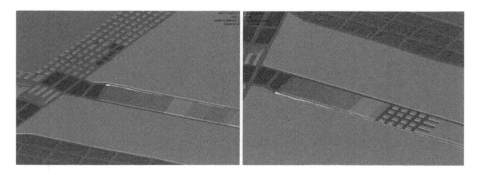

Figure 4. View of the FEM model for longwall mining system (left) and the FEM model for room-and-pillar mining system (right).

4 ANALYSIS OF SAFETY MARGIN DISTRIBUTION

It was assumed that the so-called safety margin, representing relation between load intensity and material strength, may serve as the basic indicator of possible roof strata instability.

The safety margin have been estimated by comparing the actually computed values of components of stress-strain tensors with ultimate values characteristic for a given material determined in laboratory or in the field. This may be done using some functions of load intensity formulated in the form of various strength theories. A measure of the goodness of fit of a given strength theory to the actual conditions is simply the percentage of the instabilities which occurred in the area of interest that are explained by the theory. The Coulomb strength theory has been chosen (see above) as the representative for the geology encountered in the pilot panel's site. The mine workings basic geometry was evaluated from safety point of view with no tectonic stress. Calculated safety margin M_c contours above the immediate roof strata are shown in Figure 5-8 (longwall mining system) and Figure 9-12 (room-and-pillar mining system).

5 DISCUSSION OF RESULTS

Results of numerical analyses presented above indicate that when using longwall mining systems with roof deflection (Figures 5 - 8), the immediate roof instability due to excessive shearing (Figure 13) may occur up to 2-3 m within roof strata unsupported or supported by wooden cribs or artificial TekPaks. However, when cemented backfilling is located within the goaf area these instabilities will cease completely.

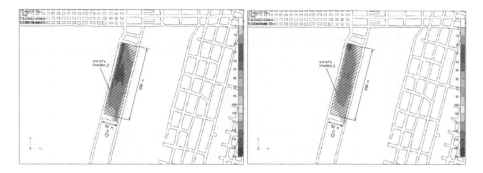

Figure 5. Safety margin distribution 0.55 m (left) and 1.7 m (right) above the immediate roof strata (longwall with roof deflection, excavation height H_p = 2.0 m, wooden cribs support or TekPak).

Table 1. Assumed geological data in the analyzed area (excavation height H_p = 2.0) – based on boreholes Gs-69 and Sr-19 H-18.

	Thickness (m)	C_o (MPa)	T_o (MPa)	$\phi^{(n)}$	$\phi^{(r)}$	$A = \frac{1+\sin\varphi}{1-\sin\varphi}$	E_s (MPa)	Poisson's ratio v	σ_{cm}**	ϕ**	A
Quaternary	390							0,3			
Lower Triassic	120	76,5	3.55	59	50.5	7,75	20000	0,15	11,5	32,4	3,31
Clay shale + gypsum	32	22,5	1.7	60.5	51.8	8,3	13500	0,18	1,6	24,3	2,40
Main Anhydrite	26	88,5	6.25	64	54.8	9,9	55500	0,25	17,0	34,4	3,602
Main Dolomite	5	124	7.55	63.5	54.4	9,7	62000	0,25	40,7	40,9	4,78
Upper Anhydrite	157	88,5	6.25	64	54.8	9,9	55500	0,24	17,0	34,4	3,00
Calcareous dolomite + gypsum (II)	8	52,8	4.4	59.9	51.3	8,1	29000	0,24	4,09	28,7	2,84
Calcareous dolomite (I)	8,9	147,8	9.3	64.7	55.4	10,3	67440	0,25	62,5	45,5	5,99
Calcareous dolomite (III)	1,2	148,77	8.95	64.8	55.5	10,4	56920	0,24	63,5	45,7	6,05
Calcareous dolomite (IV)	1,1	134,65	9.0	64	54.8	9,9	48100	0,25	49,8	42,9	5,27
Streak dolomite + shale	1,2	98,96	6.68				29610	0,23			
Grey sandstone	0,8	74,0	4.4				27800	0,20			
Quartz sandstone	1,5	41,4	1.8				12100	0,16			
Clay sandstone	4,4	19,6	0.91				8800	0,13			
Quartz red sandstone	200	13,4	0.64				5400	0,12			

Laboratory data

PN-G-05020
PN-B-03020

Longwall cutting height C_0 = 89,0 MPa

Values scaled down acc. to Hoek (2007)

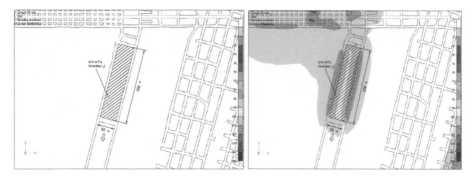

Figure 6. Safety margin distribution 7.75 m (left) and 15.2 m (right) above the immediate roof strata (longwall with roof deflection, excavation height H_p = 2.0 m, wooden cribs support or Tekpak).

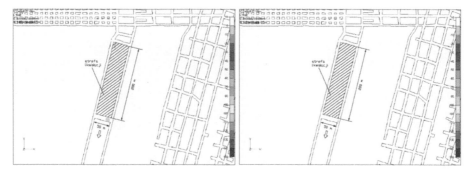

Figure 7. Safety margin distribution 0.55 m (left) and 1.7 m (right) above the immediate roof strata (longwall with cemented backfilling, excavation height H_p = 2.0 m).

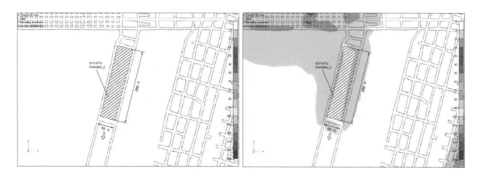

Figure 8. Safety margin distribution 7.75 m (left) and 15.2 m (right) above the immediate roof strata (longwall with cemented backfilling, excavation height H_p = 2.0 m).

Analysis of the behavior of the room-and-pillar mining system showed also (see Figure 14) that independently on the rock excavation technology, main roof instability may occur about 10 m above the immediate roof strata due to possible excessive value of σ_1 principal stress resulting in rupture of the upper rock mass.

Based on the former experiences in the field of numerical analyses (Pytel, 2010; Pytel and Butra, 2010) one may conclude, that instability hazard associated with loosening of roof rock layers (due to the high tangential stresses) and their crosswise shear, which are represented by margin of safety M_c formulated based on Coulomb's theory of failure is particularly

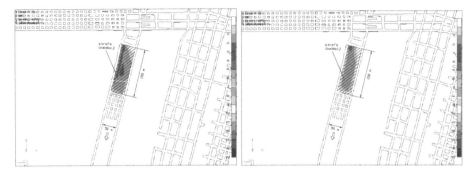

Figure 9. Safety margin distribution 0.55 m (left) and 1.7 m (right) above the immediate roof strata (room-and-pillar with roof deflection, excavation height H_p = 2.0 m, wooden cribs support or Tekpak).

Figure 10. Safety margin distribution 7.75 m (left) and 15.2 m (right) above the immediate roof strata (room-and-pillar with roof deflection, excavation height H_p = 2.0 m, wooden cribs support or Tekpak).

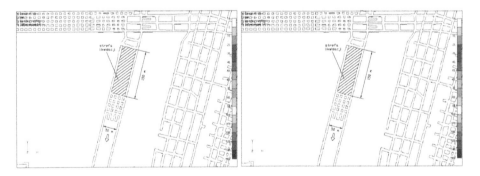

Figure 11. Safety margin distribution 0.55 m (left) and 1.7 m (right) above the immediate roof strata (room-and-pillar with cemented backfilling, excavation height H_p = 2.0 m).

important on all levels of roof layer for most real cases of mining analyzed by the presented above numerical methods.

Therefore, margin of safety M_c is one of the key indicator of presented analysis. Installation of wooden cribs or Tekpak support within the mined out areas does not provide the expected, beneficial results in terms of safety in regard of roof falls and

296

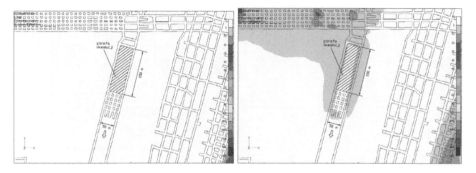

Figure 12. Safety margin distribution 7.75 m (left) and 15.2 m (right) above the immediate roof strata (room-and-pillar with cemented backfilling, excavation height H_p = 2.0 m).

Figure 13. Zone of possible shearing type damage allowing immediate roof strata falling down.

rockburst. Their effectiveness is comparable to the effect of mechanized support, however, a significant role of the last one as an element protecting the working zone against the local roof falls cannot be excluded. Wooden cribs and Tekpak support may be treated as an element of roof falls prevention, up to 8÷10 m within the roof strata (see Figure 14). This value depends on the opening height.

The effect of density of artificial support location within the goaf area (Figure 15, Variants I-IV) on safety margin values M_c is illustrated on the example of the same pilot panel which

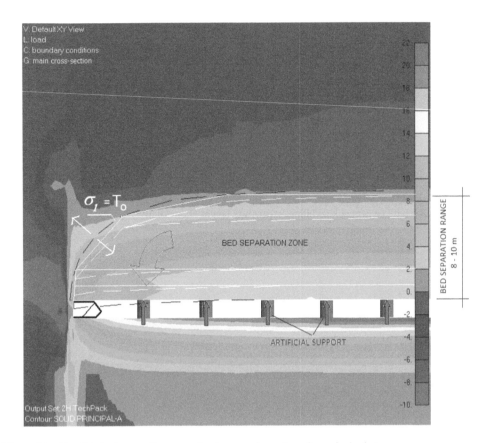

Figure 14.   Possible bed separation zone due to excessive value of σ_1 principal stress.

was analyzed above. Selected computation results indicate that from point of view of shear type instability, the rock strata behavior is not sensitive on the density of Tekpak location within the goaf area (Figure 16). This means for the mining practice, that supports of the Tekpak type are not able to resist effectively against the horizontal bed separation within the immediate roof strata due to the eccessive shearing stress. Moreover, this kind of roof suport does not secure roof strata against the rupture at the mutual contact surfaces. Therefore the artificial Tekpak suport may successfully carry out only the weight of the separated strata unless it is less than the support load capacity.

In presented numerical analyses there were utilized extremely unnfavouring values of rock strength parameters scaled down to in-situ rock mass conditions acc. to Hoek's approach. This way, the lower bound of safety has been determined, while the upper safety bound has been assessed using the laboratory obtained strength parameters reduced by the material design coefficient $\gamma_m = 0,9$ and the rock cleavage factor $k_1 = 1/1,05$. Selected calculation results are presented in Figure 17.

The obtained results also indicate significant safety improvement when laboratory determined rock mass strength parameters have been used. This may prove that the actual values of safety margin distribution along the line A-A (Figure 1) may be located somewhere between these two curves describing the upper and lower bounds of safety. Concluding that one may finally assume that the zone of possible roof strata separation moves definitely to goaf area.

Based on numerical analyses of considered mining parcels one may conclude, that support of the roof strata by the use of cemented backfilling guarantee an appropriate level of safety,

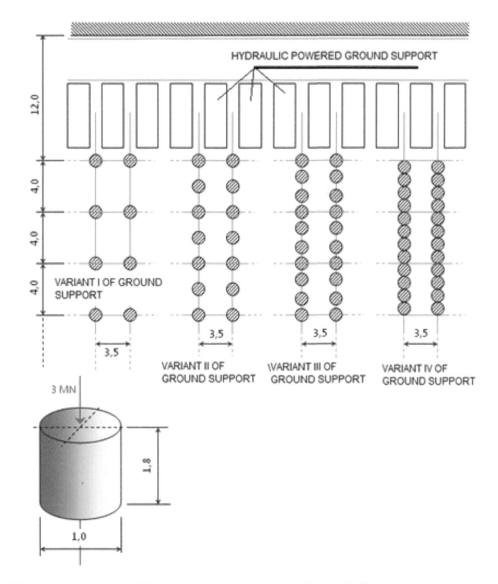

HYDRAULIC POWERED GROUND SUPPORT

12,0

4,0

4,0

4,0

VARIANT I OF GROUND
SUPPORT

3,5

VARIANT II OF
GROUND SUPPORT

3,5

VARIANT III OF
GROUND SUPPORT

3,5

VARIANT IV OF
GROUND SUPPORT

3 MN

1,8

1,0

Figure 15.   Variants of artificial supports' location within the pilot panel adjacent goaf area.

regardless on the size of opening and mining-geological conditions what is presented on Figures 18 - 19. Yielding of the rock mass on the mining front, which is typical for a classical room-and-pillar mining system leads to substantial rockburst and roof falls hazard reduction. Applying of a lighter support may be effective only when geology of the immediate roof would not be prone to cracks formation and delamination in the roof layers located deeper that 6÷8 m in the roof outer surface.

Application of mechanical excavating system at a depth greater than 700 m where the dynamic pressure of the rock mass is higher, is associated with significant difficulties with roof control, especially when the technology with possible roof falls is used or lighter support like wooden cribs or Tekpaks are used. This stipulation has been proved in reality few weeks after the mechanized mining system was commenced in the pilot

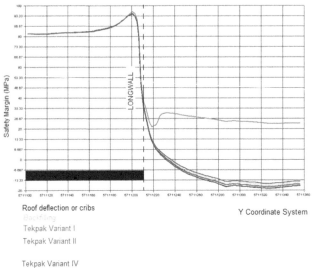

Roof deflection or cribs
Blackfilling
Tekpak Variant I
Tekpak Variant II

Tekpak Variant IV

Figure 16. The effect of roof suport in goaf area on values of safety margins M_c along the line A-A (Figure 1) at the level 1.7 m above the immediate roof strata.

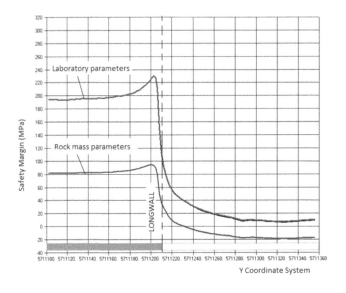

Figure 17. The effect of rock mass parameters assessment: values of safety margins M_c along the line A-A (Figure 1) at the level 1.7 m above the immediate roof strata.

panel, when extremely large (36 × 50 × 2 m) slab of immediate roof stratum felt down, fortunately without any casualties or severe loses (see Figure 20). Wooden cribs placed in goaf area have disappeared completely, crushed there between immediate roof and floor strata.

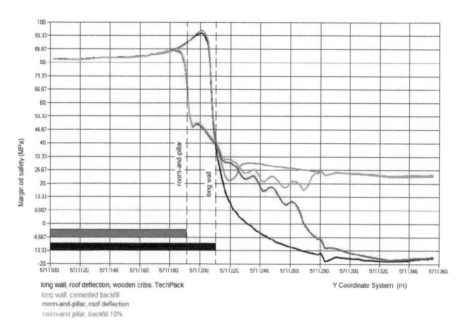

Figure 18. Values of safety margin M_c along the line A-A in the pilot panel (see Figure 1) in dolomite III, 1.7 m above the immediate roof strata.

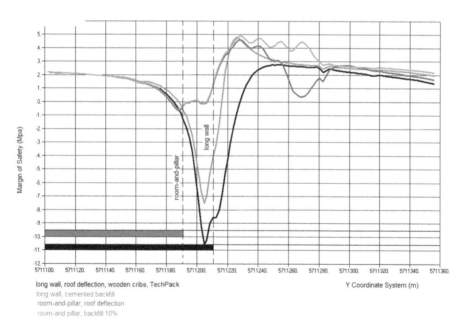

Figure 19. Values of safety margin M_c along the line A-A in the pilot panel (see Figure 1) in dolomite II, 15.6 m above the immediate roof strata.

Figure 20.   Massive roof fall observed in the pilot panel at the beginning of the ore extraction.

5   CONCLUDING REMARKS

The above presented results of numerical modeling of rock mass behaviour when exploited in the pilot panel using (a) the mechanized longwall system and/or (b) room-and-pillar mining, are the base of the following conclusions to be drawn:

A. It is confirmed that increasing the excavation height from 1.2 to 2.0 m induced unfavorable effect on geomechanical safety, however the scale of this phenomenon is not so large as one could expect.
B. It was stated, that increasing the depth of exploitation has unfavorable impact on safety in the vicinity of mining front. The presented results of performed calculations have proved, that the value of M_c parameter is very high on larger depth. It means that the transition from the safe to danger zone occurs very rapidly, even at 10 m distance. On smaller depth this distance reach the distance of about 20-30 m.
C. The experience to date in the field of numerical analysis shows that the risk of instability phenomena associated with the possibility of loosening the rock layers (due to excessive tangential stress) and their lateral shearing, represented by the value of the M_c safety reserve formulated on the basis of Coulomb's strength hypothesis, is relevant for all levels of the roof strata, for most real cases of exploitation analyzed by numerical methods. Therefore, the safety margin M_c is the leading indicator of the risk analysis performed in this study.

D. It is assumed that the negative values of Mc safety margins determined for the level of 1.15 m above the roof indicate a local rock fall threat. However, in the case of locations above 8.0 m above the ceiling, for which the value of the safety margin is less than zero, one can conclude about a significant rock burst hazard.

E. Rockburst hazard estimated on the basis of calculated spatial distribution of safety margins, which are a practical indicator of the degree of rock mass effort and the distance of appropriately formulated quantifiers of the current state of stress from the boundary surface of destruction, is in essence a relative assessment, allowing to identify the best solution from a number of proposed solutions from a risk point of view.

F. The analysis of the calculation results also shows that the proposed calculation procedure and the destruction criteria used to assess the hazard fulfill their role properly, without leading to conclusions contrary to the experience and results of previous observations of the rock mass behavior.

G. For the assumed mining-geological conditions, calculated safety margin distributions indicates a real rockburst hazard when the mechanized longwall technology is applied, without the highly effective backfilling. This situation may be significantly changed by including into the analysis the presence of high horizontal stresses, which haven't been considered due to the lack of the reliable data.

H. Because the presented results of numerical calculations were obtained by applying a drastic reduction of rock strength parameters according to the Hoek' approach, all the conclusions and reservations formulated above have extremely conservative character in the sense of the assumed safety margin.

REFERENCES

Amick, M., Mazzoca, J., and Vosefski, D. 1993. *The use of foamed cement cribs at American Electric Power Fuel Supply Meigs Division*, in Proceedings of 12th International Conference on Ground Control in Mining, S.S. Peng, ed., West Virginia University, pp. 55–58.

Barczak, T.M., Chen, J., and Bower, J. 2003. *Pumpable roof supports: developing design criteria by measurement of the ground reaction curve*, in Proceedings of 22nd International Conference on Ground Control in Mining, S.S. Peng, ed., West Virginia University, pp. 283–293.

Butra J., Debkowski R. and Pytel, W. 2001. *Bump Hazard Control in Deep Copper Mines in Poland*. Proc. of the 10th Int. Symp. On Mineral Planning and Equipment Selection, New Delhi, 2001, pp. 761–768.

Hoek, E. 2007. *Practical Rock Engineering*. Retrieved from https://www.rocscience.com.

Hustrulid, W.A. 1982. *Underground Mining Methods Handbook*. New York: SME-AIME.

Hustrulid, W.A. and Bullock R.L. 2001. *Underground Mining Methods: Engineering Fundamentals and International Case Studies*. Society of Mining Engineers. ISBN 978-0-87335-193-5, 728 pp.

Jaeger, J.E., Cook N.G.W. and Zimmerman, R. 2007. *Fundamentals of Rock Mechanics*. Wiley and Sons, 488 pp.

KGHM Guideline concerning the roof deflection method of mining in underground copper mines (in Polish), Lubin, 1994.

Kijewski P. and Lis, J. 1994. *Geomechaniczne właściwości skał z obszaru LGOM – próba systematyki*. Sci. Rep. of Institute of Geotechnics and Hydrotechnics, Wroclaw University of Science and Technology, pp. 103–110.

Mark, C. 1990. *Pillar design methods for longwall mining*. Pittsburgh, PA: U.S. Department of the Interior, Bureau of Mines, IC 9247.

Minova. 2017. *TekpakTM Pumpable Crib. The Advanced Pumpable crib System*. www.minovaglobal.com

NEi/Nastran (Version 9.2) [Computer software]. Westminster, CA: Noran Engineering, Inc.

Orlecka-Sikora, B. and Pytel, W. 2013. *Integration on geomechanical and geophysical analysis methods for the prediction of seismic events in underground mining*. In: Rock Mechanics for Resources, Energy and Environment – Kwasniewski & Łydżba (eds), Taylor & Francis Group, London.

Pariseau, W. G. 2009. *Design Analysis in Rock Mechanics*. Taylor and Francis, 499 pp.

Pytel W.M. and Chugh, Y.P. 1992. *Design of Partial Extraction Coal Mine Layouts for Weak Floor Strata Conditions*. Proc. of the Workshop on Coal Pillar Mechanics and Design, Information Circular 9315, U.S. Bureau of Mines 1992, pp. 32–49.

Pytel, W. 2003. *Rock Mass – Mine workings interaction model for polish copper mine conditions*. International Journal of Rock Mechanics & Mining Sciences, 40, pp. 497–526.

Pytel, W. 2010. *Room-and-pillar mine workings design in high level horizontal stress conditions*. Case of study from the Polish underground copper mines, *Rock stress and Earthquakes* (ed. Xie), © 2010 Taylor & Francis Group, London, ISBN 978-0-415–60165–8.

Pytel, W. and Butra, J. 2010. *Mine workings design in regional pillar mining conditions – a case study from a Polish copper mine*. Deep Mining 2010 – (eds. M. van Sint Jan and Y. Potvin), © 2010 Australian Centre for Geomechanics, Perth, ISBN 978-0-9806154-5-6

Pytel, W. and Palac-Walko, B. 2015. *Geomechanical safety assessment for transversely isotropic rock mass subjected to deep mining operations*. Canadian Geotechnical Journal. Oct. 2015, vol. 52 Issue 10, pp. 1477–1489.

Mark, C. 1990. *Pillar design methods for longwall mining*. Pittsburgh, PA: U.S. Department of the Interior, Bureau of Mines, IC 9247.

Spisak J. and Zelko, M. 2015. *Environmental aspects of raw materials SMART processing*. In: Environmental Engineering and Computer Application – Chan (Ed.), Taylor & Francis Group, London.

Spisak J. and Zelko, M. 2010. *The Advanced Technologies Development Trends for the Raw Material Extraction and Treatment Area*. Products and Services: from R&D to Final Solutions, SCIYO, Croatia.

Legal, safety, and environmental drivers for backfill

Minefill 2020-2021 – Hassani et al (eds)
© *2021 Taylor & Francis Group, London, ISBN 978-1-032-07203-6*

Filling underground voids to prevent water hazards in active and decommissioned hard coal mines

Grzegorz Strozik
Silesian University of Technology, Gliwice, Poland

SUMMARY: In hard coal mines in Poland, the use of filling of caving is widespread. Mixtures of water and fly ash from coal combustion in power plants are used for this purpose. In addition to many other benefits, attention has been paid to this technology's usefulness in reducing water hazards. For this purpose, laboratory tests have been undertaken on fly ash-water mixtures samples conductivity using a falling-head permeameter. The results showed that the mixtures achieve permeability values range from $8x10^{-7}$ m/s to $1,2x10^{-9}$ m/s. Additionally, caved zone structure and flow of fill mixtures in longwall cavings have been discussed.

Keywords: water hazard, mining with caving, filling of cavings, permeability of fill materials, insulation of gob area

1 INTRODUCTION

Underground mining is associated with numerous impacts on both surface and subsurface environments. When operating with a roof collapse, the most considerable impact is rock masses' movement, resulting in deformations of the rock mass and the ground surface. These deformations have the nature of continuous displacements - without disturbing the structure of the rock and discontinuities, which includes, among other things, the formation of underground voids and fractures, the expansion and development of existing fissures, which accompanied the transformation of continuous rock layers into separate rock blocks and blocks.

For groundwater flow and water hazard considerations, it is convenient to adopt the model of the rock mass fracturing process proposed by Palchik (2003). It assumes the creation of zones of increased vertical and horizontal interconnections between rock strata in the roof of a coal seam extracted using longwall with caving, Figure 1. This model is still in use, also in currently appearing works (Guo et al. 2019; Ning et al. 2019).

Above the gob area (caved zone), a fracture zone occurs, divided into three sub-zones, as illustrated in Figure 1. The lower subzone consists of separated blocks separated by vertical and horizontal fractures, and the vertical displacement of the rock blocks may occur. However, in the middle zone, flat and interconnected vertical fractures exist of smaller dimensions and without individual rock blocks. Finally, there are only interlayer bed separations in the upper area. Above this structure, a continuous subsidence trough extends up to the carbon-iferous formation roof and further upward to the ground surface (Guo et al. 2019; Ning et al. 2019; Palchik 2003).

Similarly, the process of creating a fracture zone proceeds in the case of longwall mining with backfill. However, in this case, the fracture zone's extent is reduced, and the dimension of the fractures is smaller (Figure 2). Separate rock blocks are larger and their convergence smaller than in the case of mining with caving.

DOI: 10.1201/9781003205906-27

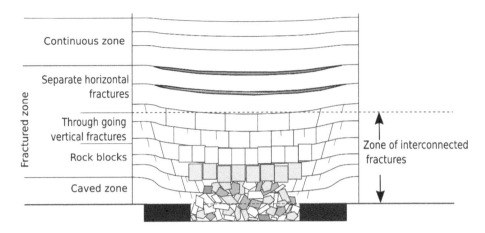

Figure 1.   Zones of fractures in the roof rock strata resulted from the longwall mining with caving.
Source: own elaboration after Palchik, 2003.

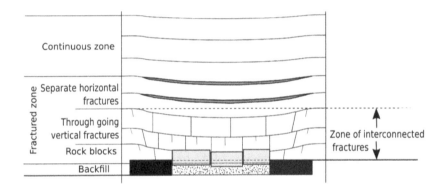

Figure 2.   Zones of fractures in the roof rock strata resulted from the longwall mining with backfill.
Source: own elaboration after Palchik, 2003.

The number of fractures and their length per specific volume of roof rocks is much smaller than in the case of caving. It depends primarily on backfill material compressibility, roof convergence at the distance of the open space of the longwall, and quality of backfill operations. Also, the eventual presence of the "backfill zero" (space below the roof that cannot be filled up, e.g., due to the disadvantageous geometry of the coal seam) is an essential factor. Of course, the use of backfill reduces the values of deformation parameters of subsidence troughs, which is the primary purpose of using backfill.

However, any mining operations cause water relations changes in the rock mass (Rogoż 2004; Sztelak 1998; Trembecki 1972). Minimal increase in water hazards and ground waters migration can be achieved using backfill. Still, due to the low profitability of hard coal mining in Upper Silesia Coal Basin conditions, the total output comes from mining with caving. Therefore, the paper attempts to discuss the impact of filling of cavings on the reduction of water risk and groundwater migration conditions in the rock mass, which arises as a result of mining with caving.

The purpose of this paper is to present the process of the filling of cavings in the aspect of limiting water hazard and creating opportunities to improve hydrogeological conditions in the active and decommissioned mines through the planned creation of insulating layers in the gob area of extracted coal seams.

2   GROUNDWATER THREAT MITIGATION IN HARD COAL MINES

In the conditions of the existing water hazard in mines, especially when mining with caving is the only or dominating mining system, an increase in water threat by creating new fracture zones and gob areas in extracted coals seams may occur. This increase in risk is intensified in multi-seam mining operations when the extraction of lower and lower coal seams promotes an increasingly higher zone of hydraulically interconnected fractures and voids (Ning et al. 2019; Sztelak 1998).

In the mine liquidation phase, previously drained rock mass areas become zones of accumulation and flow of groundwater (Mzyk 2016; Trembecki 1072). When a decommissioned mine lies adjacent to an active mine, it may be advantageous to limit groundwater migration from a flooded to an active mining area. That target can be achieved, e.g., by isolating specific parts of the rock mass and creating convenient groundwater runoff routes to other parts of it. In practice, this goal may be achieved by planned closure activities (e.g., by tight isolating or damming of working) or by maintaining selected main workings as water collectors (Mzyk 2016). In this way, it is possible to divert groundwater to drainage pumps of the active mines' protection system.

The caved zone formed by the gob area and the separated rock blocks creates space for retention of large volumes of groundwater, and the large fractures allow its unrestricted flow (Figure 3). The vertical flow becomes more limited in the higher layers, while the horizontal movement of water and their retention are still high. Only in the zone of separated horizontal layers, the horizontal flow becomes moderately more extensive than in initial conditions. Finally, in the area of continuous deformations, groundwater flow and retention dominate, governed by natural hydrogeological conditions of the rock mass (Figure 3).

In mining with backfill, the fracture system is weakly developed, which means that the increase in groundwater flow dynamics and the disturbed rock mass strata's retention capacity are much smaller (Figure 4).

In both cases, further vertical run-off of groundwater depends on the rock material's insulating properties lying on the extracted coal seam floor. In the case of mining with backfill,

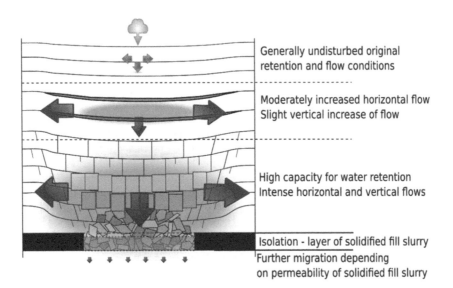

Generally undisturbed original retention and flow conditions

Moderately increased horizontal flow
Slight vertical increase of flow

High capacity for water retention
Intense horizontal and vertical flows

Isolation - layer of solidified fill slurry
Further migration depending on permeability of solidified fill slurry

Figure 3.   Schematic diagram of the changes in groundwater dynamics and retention in the rock mass caused by longwall mining with caving.

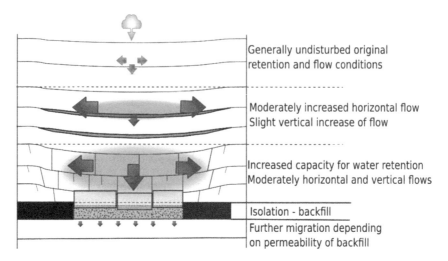

Figure 4. Schematic diagram of the changes in groundwater dynamics and retention in the rock mass caused by longwall mining with backfill.

the insulation level depends primarily on the water permeability of the backfill material. On average, the permeability of the hydraulic backfill made from sand is 2×10^{-4} m/s, which is a small value considering that the permeability of coal in an undisturbed seam equals about 5×10^{-6} m/s (Sztelak 1998; Trembecki 1972). Therefore, the main effect of restricting the free flow of groundwater in the rock mass through the backfill occurs by limiting the size of the fracture system in layers above the coal seam.

As already mentioned, the conditions of free groundwater movement and extensive retention possibilities can significantly increase the level of a water hazard in mining with caving. Filling cavings on the entire surface of the extracted coal seam with solidifying mixtures may substantially reduce this water hazard. (Figure 3).

3 USE OF FINE-GRAINED FILL MIXTURES IN LONGWALL MINING WITH CAVING

The primary purpose of filling cavings in mines is to improve ventilation and protect against spontaneous ignition of coal residues in the gob areas. As a material for filling voids, fly ash from the combustion of hard coal mixed with water is used, mainly due to these mixtures' ability to solidify and convenient rheological properties that allow their easy hydraulic transport in pipelines and placement into the cavings (Palarski et. al. 2014). Various aspects of the technology of filling of cavings have been discussed in numerous works devoted to this issue, so there is no need to discuss them in this paper.

From the point of view of this paper's subject, important is the ability of the fly ash – water mixture to efficiently penetrate the fractures and voids in the gob area. Large fractures make the filling process more effective, so the filling procedure is performed at a short distance behind the front of the longwall. The most successful filling of cavings may be achieved for (Palarski et. al. 2006):

– Seams with inclination $5° \div 10°$ and thickness 2,0 m $\div$ 2,5 m,
– Roof strata with advantageous fragmentation and regular fracturing,
– Longwalls 200 m $\div$ 220 m wide, advancing up the dip with filling from the pipeline with multiple outlets in the longwall face or
– Shorter longwalls in more inclined seams, advancing along with the dip with filling from a single pipeline outlet located in the headgate (Figure 5).

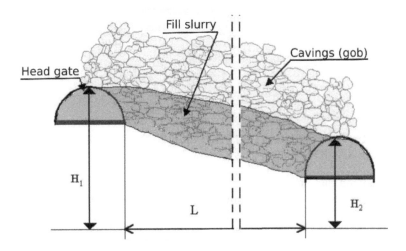

Figure 5.   Flow of fill mixture through the longwall cavings.
Source: own elaboration after Palarski et.al. 2006.

4   FLOW CONDITIONS OF FLY ASH – WATER MIXTURES THROUGH CAVINGS

In most known bibliography, mixtures of water and fly ash are considered the same as real liquids. The flow in pipelines is relatively easy to describe using well-known equations for laminar and turbulent flow with reference to their rheological properties (most often, these mixtures meet the criteria of the Bingham plastic fluid). Much less attention is paid to the conditions of the mixture spreading freely in the caving zone.

The equation of mixture flow in the caving zone can be derived by combining the Darcy equation's three-dimensional form with the continuity equation. The flow takes place as the inclination of the seam floor (hydraulic gradient I) is larger than (Palarski et. al. 2006):

$$I = \frac{H_1 - H_2}{L} > \frac{3f\tau_0(1-n)}{\rho gnd} \tag{1}$$

Where:
H_1 – the height of the pipeline outlet in the headgate,
H_2 – the height of the fill at the wall of the tailgate,
L – length of the longwall,
f – factor of the Surface roughness in the gob area,
τ_0 – the yield stress of the filling mixture,
n – the porosity of the cavings,
g – acceleration of gravity,
d – average diameter of rock blocks,
The formula for a volumetric flow rate through the cavings downwards the slope of the gob area is:

$$Q = \frac{n^3}{1-n^3} \frac{d^2}{f^2 N} \frac{\rho g}{\eta} \left(1 + 4\delta^3 - 3\delta\right) Ins \frac{H_1 - H_2}{2} \tag{2}$$

Where:
η – dynamic viscosity of fill mixture,
N – coefficient ($N = 250 \div 300$),

s – the width of flowing stream of a mixture,

In equation (2), δ is equal to:

$$\delta = \frac{3\tau_0 f}{2\rho g \, I n d} \tag{3}$$

5 ASSESSMENT OF INSULATING PROPERTIES OF SOLIDIFIED FLY ASH – WATER MIXTURES

The basic condition that must be met by the material constituting the insulation barrier is the satisfactory value of tightness. The tightness is characterized by the filtration coefficient K. Under natural conditions, rocks achieve a filtration coefficient down to about $10^{-10} \div 10^{-12}$ m/s (Freeze and Cherry 1979). Such almost impervious behavior is represented by unweathered clays (among unconsolidated rocks) and unfractured metamorphic and igneous rocks and shales – perpendicular to the layering (among consolidated hard rocks). Natural materials for constructing impermeable barriers for waste depositories should exhibit a filtration coefficient no less than 1×10^{-9} m/s (Plewa et. al. 2006; Plewa et. al. 2009). Most of the rock formations in Carboniferous rock mass and its younger, Quaternary overburden represent significantly higher permeability and the ability for water retention (aquifers). Hydraulic conductivity of highly fractured rocks and gravels may range from over 1 m/s down to 10^{-3} m/s. Sandstones, mudstones, hard coal (unfractured), as well as fine sands and silts, may achieve conductivity in the range from 10^{-4} m/s to 10^{-7} m/s. Such rocks may be considered semi-permeable but not capable of containing reach aquifers (Rogoż 2004).

5.1 Measurement method and equipment

There are two laboratory methods for measuring hydraulic conductivity: the constant-head method and the falling-head method (Amoozegar and Wilson 1999). In the constant-head method, water moves through the specimen under a steady-state head condition, while the volume of water flowing through the soil specimen is measured over a period of time. By knowing the volume ΔV of water measured in a time while the volume of water flowing through the soil specimen is measured over a period of time over a specimen of length l and cross-sectional area A, as well as the head h, the hydraulic conductivity, K, can be derived simply rearranging Darcy's law (Amoozegar and Wilson 1999; Laksmahan 2011):

$$K = \frac{\Delta V}{\Delta t} \frac{l}{Ah} \tag{4}$$

In the falling-head method, the soil sample is first saturated under a specific head condition. The water is then allowed to flow through the soil without adding any water, so the pressure head declines as water passes through the sample. If the head drops from h_1 to h_2 in a time Δt, then the hydraulic conductivity is equal (Amoozegar and Wilson 1999); Laksmahan 2011):

$$K = \frac{l}{\Delta t} \ln \frac{h_f}{h_i} \tag{5}$$

Both measurement methods are generally supposed to be used to measure soils as relatively highly permeably grained materials (Sandoval et. al. 2017). For rocks with low permeability, the most appropriate measurement method is the flow pump method. The water flow rate through the sample is kept constant; however, this method requires a special laboratory stand.

The falling-head method has been selected to measure solidified fill materials' conductivity obtained from fly ash – water mixtures. The method has been selected because of the simplicity of the measuring system and the ability to perform measurements in the range of very low conductivities (Plewa et al. 2006; Plewa et al. 2008; Wilson et al. 2000).

Measurements were carried out in a permeameter, the schematic diagram of which is shown in Figure 6. Permeameter has been adjusted to measure small conductivity values. It means small diameter and height of specimens, high water tube – and what is important during long time measurements – a control tube was placed to measure the effect of water evaporation on the downward rate of water in the cylinder.

Hydraulic conductivity is related to the permeability with relationship (Sztelak 1998):

$$T = Km \ [\text{m}^2/\text{s}] \tag{6}$$

Where:

T – Permeability index,

m – thickness of the water-bearing layer,

Water permeability index T describes the volume of water flowing in the unit of time through a water-bearing layer of thickness m and 1 m wide.

5.2 *Materials used in conductivity measurements*

Hard coal-powered power plants use different combustion vessel constructions, flue gas desulphurization technologies, etc. These produce different physical and chemical properties of fly ash, affecting their mixtures' properties with water.

For the conductivity measurement, three types of fly ash have been used: ordinary fly ash from vessels without desulphurization by-products (OFA), fly ash from a vessel with semi-dry flue gas desulphurization process (SDD) and fly ash from fluidized bed vessel (FLB). Figure 7 shows the grain-size distribution of these kinds of fly ash.

Tap water has been used to prepare mixture samples, and the temperature was kept constant at 20°C throughout the measurements. Three samples of each mixture have been prepared, and average values from three measurements have been considered as results.

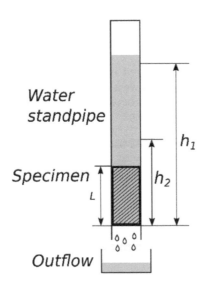

Figure 6.   Falling-head permeameter used in measurements.

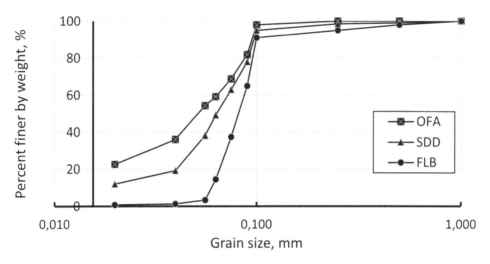

Figure 7.　Grain-size distribution of fly ash samples used in measurements.

Source: own measurements.

The mixtures' composition has been selected so that mixtures from different types of fly ash exhibit similar rheological properties. The fly ash's general properties –water mixtures used in conductivity measurements have been shown in Table 1.

5.3　*Hydraulic conductivity measurement results and analysis*

The specimens of fly ash – water mixtures after preparation have been cured inside permeameter tubes constantly for a period of 98 days (14 weeks). The water levels in permeameter tubes have been measured daily for the first week of curing and then once a week. At the early stage of stabilization of the specimens, the hydraulic conductivity values decreased rapidly from about $(3 \div 7)x10^{-5}$ m/s down to $6x10^{-6}$ m/s for OFA fly ash, $1,2x10^{-6}$ m/s for SDD fly ash, and $8,5x10^{-8}$ m/s for FLB fly ash – Figure 8. In the next phase of sample curing, lasting up to about 11 weeks, a slight decrease in all samples' hydraulic conductivity was observed, followed by no further significant changes in hydraulic conductivity. Specimens of fly ash – water mixtures made from OFA, SDD, and FLB fly ash after 98 days cure time achieved final values of the hydraulic conductivity equal $8x10^{-7}$ m/s, $1x10^{-8}$ m/s, and $1,2x10^{-9}$ m/s respectively, Figure 8.

It can be concluded that the stabilized (solidified) fly ash – water mixtures about three months after placement in cavings achieve a relatively very high level of impermeability compared to typical rocks found in the Carboniferous layers. A similar range of hydraulic conductivity exhibits unfractured sandstones (Freeze and Sherry 1979). The highest value of

Table 1.　Physical properties of fly ash – water mixtures used in hydraulic conductivity tests.

Parameter	Type of fly ash		
	OFA	SDD	FLB
Density of fly ash [kg/m^3]	2060	2340	2620
Bulk density [kg/m^3]	1040	1120	1340
Solids/water ratio by mass S:W	2,78	1,88	0,95
Density of mixture [kg/m^3]	1555	1528	1380
Table spread [mm]	180	180	180

Source: own measurements.

314

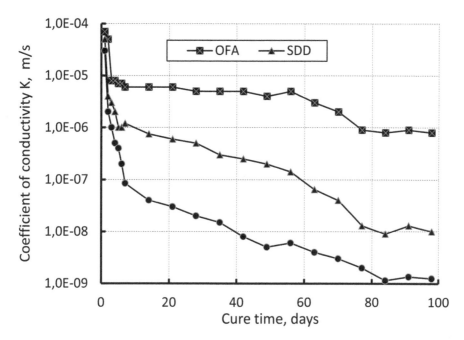

Figure 8. Hydraulic conductivity measurement results.
Source: own measurements.

hydraulic conductivity shows mixtures made from ordinary fly ash from combustion vessels without any desulphurization process (OFA); however, this type of combustion vessel becomes obsolete. The best result has been achieved for a mixture made from fly ash from a fluidized bed combustion vessel (FLB). This type of ash contains a significant amount of calcium oxide (CaO), which gives it strong binding properties, which are probably responsible for the low conductivity of its solidified mixture with water.

6   CONCLUSION

Filling cavings is a technology widely used in the mining industry, primarily aimed at improving ventilation conditions and protecting against the formation of endogenous fires in the remains of seams (gob areas) during longwall mining with caving.

Mining with caving is associated with significant damage to the roof rock structure, which in the conditions of presence of static or flowing groundwater in mines may cause an increase in the water hazard.

Laboratory tests of the permeability of ash – water mixtures used for filling cavings have shown that this procedure can promote the creation of an insulating protective layer limiting the flow of water to lower parts of the coal bed.

The hydraulic conductivity of tested fly ash – water mixtures ranges from 8 x 10^{-7} m/s to 1,2 x 10^{-9} m/s, which allows them to be classified as low- or impermeable materials.

The creation of impermeable screens can also protect adjacent active mines and regulate the direction of groundwater flow in decommissioned mines.

BIBLIOGRAPHY

Amoozegar, A. and Wilson, G.V. 1999. *Methods for Measuring Hydraulic Conductivity and Drainable Porosity.* [In:] Skaggs, R.W. and Schilfgaarde J. van eds. *Agricultural Drainage*, chapter 37. Society of

Agronomy, Crop Science Society of America and Soil Science Society of America, Madison, Wi, USA, pp. 1149–1205.

Freeze, R.A. and Cherry, J.A. 1979. *Groundwater*. Englewood Cliffs: Prentice Hall Inc., 624 pp.

Guo et.al. 2019 – Guo, W., Zhao, G., Lou, G. and Wang S. 2019. Height of fractured zone inside overlying strata under high-intensity mining in China. *International Journal of Mining Science and Technology* 29(1), pp. 45–49.

Lakshmahan, E. ed. 2011. *Hydraulic Conductivity – Issues, Determination and Application*. Rijeka: Intech, 446 p.

Mzyk, T. 2016. Forecast of flooding a decommissioned mine in north-eastern part of USCB. Assisted systems in Production Engineering 5(17), pp. 147–163 *(in Polish)*

Ning et. al. 2019 – Ning, J., Wang, J., Tan, Y. and Xu Q. 2019. Mechanical mechanism of overlying strata breaking and development of fractured zone during close-distance coal seam group mining. . *International Journal of Mining Science and Technology*. [Online] https://www.sciencedirect.com/science/article/pii/S2095268617304391?via%3Dihub

Palarski J., Plewa F., Babczyński W., 2002. Factors affecting the underground waste disposal – hydrotransport and sedimentation. 11th International Conference on Transport & Sedimentation, pp. 159–173, Ghent, Belgium, September 2002.

Palarski et. al. 2006 – Palarski J., Plewa F., Strozik, G. and Jendruś R. 2006. Problems of migration and sedimentation of fly ash - water mixtures in gob areas in the light of laboratory experiments. [In:] *Proceedings of the 13-th Conference Transport and Sedimentation of Solid Particles*. Tbilisi, Georgia, 18–20 September 2006, pp. 240–248.

Palarski et. al. 2014 – Palarski J., Plewa F. and Strozik, G. 2014. Filling of voids in coal longwall mining with caving - technical, environmental and safety aspects. [In:] Potvin, Y. and Grice, T. eds. *Proceedings of the 11-th International Symposium on Mining with Backfill*. Perth, Australia, 20–22 May 2014. Nedlands: Australian Centre for Geomechanics, pp. 483–492.

Palchik, V. 2003. Formation of fractured zones in overburden due to longwall mining. *Environmental Geology*, 44, p. 28–38.

Plewa et. al. 2006 – Plewa, F., Mysłek, Z. and Piontek, P. 2006. Study on the impact of the addition of water glass and lime on the mechanical and filtration properties of fly ash without desulphurization products from power plant X in terms of the possibility of their use for the construction of insulation barriers in landfills. [In:] *Proceedings of the XII International Symposium Geotechnics 2006*. Gliwice-Ustroń, 17–20 October 2006. Gliwice: Publishing House of Silesian University of Technology, pp. 293–304 *(in Polish)*

Plewa et. al. 2008 – Plewa, F., Pierzyna, P. and Filrit I. 2008. Investigation of the effect of the addition of lime and water glass on the mechanical and filtration properties of solidified hydraulic mixtures of fly ash from semi-dry flue gas desulphurization. [In:] *Proceedings of the XIII International Symposium Geotechnics 2008*. Gliwice-Ustroń, 14–17 October 2008. Gliwice: Publishing House of Silesian University of Technology, pp. 331–342 *(in Polish)*

Plewa et. al. 2009 – Plewa, F., Pierzyna, P. and Kanafek, J. 2009. Assessment of porosimetric and filtration properties of hydraulic mixtures of fly ash for construction of isolation barriers in underground waste depositories. *Energy Policy* 12(22), pp. 475–483 *(in Polish)*

Rogoż, M. 2004. *Mine hydrogeology with basics of general hydrogeology*. Katowice: Publishing House of GIG, 480 pp. *(in Polish)*

Sandoval et. Al. 2017 – Sandoval, G.V.B., Galobardes, I., Teixeira, R.S. and Toralles, B.M. 2017. Comparison between the falling head and the constant head permeability tests to assess the permeability coefficient of sustainable Pervious Concretes. *Case Studies in Construction Materials* 7, pp. 317–328.

Sztelak, J. 1998. *Mine hydrogeology and methods of combating water threats in underground mines*. Gliwice: Publishing house of Silesian University of Technology, 500 pp. *(in Polish)*

Trembecki, A. 1972. *Water hazards in mining*. Katowice: Publishing house "Śląsk", 458 pp. *(in Polish)*

Wilson et. al. 2000 – Wilson, M.A., Hoff, W.D., Brown, R.J.E. and Carter M.A. 2000. A falling head permeameter for the measurements of the hydraulic conductivity of granular solids. *Review of Scientific Instruments* 71, 3492.

Minefill 2020-2021 – Hassani et al (eds)
© *2021 Taylor & Francis Group, London, ISBN 978-1-032-07203-6*

Impact of the method of managing opencast excavations by filling with mining waste on the quality of leachate entering the surface water

Sławomir Rzepecki
CTL Maczki-Bór S.A., Poland

Aneta Grodzicka & Katarzyna Moraczewska-Majkut
Silesian University of Technology, Gliwice, Poland

SUMMARY: For years, environmental protection has been struggling with the problem of water contamination in Poland, especially salinity in surface water. The main source of water salinity in Poland is mining, so the problem of water salinity most affects the region of Silesia, which is characterized by a large number of hard coal mines and industrial plants. The hard coal mines located in Upper Silesia, during mining operations, pump large quantities of highly saline waters from the drainage of mine workings to the surface. Therefore, the problem of environmental protection is nowhere else on such a large scale. However, due to the fact that hard coal mines in our country constitute the main source of energy resources, solving the problem of saline water is difficult. The first part of the article presents the negative impact of salinity on water quality and resources in Poland and and it also indicates the economic effects associated with excessive salinity of rivers, which mainly include corrosive effects. The second part of the article will indicate the method of reclamation of the minefilling sand excavation, from which the leachate is discharged into surface water. As the reclamation takes place by filling the excavation with post-mining waste, there is a risk that saltwater may enter the surface water. This article also presents the example how to properly reclaim areas with post-mining waste and maintain good water status, to which leachate from the reclaimed area is discharged. The article specifies how much water from post-mining waste can be safely discharged into surface water, without creating the risk of excessive salinity. In the event of an imminent threat, there is an obligation to immediately take preventive measures, including immediate control, containment, removal or reduction of pollution or other harmful factors e.g. by controlling leachate effluent flows.

Keywords: opencast excavation, mining waste, surface water salinity

1 INTRODUCTION

Poland has relatively small water resources, which due to the geographical location, are supplemented by precipitation to a much lesser extent than in other European countries. In addition, high temperatures occurring in summer, lead to high evaporation losses. The greatest water reclamation capability is characterized by mountain areas, despite the fact that they do not have high retention capacities of the geological base (Chełmicki 2002).

The negative impact on water quality and resources related to human activities grew with the socio-economic development of the country. The strong development of industry caused an increasingly adverse impact on the water environment, including the salinity of waters. As

DOI: 10.1201/9781003205906-28

of today, the question of the salinity of flowing waters is a significant problem in our country, whose solution has become a necessity. Undoubtedly, the greatest impact on the increase in the level of salinity of flowing water are adequate:

- discharges of underground water from hard coal mine,
- discharges of saline industrial sewage,
- municipal economy and agriculture.

Due to the occurrence of the largest hard coal mines in the Upper Silesian Industrial District, in connection with the sources of the two largest rivers in Poland - the Vistula and the Odra, the problem of salinity of waters flowing through mine water from mines occurs in Poland on a much larger scale than in other countries (Lipiński 1987).

The problem of water salinity has been known in Poland for many years. The first reports on this problem took place in the early 1960s. As previously mentioned, mining is the main source of water salinity in Poland. Upper coal mines located in Upper Silesia, in connection with mining activities, pump large quantities of highly saline waters from the drainage of mine workings to the surface. Additionally, an unfavourable aspect resulting from this fact is that hard coal mines are located in the upper section of the Vistula and Odra rivers, contributing to water pollution in other regions of the country (Miller 2004). Excessive salinity of waters causes degradation of biological life in rivers and the inability to use water resources for economic and industrial purposes. In addition, too high salt content can have a negative impact on the life and efficiency of hydrotechnical equipment, thus generating additional, high costs for overhauls and current repairs (Miller 2004).

2 THE WATER SALINITY IN POLAND

The water salinity in Poland is a very important problem in ecological and economic terms. The basic adverse effects of discharges, mainly through hard coal mine, excessively saline waters to rivers are (Grynkiewicz-Bylina & Majewski 2000):

- degradation of the natural water environment,
- harmful and toxic effects on river flora and fauna,
- limiting or preventing the economic use of water resources,
- increase in costs related to saltwater treatment for municipal, industrial and agricultural purposes,
- corrosion of hydrotechnical equipment structures,
- the need to provide water for communal purposes from other regions of Poland.

The Silesian Voivodship, which is characterized by a large number of hard coal mines and industrial plants, is the most vulnerable region in Poland in the adverse impact of excessively saline flowing waters. Therefore, the problem of environmental protection is nowhere else on such a large scale (Nocoń & Nocoń 2011). However, due to the fact that hard coal mines in our country constitute the main source of energy resources, for this reason solving the problem of saline water is difficult (Miller 2004). Currently, in Poland, hard coal is exploited from about 20 mines and they are the main source of water salinity.

Hard coal mining has been systematically declining - in the last 3 years it has not exceeded 67 million tonnes, as illustrated in Figure 1 (Państwowy Instytut Geologiczny 2018). However, it should be taken into account that every ton of extraction is associated with generated waste and mine waters.Source: Państwowy Instytut Geologiczny, 2018

Mines constitute the main source of excessive salinity of the Vistula and Odra rivers by pumping out and introducing huge amounts of chloride and sulphates charges into rivers and streams. The level of the components of the sum of chlorides and sulphates in the flowing waters of the Silesian agglomeration is notoriously exceeded. The vast majority of mines in mining plants discharges mine water directly to the receiver. Most, therefore, rivers are contaminated, to which the saline waters are directly introduced (Policht-Latawiec & Kapica 2013). According to the statistical year of the Central Statistical Office, in 2016 the amount of

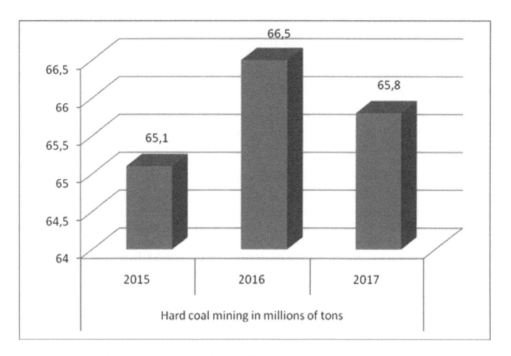

Figure 1.   Hard coal mining in the period 2015-2017.

saline water discharged in the province Silesian for surface water is 112 660 cubic decametres, and the amount of chloride and sulphate load in saline waters is 1 598 119 t/year, and discharged into inland waters is as high as 1 228 924 t/year, while only 369 195 t/year was utilized (GUS 2017).

According to data from the WIOŚ in Katowice, in 2017, in the Silesia Voivodship, the majority of industrial waste (96.9% of all waste generated in the voivodship) was generated by plants operating in the mining and quarrying (70.8%). Among the generated industrial waste, waste from the rinsing and purification process of minerals predominated – 20 703 900 Mg (65.4% of generated waste), fly ash from coal – 1 152 100 Mg (3.6%) and waste from flotation enrichment of coal – 1 087 600 Mg (3.4%) (Inspection of Environmental Protection, Provincial Inspectorate of Environmental Protection in Katowice 2018). Among the waste generated as a result of the exploitation of underground hard coal deposits, two basic types of mining waste should be distinguished: mining and processing waste. Mining wastes come directly from mining preparatory works and are characterized by high variability of petrographic composition. Their grain composition ranges from 0-500 mm. Processing wastes come from various processes of enriching raw material and are characterized by greater stability of the petrographic composition. Depending on the enrichment equipment used and the technology, the tailings are divided into: coarse-grained, fine-grained and flotation and other mud. The impact that mining waste may have on the aquatic environment is determined by many factors, including their physico-chemical properties, hydrological, geological and climatic conditions prevailing at their place of use, as well as the construction of the deposit. During the deposit of waste they are exposed to atmospheric conditions and take part in the natural cycle of water circulation. They also depend on the methods of use and securing the excavation and on the ability of the environment to self-purify. In the case of utilization of post-mining waste excavations for reclamation, the potential environmental pollution is mainly related to the presence of chloride and iron sulphides in the material, which as a result of the development of oxidation processes are a source of sulphate and chloride ions (Kugiel & Piekło

2012). Taking into account the identification of the causes of the salinity of rivers through-out the country, it can be noticed that this problem is mainly of a local nature, mainly related to the region of the Upper Silesian Industrial District. The main source of saline water is the water discharged from the mines, but the article analyses the results of moni-toring of surface waters discharged from the sites deposited post-mining waste.

3 RECLAMATION OF THE OPENCAST EXCAVATIONS ON THE EXAMPLE OF MACZKI-BÓR

3.1 *The method of reclamation of opencast excavations by filling with mining waste*

Mining activity of CTL Maczki-Bór S.A. in Sosnowiec is consists in the exploitation of sand deposits and the parallel implementation of the process of reclamation of post-mining excava-tions. The reclamation works are filled with mining waste from hard coal mines. They affect surface waters due to the location of backfilling sand deposits. The deposit lies in the fork of the Biała Przemsza River and its right-bank tributary - the Bobrek River. The deposit covers an area with a total area of over 558 ha, which has been divided into exploitation fields - the Bór Zachód field and the Bór Wschód field. Mine area dewatering is conducted in a gravitational manner with the participation of an extensive system of drainage ditches, which converge in the central part to the main channel. By this channel, the water is dis-charged into the sump of the main drainage, and from there to the Biała Przemsza River.

In the southern part of the Bór Zachód excavation, works related to the exploitation of sand are underway, and the remaining ones are subject to reclamation works aimed at leveling the basin of the excavation to the elevation of the surrounding area. In the area of about 200 ha, technical reclamation was completed. Land reclamation of the Bór Zachód field is intended for investments or as green areas. The remaining part of the degraded land surface exploitation of sand is covered by the reclamation process.

In the northern part of the Bór Wschód sand pit, the residual sand exploitation takes place. The area covered by the area of 220 hectares was covered by technical reclamation (from December 2004) aiming at the construction of a water reservoir with an area of 86.7 ha for recreational, retention or mixed purposes. Currently, the intended destination of the area is an industrial and investment objective with an accompanying green zone in the southern part.

Until October 2006, non-hazardous waste was delivered to the area of sandblast excava-tions: post-mining, energy (used for fire prevention) and neutral. Only mining waste from coal pre-treatment and enrichment as well as construction debris waste has been used for reclam-ation of sandblasting pits since October 2006. All used waste is non-hazardous waste. Before accepting waste for use in the reclamation of CTL Maczki-Bór S.A. requires suppliers to pro-vide information on their physicochemical properties.

During filling the basin of the Bór Zachód excavation, dozens of hard coal mines were sup-pliers of waste including hard coal mines:

- Zagłębie coal mines: Niwka Modrzewów Sosnowiec, Jan Kanty Jaworzno and Siersza;
- Katowice: Kleofas, Wujek, Staszic, Wesoła;
- Rybnik coal mines: Śląsk, Polska Wirek, Zabrze Bielszowice, Pokój;
- Vistula coal mines: Ziemowit, Piast, Czeczot, Janina,
- mines from the area of Ruda Śląska: Silesia, Poland, Wirek, Zabrze, Bielszowice, Pokój.

In the second stage of operation, waste from the mines was used to fill the excavation:

- Katowicki Holding Węglowy: Staszic, Wujek, Wesoła, Murcki,
- Kompania Węglowa: Piast, Ziemowit,
- Południowy Koncern Energetyczny: Janina, Sobieski.

These wastes are gangue rocks composed of clays and siltstones, and in a small part sand-stones with overgrown coal. In addition, the qualified "Haldex" aggregate from the mining waste decarburization processes carried out at the Extraction Waste Treatment Plant (ZPOW), located at CTL Maczki-Bór SA, is also used.

3.2 *Impact of leachate from the excavation fill on surface water*

The Biała Przemsza River flows along the southern border of the Mines Mining Plant CTL Maczki-Bór S.A. in the latitudinal direction from east to west. It is a tributary of the Czarna Przemsza, with which it connects in the "Jęzor" district on the site of the former Niwka-Modrzejów mine. From this point the river takes the name of Przemsza and flows to the south-east along the border of the "Jaworzno" OG. Drainage of headings CTL Maczki-Bór S. A. is guided by a system of ditches, converging in the middle part to the east-west main duct. With this channel, the water flowing into the outcrop is discharged to the area of BórZachód, and from there pumped to the Biała Przemsza River. The flow of water at the mouth of the channel to the river varies at the level of 0.38 m³/s.

In order to demonstrate the impact of water discharged from the "Maczki - Bór" CTL on the salinity of the Biała Przemsza River, the results of monitoring studies carried out in 5 research periods presented. The research included water sampling at the place of discharge from the grit chambers to Biała Przemsza and two points located about 400 m above and below the discharge from the CTL "Maczki - Bór" to Biała Przemsza. Table 1 presents the results of analyses.

With regard to limit values of indicators specified in the ordinance of the Minister of the Environment regarding the method of classifying the status of uniform surface water bodies and environmental quality standards for priority substances (defined in the Journal of Law from 2014 item 1482), in the drainage waters of the "Bór-Zachód" field, increased conductivity and suspension were observed, which was influenced by concentrations of chlorides and sulphates in the discharged water.

In terms of sulphates, the water discharged from the grit chamber to the river contributed meaningless changes in sulfate concentrations after the discharge. In 1 period, they indicated even smaller values than before the discharge, which resulted in an additional improvement in water quality. In terms of chlorides, water discharged from the grit chamber into the river also contributed small increase in chloride concentrations, without affecting the change in the water quality class. The water of the river, both before and after discharge, usually belong to the waters below class II. Only two single measurements in 2 and 3 periods indicated the quality of first class water before discharge, and after discharge these waters were of a quality below class II, so the discharge from the grit chamber deteriorated water quality.

It is important that the pollution load does not limit the process of self-purification of surface water, as the receiver is the next stage in the treatment of polluted waters. Most models used to analyze the impact of pollutant discharge on water quality assume complete mixing at or near the discharge point. This mixing process is rare, however, for the purposes of describing the state of the river, the assumption of complete mixing is sufficient. The speed of mixing wastewater with river water depends on its amount and the size and speed of water flow, as well as the width and depth of the watercourse. It is also very important to maintain oxygen conditions throughout the river, appropriate for living organisms receiving the receiver and an efficient self-purification process. Therefore, it is important not only to monitor the concentration of pollutants in treated wastewater, but also to determine their impact on the receiving water. In the event that grit chamber outlet worsen the quality of the receiving waters and the results indicate a reduction in water class (as in periods 2 and 3), it should be recommended to reduce the leachate flow. Using the formula for concentration of pollutants in the river after entering the outlet assuming complete mixing:

$$C = \frac{Q_r C_r + q_O C_O}{Q_r q_O}$$

Table 1. Results of analyses water above and below the grit chambers and discharged from the grit chambers to Biała Przemsza.

Analyses period	Place of sampling	Total suspension mg/l	Chloride mg/l	Sulphate mg/l	pH	Electric conductivity µS/cm
1	Biała Przemsza above the grit chamber	77	44	183	8,3	580
		8,2	46	233	7,4	792
	Water discharge from the grit chamber to Biała Przemsza	23	227	263	8,1	1641
		6	310	184	7,6	1943
	Biała Przemsza below the grit chamber	48	43	167	8,3	605
		8,4	72	202	7,7	896
2	Biała Przemsza above the grit chamber	86	19	50	7,6	282
		9,2	43	199	8,1	737
	Water discharge from the grit chamber to Biała Przemsza	20,4	261	219	7,6	1920
		5,6	231	266	7,7–8,3	1963
	Biała Przemsza below the grit chamber	108	28	58,8	7,6	283
		9,6	69	221	8,1	849
3	Biała Przemsza above the grit chamber	28,5	43	181	7,9	790
		19,4	39	178	8,0	772
	Water discharge from the grit chamber to Biała Przemsza	24,8	214	270	no data	1530
		7,8	300	180	no data	1910
	Biała Przemsza below the grit chamber	38,5	65	209	7,8	890
		30,3	72	230	8,0	860
4	Biała Przemsza above the grit chamber	36,5	52	200	7,9	797
		5	52	226	7,8	855
	Water discharge from the grit chamber to Biała Przemsza	6	262	228	no data	1510
		4	240	245	no data	1710
	Biała Przemsza below the grit chamber	28	41	201	7,8	884
		15	85	234	7,6	820
5	Biała Przemsza above the grit chamber	27	40	213	7,5	750
		12,4	40	227	7,9	870
	Water discharge from the grit chamber to Biała Przemsza	8	277	222	no data	1670
		7,2	220	320	no data	1620
	Biała Przemsza below the grit chamber	28	69	227	7,8	900
		14	56	227	8,0	968

Source: CTL Maczki-Bór S.A..

which: Q_r and q_O - river and outlet water flow [$m^3 \cdot s^{-1}$] C_r and C_O - concentration of a given component in the river water above the outlet and in the outlet water [$g \cdot m^{-3}$],
it is possible calculate the appropriate amount of leachate discharged from the grit chamber:

$$q_O = \frac{C_r}{C - \frac{C_O}{Q_r}}$$

Although the water discharged from the excavation drainage system meets the requirements of the water permit, it seems reasonable to increase the protection of the receiving waters, i.e. the river, by using an intelligent water drainage system.

Table 2 presents the results of physicochemical analyzes of water discharged to the Biała Przemsza in relation to the permissible values of pollutants contained in the Regulation of the Minister of Environment of November 7, 2019 on conditions to be met when introducing sewage into waters or to land, and on substances particularly harmful to the aquatic environment (Journal of Laws of 2019, item 2149). Analyses included determination of reaction, dry residue, total suspended solids, conductivity, chlorides, sulphates, bicarbonates, calcium, magnesium, general hardness (total calcium and magnesium), sodium, potassium, general iron, manganese, zinc and lead.

The quality of wastewater discharged from the grit chamber to Biała Przemsza, in terms of all analyzed indicators, meets the requirements of the Regulation of the Minister of the Environment of November 7, 2019, on conditions to be met when introducing wastewater into waters or soil, and on particularly harmful substances for the water environment (Journal of Laws of 2019, item 1311), for treated industrial wastewater. In this case, there are no indications to take any additional action.

Table 2. Results of analyses physicochemical parameters in water at the outlet of the grit chamber and limit values in treated industrial wastewater (according Journal of Laws of 2019, item 1311).

Parameters [unit]	Period 1	Period 2	Period 3	Period 4	Period 5	The highest permissible values of pollution
pH	8,1 7,6	7,6 7,7-8,3	No data No data	No data No data	No data No data	6,5-9,0
Electric conductivity [μS/cm]	1641 1943	1920 1963	1530 1910	1510 1460	1670 1620	No data
Dry residue [mg/l]	1020 1005	1408 1332	1384 1580	1200 1372	1500 1280	No data
COD-Cr [mgO$_2$/l]	27 27	57 <25	32 38	32 24	18 20	125
BOD-5 [mgO$_2$/l]	6,78 4,5	9,5 3	8 12,6	6,4 4,8	4,5 5,8	25
Ammonium nitrogen [mgN-NH$_4$/l]	0,24 0,21	0,4 0,84	0,46 0,7	1,0 1,09	1,06 2,61	10
Nitrate nitrogen [mgN-NO$_3$/l]	0,34 0,44	0,58 0,43	0,45 0,28	0,76 0,14	0,25 0,36	30
Total nitrogen N$_{total}$ [mgN/l]	1,6 4,2	3,6 21	2,06 3,9	2,6 8,8	4,96 4,62	30
Total phosphorus [mgP/l]	0,14 0,35	0,41 0,06	0,13 <0,10	<0,10 <0,10	<0,10 <0,10	2
Chloride [mgCl/l]	227 310	261 231	214 300	262 218	277 220	1000
Sulphate [mgSO$_4$/l]	263 184	219 266	270 180	228 266	222 320	500
Volatile phenols [mg/l]	0,01 <0,002	<0,002 <0,002	<0,002 <0,002	0,008 <0,002	<0,002 0,011	0,1
Fluoride [mg/l]	<0,10 7,6	<0,10 <0,10	<0,10 <0,10	<0,10 <0,10	<0,10 <0,10	25
Bicarbonates [mgHCO$_3$/l]	271 <0,10 317	268 323 346	311 280 337	305 366 492	342 336 407	No data
Calcium [mgCa/l]	144 486	148 85	148 88	146 137	70 116	No data
Magnesium [mgMg/l]	33,8 138	44,1 32,5	43,3 28,6	39,8 36,3	28 28,6	No data

(Continued)

Table 2. (Continued)

Parameters [unit]	Period 1	Period 2	Period 3	Period 4	Period 5	The highest permissible values of pollution
Sodium [mgNa/l]	179	199	222	202	220	800
	34,5	129	180	188	140	
Potassium [mgK/l]	8,86	11,2	10,4	15,2	5,78	80
	168	7,4	9,2	12,3	16	
Total iron [mgFe/l]	0,25	<0,010	<0,010	0,77	0,18	10
	7,82	<0,010	<0,010	0,85	0,026	
Manganese [mgMn/l]	0,7	0,57	0,43	0,71	0,28	No data
	0,58	0,13	<0,005	0,64	0,46	
Bar [mgBa/l]	0,14	0,12	0,11	0,13	0,05	2,0
	<0,010	0,1	0,06	0,12	0,18	
Total chrome [mgCr/l]	<0,005	<0,005	<0,005	<0,005	<0,005	0,5
	<0,005	<0,005	<0,005	<0,005	<0,005	
Zinc [mgZn/l]	0,059	<0,020	<0,020	<0,02	<0,020	2,0
	0,024	<0,020	<0,020	<0,020	<0,020	
Cadmium [mgCd/l]	<0,0010	<0,001	<0,0010	<0,0010	<0,001	0,4
	<0,001	<0,001	<0,001	<0,001	<0,0010	
Copper [mgCu/l]	<0,005	<0,005	<0,005	<0,005	0,0078	0,5
	0,018	<0,005	<0,005	<0,005	<0,005	
Lead [mgPb/l]	<0,010	<0,010	<0,010	<0,010	<0,010	0,5
	<0,010	<0,010	<0,0020	<0,010	<0,010	
Arsenic [mgAs/l]	<0,020	<0,020	<0,020	<0,020	<0,020	0,1
	<0,020	<0,020	<0,020	<0,020	<0,020	
Nickel [mgNi/l]	<0,010	<0,010	<0,010	<0,010	<0,010	0,5
	<0,010	<0,010	<0,010	<0,010	<0,010	
Petroleum compounds [mg/l]	<0,10	0,27	<0,10	0,18	0,13	15,0
	No data	0,14	<0,10	<0,10	<0,10	
Suspension [mg/l]	23	20,4	24,8	6	8	35

Source: CTL Maczki-Bór S.A..

4 CONCLUSIONS

Due to the poor water resources of the country and the poor quality of water in Poland, it is necessary to systematically strive to improve the condition of rivers in Poland. The degree of salinity of flowing waters in Poland presented in the article, illustrates the large scale of the problem that is the over-normative salinity of Polish rivers. The impact resulting from the discharging of salt water from hard coal mines to rivers in this region, causes contamination of other parts of the country, caused by a large amount of chloride and sulfate discharges along with the Vistula and Odra rivers. The high salinity of the Odra waters, which is the border between Poland and Germany, causes that the problem of over-normative salinity of waters flowing in Poland takes on an international character. Knowledge of the degree of water salinity is therefore very important for political, ecological and economic reasons. Bad water quality makes it impossible to use them for industrial, economic or recreational purposes. Determining the sources of water salinity is important due to the indication of appropriate remedies that can partially or completely eliminate the negative impact of saline waters on the aquatic environment. This is particularly important in the case of underground water discharging from hard coal mines, which are the main source of river salinity. Considering the large amount of salt water discharged by mines, for technical and economic reasons it is not possible to completely eliminate this problem, however, it is possible to partially avoid this problem by applying appropriate methods of management and utilization of these waters. In line with EU requirements, Polish mines are gradually implementing environmental protection systems. When planning a new investment, they use and implement possible environmental technologies, contributing to reducing the amount of pollutants introduced into the environment. In addition, the introduction of appropriate remedial actions related to water protection in many industrial plants, which, in addition to the mine, also have an impact on the salinity of water, may contribute to reducing the degree of water salinity in Poland. The example of water discharged from the excavation filled with mining waste given in the article indicates a low degree of deterioration of surface.

The most important aspect of depositing mine waste is minimizing the negative impact on the natural environment. One of the basic directives in force in the field of environmental protection is the precautionary principle that obliges entities that may have a negative impact on the environment to prevent the emergence of such impacts. In practice, it is not always possible to implement it, therefore, in the current system of environmental protection introduced a whole range of legal instruments intended to bring the environment to the proper state in case of negative changes in it.

Reclamation of excavations with mining waste causes impact on surface waters, bringing in sulphates, chlorides, iron and manganese, but generally water pumped from the Mielnik Maczki-Bór Mining Plant and discharged through the sandstone to the Biała Przemsza River in the light of the ordinance of the Minister of Environment of 7 November 2019 on the method of classification of the surface water bodies and environmental quality standards for priority substances slightly affect the quality of water in the Biała Przemsza River, allowing it, in the case of the majority of analyzed parameters, to preserve the chemical status of water as before their introduction to river.

In the event of an imminent threat, there is an obligation to immediately take preventive measures, including immediate control, containment, removal or reduction of pollution or other harmful factors e.g. by controlling leachate effluent flows.

ACKNOWLEDGEMENTS

The results of the research come from the internal data of the company CTL Maczki-Bór S. A., carried out in accordance with the decisions of the Marshal of the Silesian Voivodeship and Polish legislation. For the purposes of the article, the results of the research were compiled and gave rise to the conclusions.

BIBLIOGRAPHY

Chełmicki W., 2002. *Water. Resources, Degradation, Protection (Woda. Zasoby, Degradacja, Ochrona)*, Warszawa: Wydawnictwo Naukowe PWN *(in Polish)*.
Environmental monitoring in the area of the "Bór-Zachód" and "Bór-Wschód" grit chamber CTL "Maczki Bór" 2012 *(Monitoring środowiska w rejonie wyrobiska piasku „Bór-Zachód" i „ Bór-Wschód" CTL „Maczki Bór")* *(in Polish)*.
Environmental monitoring in the area of the "Bór-Zachód" and "Bór-Wschód" grit chamber CTL "Maczki Bór" 2013 *(Monitoring środowiska w rejonie wyrobiska piasku „Bór-Zachód" i „Bór-Wschód" CTL „Maczki Bór")* *(in Polish)*.
Environmental monitoring in the area of the "Bór-Zachód" and "Bór-Wschód" grit chamber CTL "Maczki Bór" 2014 *(Monitoring środowiska w rejonie wyrobiska piasku „Bór-Zachód" i „ Bór-Wschód" CTL „Maczki Bór")* *(in Polish)*.
Environmental monitoring in the area of the "Bór-Zachód" and "Bór-Wschód" grit chamber CTL "Maczki Bór" 2015 *(Monitoring środowiska w rejonie wyrobiska piasku „Bór-Zachód" i „ Bór-Wschód" CTL „Maczki Bór")* *(in Polish)*.
Environmental monitoring in the area of the "Bór-Zachód" and "Bór-Wschód" grit chamber CTL "Maczki Bór" 2016 *(Monitoring środowiska w rejonie wyrobiska piasku „Bór-Zachód" i „ Bór-Wschód" CTL „Maczki Bór")* *(in Polish)*.
Główny Urząd Statystyczny, 2018. *Environment 2017 (Ochrona środowiska 2017)*. online: http://stat.gov. pl/files/gfx/portalinformacyjny/pl/defaultaktualnosci/5484/1/18/1/ochrona_srodowiska_2017.pdf [access 15.11.2018].

Grynkiewicz-Bylina B., Majewski M., 2000. Monitoring of mine waters salinity *(Monitoring zasolenia wód kopalnianych)*, KARBO No. 2, pp. 60–65 *(in Polish)*.

Inspekcja Ochrony Środowiska Wojewódzki Inspektorat Ochrony Środowiska w Katowicach, 2018. *State of the environment in Śląskie Voivodeship in 2017 (Stan środowiska w województwie śląskim w 2017 roku)*, Katowice: Biblioteka Monitoringu Środowiska *(in Polish)*.

Kugiel M., Piekło R., 2012. *Directions of mining waste management in Haldex S.A., (Kierunki zagospodarowania odpadów wydobywczych w Haldex S.A.)*, Górnictwo i Geologia, T. 7 z. 1, pp. 133–145.

Lipiński K., 1987. *Protection of wateragainstsalinity (Ochrona wód przed zasoleniem)*, Ochrona Środowiska, Nr 521/1(31), pp. 7–11 *(in Polish)*.

Miller P., 2004. *Salinity of waters in Poland (Zasolenie wód w Polsce)*, Przyroda Polska, No. 10, pp. 24–25 *(in Polish)*.

Nocoń K., Nocoń W., 2011. *Flowing surface water of the Silesian agglomeration - problems and challenges (Płynące wody powierzchniowe aglomeracji śląskiej – problemy I wyzwania)*, Ochrona Środowiska, No. 3, pp. 26–31 *(in Polish)*.

Państwowy Instytut Geologiczny, 2018. *Balance of mineral resources in Poland at 31.12.2017 (Bilans zasobów złóż kopalin w Polsce według stanu na dzień 31.12.2017)*, online: https://www.lw.com.pl/pl,2,d1097, gospodarka_odpadami.html [access: 15.11.2018].

Minefill 2020-2021 – Hassani et al (eds)
© *2021 Taylor & Francis Group, London, ISBN 978-1-032-07203-6*

Properties of cemented backfill prepared on the basis of selected Coal Combustion Products (CCPs)

Piotr Pierzyna
Silesian University of Technology, Gliwice, Poland

SUMMARY: The article presents the results of research on the physico-mechanical properties of Cemented Paste Backfill mixtures (CPB) produced from low-calcium fly ash from a conventional boiler and from slag taken from a fluidized bed boiler with the addition of metallurgical cement in the amount of 5-15 wt%. Fly ash and slag were mixed in big quantities in various proportions. The tested mixtures with a constant concentration of solids to water of 62 wt% were characterized by a wide range of consistency: from paste to very liquid. The conducted research demonstrates that the best results were obtained for the mixtures with ash to slag ratio of 75:25 wt%.

Keywords: Cemented paste backfill, Coal Combustion Products, Mechanical property

1 INTRODUCTION

Wastes generated by power industry or mining industry have been widely used in civil engineering for many years, i.e. in road engineering, construction, land reclamation, as well as in mining. Wastes from the energy sector are of particular technological importance for Polish hard coal mining industry. They are most often used in the backfilling technology of goafs and to eliminate inactive workings, usually dog headings. And post-mining waste in the form of crushed gangue is used by Polish hard coal mines only in traditional hydraulic backfill, which is applied infrequently. In contrast, in ore mining industry they use both typical hydraulic backfill (for example KGHM Polska Miedź) and solidified backfill (for example ZG "Trzebionka"). In room-and-pillar systems used by ore mining industry, massive pillars, which were left behind to ensure stability, limit mining operations. The use of solidified backfill allows for safe extraction of ore remaining in the pillars and ensures long-term stability of the mined-out areas (Jung & Biswas, 2002). A typical hydraulic backfilling is made of sand. And the solidified backfill is usually made of dehydrated post-flotation waste and cement, forming the so-called cemented paste backfill (CPB). Other additives that are used in that type of backfilling are: lime, fly ash and metallurgical slag (Belem & Benzaazoua, 2004). The paste-type solidified backfills are usually pumped over large horizontal distances (1 km) using concrete pumps. However, more frequently the feeding of the mixtures to the void to be filled up is done by gravity, using mixtures of liquid consistency (Brackebusch, 1995).

DOI: 10.1201/9781003205906-29

2 MATERIAL AND METHODS

The following wastes from power-generation plants were used to produce CPB mixtures:

- fly ash from conventional boilers (code 10 01 02),
- slag from fluidized bed boilers (code 10 01 24).

The used wastes were from selected power plants located in southern Poland where hard coal is burned.

The metallurgical cement CEM III/A 42.5N - LH/HSR/NA was used as the binder. It is characterized by low content of alkali (NA) and high resistance to corrosive agents (HSR) (Shi, 2003). Its application states to reason due to the use of slag from fluidized bed boilers, which is generally characterized by a high content of sulfates (Qian et al., 2008), (Z. Zhang et al., 2012). Another aspect of using this type of cement involves the presence of acid and saline waters in the mines (PluTa, 2006), (Singh, 1988)(Skoczyńska-Gajda & Labus, 2012), which is a highly corrosive environment.

Fly ash and slag were mixed in large volume in various proportions, replacing one component with another with a step of 25 wt%. To the wastes and mixtures thereof, cement was added in the amount of 5, 10 and 15 wt%. In this way, ash-cement, ash-slag-cement and slag-cement mixtures were produced. In total, 15 CPB mixtures were investigated. Table 1 presents the proportioning of the mix for all CPB samples.

Dry components (fly ash, slag, cement) were homogenized in a mixer for 3 min before being combined with water. Then water was added and the whole was further stirred for 2 min. The prepared mixtures had a constant solids/water concentration of 62%. The proportions of water and solids (W/S ratio = 0.38) were selected so that the resulting mixtures had mainly a liquid consistency, as well as that of pastes. Liquid consistency results from the fact that the transport of mixtures from the surface to an underground working, in line with the conditions of Polish mining, takes place in a gravitational way (Palarski et al., 2014). The fluidity of the mixture determines, among others, their transport and penetration capabilities (Piotrowski & Mazurkiewicz, 2006) (Palarski et al., 2005).

The tests on the chemical composition in oxide form of ash and slag were carried out using the X-ray fluorescence (XRF) method, with the application of the spectrometer Epsilon 1 by PANalytical.

The tests of CPB mixtures were carried out in line with the backfill Standard PN-G-11011:1998 in force in Poland "Materials for solidified backfill and caulking of goafs.

Table 1. Cemented paste backfill (CPB) mixture formulations.

Proporcja Fly ash/Slag	Fly ash [wt%]	Slag [wt%]	CEM III [wt%]
-	0	95	5
1:3	23,75	71,25	5
2:2	47,5	47,5	5
3:1	71,25	23,75	5
-	95	0	5
-	0	90	10
1:3	22,5	67,5	10
2:2	45	45	10
3:1	67,5	22,5	10
-	90	0	10
-	0	85	15
1:3	21,25	63,75	15
2:2	42,5	42,5	15
3:1	63,75	21,25	15
-	85	0	15

Source: own study.

Requirements and tests" (Norma PN-G-11011:1998). The above Standard should meet the following requirements:

1. flowability test - minimum 90 mm,
2. density - minimum 1.2 Mg/m^3,
3. volume of supernatant water - maximum 7%,
4. setting time - individually determined for local conditions of using solidified backfill,
5. compressive strength - minimum 0.5 MPa,
6. slaking - maximum 20%.

The paper presents the basic parameters such as melt flowability and compressive strength. In order to map typical climatic conditions occurring in underground mine workings, the prepared samples of mixtures were seasoned in the LTB 650 RV climate chamber by Elbanton, the Netherlands. The seasoning conditions in the chamber were as follows: temperature 25°C± 1°C, humidity 90% ± 2%.

The consistency of tested mixtures was determined by flowability tests. To measure the consistency of paste-type backfill mixtures, the cone slump test method is often used, which consists in measuring the decrease in the height of the mixture accumulated in the measuring cylinder (cone) (Clayton et al., 2003). This test is of no use for mixtures with liquid consistency. In such a case, all the mixture accumulated in the measuring cylinder flows out of it and does not create the typical slump cone, but it creates a flow cake. The diameter of the resulting flow cake is a measure of flowability test, as illustrated in Figure 1. Thus, the measurement of consistency using the cone slump method after obtaining the maximum (limit) slump height corresponding to the height of the used measuring cylinder can be carried out just by measuring the flowability. The flowability tests were carried out using a cone of the dimensions: top D = 35 mm, base D = 63 mm, height H = 60 mm, which is in line with the Polish backfill Standard PN-G-11011:1998.

A peculiar (atypical) parameter for the Polish mining area is setting time. For this reason, it requires a broader presentation. According to the Polish standard PN-G 11011:1998 (Norma PN-G-11011:1998, n.d.) setting time is the time required to obtain a load capacity of 0.51 MPa. The measurement is made with a modified Vicat apparatus. It consists in measuring the

Figure 1.   Exemplary flow cakes being a measure of flowability.
Source: own study.

immersion depth of a 1 cm^2 penetrator into the sample under the specified weight exerted by the moving part of the apparatus. The sample of the tested material is placed in an identical ring which is used in the traditional cement setting time test. The setting time is assumed to be the time after which the penetrator has immersed in the sample to a depth not lower than 3 mm with a moving part weight of 5.0 kg, which corresponds to the load capacity of 0.51 MPa.

And slaking is the percentage change in compressive strength caused by the impact of water. Usually, it refers to its loss. For this purpose, the samples of the tested CPB mixtures, after 28 days of seasoning in a humid air environment (climatic chamber) are subjected to water impact by immersing them for 24 hours. After that, the hydrated samples are subjected to a compressive strength test which is referenced to the strength obtained after 28 days of seasoning in the humid air environment. The compressive strength tests and slaking tests were tested on cylindrical samples of the height and diameter of 100 mm.

3 RESULTS AND DISCUSSION

3.1 *Chemical composition*

The chemical composition in oxide form applied in the tests of ash from a conventional boiler and slag from a fluidized bed boiler is presented in Table 2.

The ash and slag used for the tests differ primarily in the content of CaO and SO$_3$, and to a lesser extent in the content of Al$_2$O$_3$. Ash contains more aluminum. In contrast, slag contains much more lime, in particular its free form, as well as more sulfates. This is due to the fact that the slag comes from a fluidized bed boiler, where the dry flue gas desulphurization method is used. LOI is also very diverse. The content of free lime, sulfur or LOI in the ashes increases their demand for water, i.e. their water demand increases (Tsimas & Moutsatsou-Tsima, 2005). Bearing in mind the guidelines of the American ASTM C618 standard in terms of the content of silicon, aluminum and iron, fly ash can be classified into class F (LCFA) (ASTM, 2012). Taking into account the content of lime, including its free form in relation to the combined content of silicon, aluminum and iron, i.e. the hydraulic module (M$_h$) (Sitko, 2016) (Wirska-Parachoniak, 1968) amounting to 0.23, we can state that this slag has low binding properties (Pierzyna, 2017). For example, for Portland cement, the hydraulic module is in the range 1.7-2.4 (Nakano et al., 2007)

Table 2. Chemical composition of ash from a conventional boiler and slag from a fluidized bed boiler.

Chemical ingredient	Fly ash	Slag
	[% by mass]	[% by mass]
SiO$_2$	50,61	43,83
Al$_2$O$_3$	26,03	19,85
Fe$_2$O$_3$	6,30	4,79
CaO	3,20	15,68
MgO	1,77	1,74
Na$_2$O	0,93	1,09
K$_2$O	2,89	2,03
SO$_3$	0,70	9,08
TiO$_2$	1,09	0,68
P$_2$O$_5$	0,47	0,09
CaO free	0,21	3,21
LOI	5,17	0,77
Total	99,16	99,63

Source: own study.

3.2 Grain size

The grain size of ash and slag applied in the study was determined using the traditional sieve analysis as presented in Figure 2.

The grain size of fly ash is below 1 mm, with approximately 80% of the fraction being below 0.063 mm. For example, ash with similar grain size was used in the studies presented in (X. Zhang et al., 2017). And the slag used in the studies had the grain size lower than 4 mm. However, the share of fraction below 0.063mm in the slag is small compared to ash, being below 2%. Taking into account the fraction below 1mm, which is the total grain size of ash, we can observe that in the slag it is around 83%. Yet, the content of grains below 0.125 mm in the ash is at the level of 90%, while in the slag it is about 10%. According to the granulometric classification recommended by the USDA (United States Department of Agriculture) (NRCS, 1993) or WRB (World Reference Base for Soil Resources) (WRB, 2015), the slag presented in its entirety can be classified as sand fraction. The said waste in the Polish nomenclature relating to the Waste Catalog is referred to as bottom sand (code 10 01 24). And the fly ash according to the above classifications can be classified to a significant extent as dusty fraction (about 75-80%), and the rest as sandy fraction.

3.3 Flowability of mixtures

For gravity transport, a sufficiently fluid consistency is required, which can be characterized by the flowability test of the mixture. In the conditions of Polish mines, and in terms of installations and pipelines applied in them, the gravitationally transportable mixtures are the ones that reach the flowability higher than about 140 mm (Jendruś et al., 2008). It is often 300 mm or higher. The fluidity of the mixture, and thus its flowability, is related to the transport distance as well as the range of its spread in the excavation and the penetration capacity into the

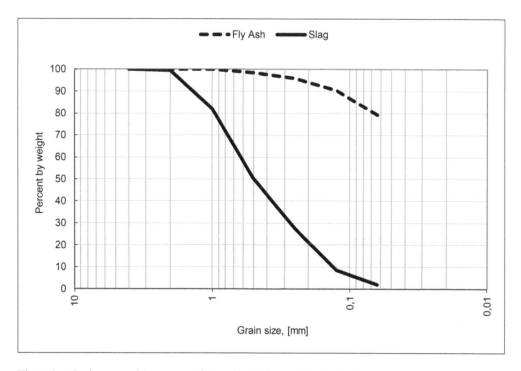

Figure 2. Grain composition curves of the ash and slag used in the study.
Source: own study.

goaf debris (Palarski et al., 2005; Piotrowki & Mazurkiewicz, 2006; Yin et al., 2012). Along with the change in the flowability of the mixtures, their rheological properties change: first of all the yield strength and density (Plewa et al., 2006; Strozik, 2010, 2018). These parameters, in turn, depend on the chemical composition, specific gravity of the particles and their size, i.e. grain size of individual components, and ultimately that of the CPB mixture (Clayton et al., 2003). The flowability of the tested CPB mixtures having the constant index w/s = 0.38 is presented in Table 3.

The flowability was determined for all mixtures except for two slag-cement mixtures with the addition of 5 and 10 wt% cement. In that case, no regular shape of spread cake was obtained, in contrast to the slag-cement mixture with the share of 15 wt% cement, as illustrated in Figure 3. The slag used in the test is characterized by grain size typical for the sandy fraction. Thus, its sedimentation is immediate, which is obvious. The share of cement in the amount of 5 and 10 wt% was insufficient to form a suspension which would not undergo sedimentation in the short time of flowability testing. The suitable suspension which could be subjected to flowability testing was ensured by the share of 15 wt% cement, with 85 wt% slag.

The conducted studies demonstrate that the flowability of the tested mixtures depends primarily on the proportion of slag and ash, and to a lower extent on the share of cement. Depending on the proportion of slag and ash, the rise of the share of cement from 5 wt% to 15wt% brings about a slight decrease or increase in flowability, by a maximum of 15 mm. However, taking into account the impact of the slag and ash ratio, we can state that the rise of ash share brings about a very significant decrease of flowability from about 215 mm to about 140 mm. In the case of ash-cement mixtures, it may be too exaggerated to state that they have liquid consistency, which is measured by flowability. The obtained spread cakes have regular shapes (Figure 4). Yet, referring to their dimensions: diameter about 100 mm and height about 20 mm, it would be more appropriate to state that these mixtures have paste-

Table 3. Flowability of the tested CPB mixtures.

Fly ash [wt%]	Slag [wt%]	Flowability [mm]		
		5% CEM	10% CEM	15%CEM
0	100	not determined	not determined	217
25	75	217	213	215
50	50	195	190	185
75	25	141	145	153
100	0	lack of flowability (94 mm) slump height=36 mm pasta	lack of flowability (100 mm) slump height=39 mm pasta	lack of flowability (102 mm) slump height=40 mm pasta

Figure 3. Obtained shapes of spread cakes of the ash-cement mixtures.
Source: own study.

Figure 4. Uzyskane kształty placków rozpływowych mieszanin popiołowo-cementowych.
Source: own study.

like consistency. In that case it would be appropriate to use the slump cone test as a measure of consistency. The cone slump height of ash-cement mixtures, depending on the cement, was changing a little only, in the range from 94 mm to 102 mm.

3.4 *Compressive strength*

The compressive strength test was carried out after 7, 14, 28 and 90 days of seasoning of the samples in the climactic chamber. The compressive strength of CPB mixtures depending on the proportions of ash and slag and the proportion of cement are presented in Figure 5.

The presented research results demonstrate that regardless of the curing time, slag-cement mixtures (without ash) are always characterized by the lowest strength, while the highest strength have the mixtures with the ash to slag ratio of 75:25 wt%. And the ash-cement mixtures (without the share of slag) have similar strengths to the slag-cement mixtures. Figure 6

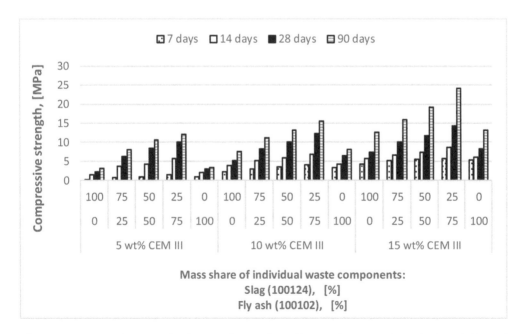

Figure 5. Compressive strength of CPB mixtures depending on the proportions of ash and slag and cement share.
Source: own study.

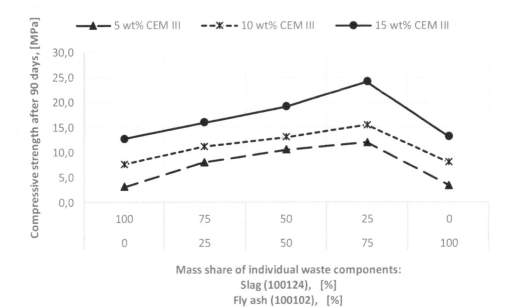

Figure 6. Compressive strength of CPB mixtures after 90 days, depending on proportion of ash to slag and the share of cement.

Source: own study.

presents the compressive strength of the investigated CPB mixtures after 90 days of curing, depending on the ash to slag ratio and the share of cement.

Having analyzed the obtained results of strength tests, depending on the proportions of ash and slag, we must say that the rise of the share of ash within the range of 0-75 wt% results in a systematic increase in strength. The rise of ash content to 100 wt% (exclusion of slag) results in a drastic decrease of strength to the level achieved by the slag-cement mixtures.

Such a behavior is characteristic for all cement shares, and it occurs in all research time periods (Figure 5). The mixtures of the ash to slag ratio of 75:25 wt% with respect to two-component mixtures (slag-cement and ash-cement) reached after 90 days about 4 times (for 5 wt% cement) and about 2 times (for 10 and 15 wt% cement) higher compressive strength.

4 CONCLUSIONS

The carried out studies demonstrate the impact of ash and slag proportions and the share of cement in the range of 5-15 wt% on the mechanical parameters of the developed CPB mixtures of the mass share of water to solids being W/S = 0.38.

As to compressive strength, the rise of the share of fly ash at the expense of slag in the range up to 75 wt% results in its significant increase. But further rise of the share of ash content to 100 wt% (exclusion of slag) results in a drastic decrease in strength to the level achieved by the slag-cement mixtures (without the share of ash).

Two-component mixtures achieve similar compressive strengths, but the slag-cement ones are characterized by almost two times higher flowability than ash-cement mixtures.

The conducted studies show that the best results were obtained for the mixtures with the ash to slag ratio of 75:25 wt%, which is consistent with the results of the work (Z. Zhang et al., 2012). They are characterized by the highest compressive strength after 90 days, reaching 24 MPa.

The requirements set by the Polish standard PN-G 11011: 1998 for the materials regarded as solidified backfill (CPB) within the scope of the described physical parameters are met by all the tested mixtures.

BIBLIOGRAPHY

ASTM, C. (2012). *Standard specification for coal fly ash and raw or calcined natural pozzolan for use in concrete*.

Belem, T., & Benzaazoua, M. (2004). An overview on the use of paste backfill technology as a ground support method in cut-and-fill mines. *Proceedings of the 5th Int. Symp. on Ground Support in Mining and Underground Construction. Villaescusa & Potvin (Eds.)*, 28–30.

Brackebusch, F. W. (1995). Basics of paste backfill systems. *International Journal of Rock Mechanics and Mining Sciences and Geomechanics Abstracts*, 3(32), 122A.

Clayton, S., Grice, T. G., & Boger, D. V. (2003). Analysis of the slump test for on-site yield stress measurement of mineral suspensions. *International Journal of Mineral Processing*, 70(1), 3–21. https://doi.org/10.1016/S0301-7516(02)00148-5

Jendruś, R., Lutyński, M., Pierzyna, S., & Strozik, G. (2008). Flow and fill properties of fly ash-water slurries for grouting of cavings in underground longwall mining. *Th International on Transport-Sedimentation of Solid Particle*, 232–240.

Jung, S. J., & Biswas, K. (2002). Review of Current High Density Paste Fill and its Technology. *Mineral Resources Engineering*, 11(2), 165. https://doi.org/10.1142/S0950609802000926

Nakano, T., Yokoyama, S., Uchida, S., & Maki, I. (2007). Badania podstawowe związane z produkcją cementu z popiołu ze spalania miejskich odpadów stałych. Cz.1: Charakterystyka popiołów i cementów specjalnych. *Cement Wapno Beton, R. 12/74, nr 4*, 187–192.

Norma PN-G-11011:1998. *Materiały do podsadzki zestalanej i doszczelniania zrobów. Wymagania i badania*.

NRCS, U. (1993). Soil survey division staff (1993) soil survey manual. Soil conservation service. *US Department of Agriculture Handbook*, 18, 315.

Palarski, J., Plewa, F., Pierzyna, P., & Zając, A. (2005). Właściwości zawiesin z materiałów odpadowych z dodatkiem środka wiążącego w aspekcie możliwości ich wykorzystania do likwidacji zawodnionych szybów. *Górnictwo i Geoinżynieria, R. 29, z. 4*, 139–144.

Palarski, J., Plewa, F., & Strozik, G. (2014). Filling of voids in coal longwall mining with caving–technical, environmental and safety aspects. *Proceedings of the Eleventh International Symposium on Mining with Backfill*, 483–491.

Pierzyna, P. (2017). Disposal of coal combustion wastes in the hydraulic backfill process. *IOP Conference Series: Materials Science and Engineering*, 268, 012011. https://doi.org/10.1088/1757-899X/268/1/012011

Piotrowski, Z., & Mazurkiewicz, M. (2006). Chłonność doszczelnianych zrobów zawałowych. *Górnictwo i Geoinżynieria, R. 30, z. 3*, 37–46.

Plewa, F., Strozik, G., & Jendruś, R. (2006). Wpływ własności reologicznych mieszanin wybranych odpadów drobnofrakcyjnych na parametry ich przepływu przez gruzowisko zawałowe w świetle wyników badań laboratoryjnych. *Zeszyty Naukowe. Górnictwo/Politechnika Śląska, z. 274*, 123–136.

PluTa, I. (2006). Wykorzystanie metod i technologii górniczych do oczyszczania wód kopalnianych. *Wiadomości Górnicze*, 57(10), 529–537.

Qian, J., Zheng, H., Song, Y., Wang, Z., & Ji, X. (2008). Special Properties of Fly Ash and Slag of Fluidized Bed Coal Combustion. *Journal of the Chinese Ceramic Society*, 10.

Shi, C. (2003). Corrosion resistance of alkali-activated slag cement. *Advances in Cement Research*, 15(2), 77–81. https://doi.org/10.1680/adcr.2003.15.2.77

Singh, G. (1988). Impact of coal mining on mine water quality. *International Journal of Mine Water*, 7(3), 49–59.

Sitko, J. (2016). Modernizacja technologii zagospodarowania odpadów hutniczych. *Systemy Wspomagania w Inżynierii Produkcji*. http://yadda.icm.edu.pl/baztech/element/bwmeta1.element.baztech-cd63ed9c-7d6c-481c-b020-087f3b4123d7/c/sitko_swzp_2016_2.pdf

Skoczyńska-Gajda, S., & Labus, K. (2012). Przegląd biernych metod oczyszczania kwaśnych wód kopalnianych. *Przegląd Górniczy*, 68.

Strozik, G. (2010). Ocena własności transportowych i migracyjnych mieszanin drobnofrakcyjnych do doszczelniania zrobów zawałowych w aspekcie ich własności reologicznych. *Cuprum : czasopismo naukowo-techniczne górnictwa rud, nr 2*, 95–109.

Strozik, Grzegorz. (2018). Influence of pipe roughness and coating built-up on pipe walls on the flow of solidifying Non-newtonian fly ash-water mixtures in hydraulic transport systems in coal mines. *IOP Conference Series: Earth and Environmental Science*, *174*, 012015. https://doi.org/10.1088/1755-1315/174/1/012015

Tsimas, S., & Moutsatsou-Tsima, A. (2005). High-calcium fly ash as the fourth constituent in concrete: Problems, solutions and perspectives. *Cement and Concrete Composites*, *27*(2), 231–237. https://doi.org/10.1016/j.cemconcomp.2004.02.012

Wirska-Parachoniak, M. (1968). Z historii wiążących materiałów budowlanych. *Ochrona Zabytków*, *4*, 17–23.

WRB, I. U. of S. S. W. G. (2015). World reference base for soil resources 2014 (update 2015), international soil classification system for naming soils and creating legends for soil maps. *World Soil Resources Reports*, *FAO, Rome*.

Yin, S., Wu, A., Hu, K., Wang, Y., & Zhang, Y. (2012). The effect of solid components on the rheological and mechanical properties of cemented paste backfill. *Minerals Engineering*, *35*, 61–66. https://doi.org/10.1016/j.mineng.2012.04.008

Zhang, X., Lin, J., Liu, J., Li, F., & Pang, Z. (2017). Investigation of Hydraulic-Mechanical Properties of Paste Backfill Containing Coal Gangue-Fly Ash and Its Application in an Underground Coal Mine. *Energies*, *10*(9), 1309. https://doi.org/10.3390/en10091309

Zhang, Z., Qian, J., You, C., & Hu, C. (2012). Use of circulating fluidized bed combustion fly ash and slag in autoclaved brick. *Construction and Building Materials*, *35*, 109–116.

Minefill 2020-2021 – Hassani et al (eds)
© *2021 Taylor & Francis Group, London, ISBN 978-1-032-07203-6*

Impact involving the sealing degree of caving goaf with fine-fraction hydraulic mixtures on the ventilation parameters of longwall headings

Marcin Popczyk & Dariusz Musioł
Silesian University of Technology

SUMMARY: The most common method to eliminate post-mining voids created after hard coal mining consists in natural filling them with rock rubble from the strata forming the roof the mining heading. Such rubble is characterized by free spaces that allow uncontrolled flow of air. The said flow is unfavorable due to disturbances in the distribution of airflow in the ventilation network, and it poses the risk of endogenous fire. One way to reduce such adverse phenomena is to fill the rubble with fine-fraction hydraulic mixtures, most frequently fly ash-water mixtures. The first part of the paper presents the method for determining the theoretical porosity of caving rubble. In the second part, the possibility of various degree of sealing of the numerically modelled caving rubble with a fine-fraction hydromixture was investigated in terms of assessing its impact on the ventilation parameters of a longwall working ventilated with the "Y" method. The presented numerical model was used to calculate the airflow distribution at various filling degrees of the rubble. The obtained knowledge involving the changes in the airflow distribution parameters, depending on the sealing degree of the rubble allows to define the possibility to slow down or to stop the development process of endogenous fire. In addition, it also allows to forecast the ventilation conditions of workings in the longwall area, limiting the development of other ventilation-related hazards, such as methane or climate hazards.

Keywords: gob grouting, fine-grained slurry, ventilation network, numerical modelling, air flow

1 INTRODUCTION

Underground coal mining is associated with the formation of post-mining voids, which often have considerable volumes. They are most commonly eliminated by means of natural filling with rock rubble from the strata forming the roof of the mining heading. The void with such rubble is characterized by the fact that it does not completely fill up the buried space, creating so-called goaf (caving) area with a certain volume of spaces. This allows an uncontrolled flow of air coming from headings with active ventilation, adjacent to the longwall panel, through the goaf. Such an airflow through the caving goaf is unfavorable for at least two reasons. Firstly, it causes disturbances in the distribution of air in the ventilation network of the longwall area, and secondly, it creates a risk of endogenous fire due to possible low-temperature oxidation of coal residues left over in the caving goaf.

One of the ways to reduce such unfavorable phenomena consists in filling up the free spaces of the caving rubble, which is technologically referred to as sealing of caving goaf, with fine-grained hydraulic mixtures produced and transported from the surface area through

DOI: 10.1201/9781003205906-30

a network of pipelines. Currently, the sealing process of caving goaf with fine-fraction power production waste material, in particular fly ash, is one of the most effective methods of their reclamation, which allows to limit fire hazard, improve ventilation conditions, as well as reduce deformation of the surface area and rock mass. The amount of fly ash that can be fed into caving goaf depends on many factors, among which the most important are: the type and porosity of caving goaf, the height of caving, the type and properties of roof layers, thickness of the mined-out bed, tightening degree and accessibility of the goafs, sealing method of the goafs and the migration properties of fly ash-water mixture.

In fact, there is no technical possibility to fill 100% of free spaces in the caving rubble with fine fraction waste due to insufficient amount of sealing material that can be daily fed into the rubble, the change in porosity of the rubble over time effected by the impact of the weight of the overlying rockmass and free uncontrolled flow of hydraulic mixture through the caving rubble. As a result, we are facing an uncontrolled airflow through the void, unfilled spaces in the caving rubble, which adversely affect the airflow distribution in the underground ventilation network, and the extent of this airflow depends on the porosity of the rubble and on its sealing degree with fine-fraction material.

2 THEORETICAL FOUNDATION FOR THE DETERMINATION OF THE POROSITY (ABSORBENCY) OF CAVING GOAFS

The presented method of theoretical determination of the absorbency of caving goafs allows with sufficient accuracy in terms of mining practice to determine the volume of free spaces that can be filled with fine-fraction material.

The volume of fine-fraction power-generation wastes, e.g. fly ash that can be introduced into caving goafs, depends on:

– the type of rocks forming the caving rubble,
– porosity of caving rubble,
– type and properties of overlying layers,
– the height of the resulting caving,
– the thickness and dip of the mined-out bed,
– the extent to which the rubble is tightened under the weight of the overlying rocks,
– physical access to goafs which may have impact on the selection of technology to carry out the process,
– migration and penetration properties of fine fraction hydromixtures,
– execution technology for the sealing process of caving goafs.

An overview illustration of the sealing process of caving goafs of a selected post-mining area of a seam deposit, depending on its dip is presented in Figures 1 and 2.

Theoretically, the volume of fine-fraction material that can be introduced into caving goafs should be equal to the volume of free spaces resulting from the porosity of the caving rubble. This porosity depends on the loosening coefficient k_r, which, depending on the rocks forming the roof layers over the mined out deposit, is changing within the range of $k_r = 1.15 - 1.35$ (Mysłek Z. 1999, Piotrowski Z., Mazurkiewicz M. 2006). The volume of free spaces which can be filled, up to the height of the mined-out deposit, can be determined from the difference between the volume of the mined-out deposit and the volume of roof rockmass forming the caving rubble:

$$V_{pz} = V_p - V_z, \ [\text{m}^3] \tag{1}$$

$$Vz = \frac{V_p}{k_r}, [\text{m}^3] \tag{2}$$

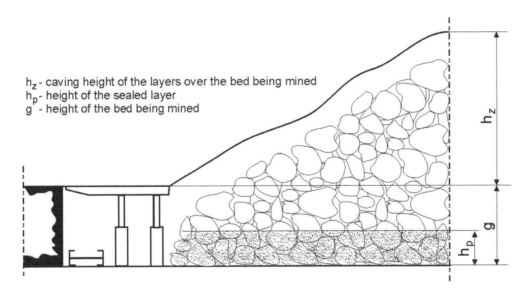

h_z - caving height of the layers over the bed being mined
h_p - height of the sealed layer
g - height of the bed being mined

Figure 1.　Sealing schematic of horizontal caving goaf with fine-fraction hydromixture.

h_z - caving height of the layers over the bed being mined
h_{pmax} - maximum height of the sealed layer
g_p - vertical height of the dipped bed
a - dip of the bed

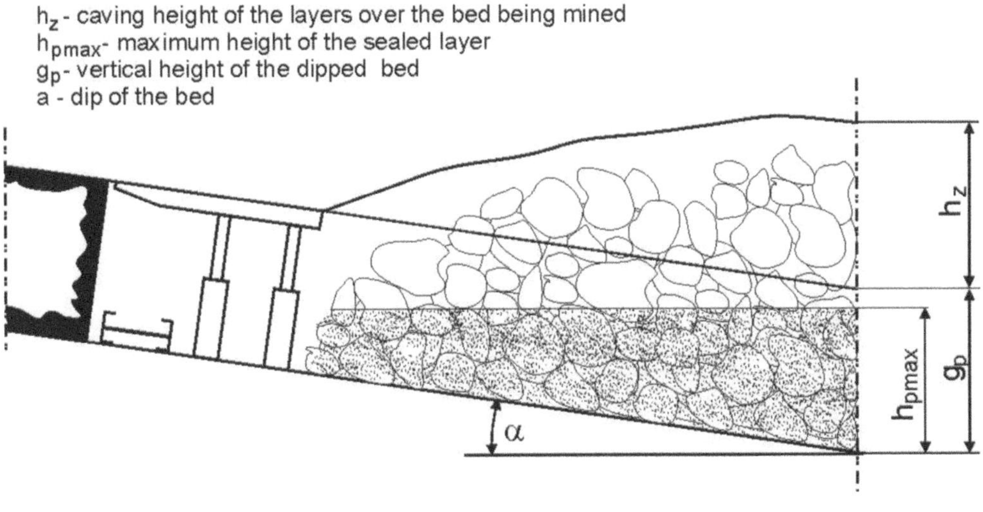

Figure 2.　Sealing schematic of dipped caving goaf with fine-fraction hydromixture.

where:
V_p – volume of the mined-out bed, m³,
V_z – volume of floor rock mass forming the caving rubble, m³,
k_r – rock loosening coefficient.
Thus, the volume of the voids that can be filled up to the height of the mined-out deposit is:

$$V_{pz} = V_p - \frac{V_p}{k_r} = V_p \left(1 - \frac{1}{k_r}\right), [\text{m}^3] \qquad (3)$$

And the dry mass of fine-fraction material that can be introduced into the free spaces of the goaf is:

$$m_p = V_{pz} \cdot \rho_p = V_p \left(1 - \frac{1}{k_r}\right)\rho_p, [\text{Mg}] \tag{4}$$

where:

ρ_p – the density of the fine-fraction material compacted after the draining of water, Mg/m^3

The relative amount of the fed fly ash in relation to the amount of extracted coal will be:

$$\frac{m_p}{m_w} = \frac{V_p \left(1 - \frac{1}{k_r}\right)\rho_p}{V_p \rho_w} = \frac{\left(1 - \frac{1}{k_r}\right)\rho_p}{\rho_w} [\text{Mg/Mg}] \tag{5}$$

The filling degree of caving goaf with fine-fraction hydromixture is strictly dependent on the feeding method of the hydromixture. Depending on the filling degree of the goaf, the amount of hydromixture that can be introduced in relation to the amount of extracted coal in line with the notation in Figure 1 will be:

$$\frac{m_p}{m_w} = \frac{h_p \left(1 - \frac{1}{k_r}\right)\rho_p}{g\rho_w} [\text{Mg/Mg}] \tag{6}$$

where:

h_p – height of the sealed layer of caving rubble, m

g – thickness of the extracted coal bed, m,

k_r – loosening coefficient of the caving,

ρ_p – bulk density of the fine-fraction material in compacted state, Mg/m^3

ρ_w – specific density of coal, Mg/m^3.

In the case when the height of the sealed caving layer is equal to the thickness of the extracted coal bed, i.e. $h_p = g$, the formula (6) has the following form:

$$\frac{m_p}{m_w} = h_p \left(1 - \frac{1}{k_r}\right)\frac{\rho_p}{\rho_w} [\text{Mg/Mg}] \tag{7}$$

However, when the hydromixture fills up all voids of the caving rubble, i.e. when

$$h_p = g + h_z \tag{8}$$

where: h_z – caving height, m, equals

$$h_z = \frac{g}{k_r - 1} [m] \tag{9}$$

the expression (6) takes the following form:

$$\frac{m_p}{m_w} = \frac{(g + h_z)\left(1 - \frac{1}{k_r}\right)\rho_p}{g\rho_w} = \frac{\left(g + \frac{g}{k_r-1}\right)\left(1 - \frac{1}{k_r}\right)\rho_p}{g\rho_w} [\text{Mg/Mg}] \tag{10}$$

Using the relations given above, we can determine the theoretical absorbency of caving goaf for the extraction of coal beds.

3 ABSORBENCY ANALYSIS OF CAVING GOAFS FOR SELECTED MINING CONDITIONS

In order to verify the derived theoretical relations, we carried out a comparative analysis of the absorbency of caving goaf of an exemplary bed of the thickness of 1.7-1.8 m, with a slight dip of bed (several degrees), and an average loosening degree of the goaf adopted at the level of 1.25 with the use of sealing technology of caving goafs. With such assumptions, we can conclude that the height of the sealed layer of the goaf will not exceed the thickness of the extracted bed layer. The calculations carried out on the basis of formula (10) demonstrate that for the given conditions, the theoretical absorbency of goaf will be about 170 kg of fly ash per 1Mg of the extracted coal.

In the case of a completely tightened goaf, i.e. when the subsidence trough becomes fully developed and the subsidence coefficient reaches the maximum value, i.e. a = 0.7÷0.8, then the amount of fly ash possible to be deposited will decrease to 20-30% of the initial value for not tightened goaf. It is estimated that with the subsidence coefficient a = 0.7, depending on the thickness of the layer being sealed and on the loosening coefficient of the caving for the analyzed operating conditions, it will be within 9 to 257 kg/Mg of the extracted coal. With the height of the layer subject to sealing equal to the thickness of the mined bed, the absorbency of the goaf will be about 51 kg/Mg. However, for the subsidence coefficient a = 0.8, the absorbency of the goaf will be changing within the range from 6 to 171 kg/Mg.

With the 50% tightening of caving goaf and with the subsidence coefficient within the range of a=0.7÷0.8, the maximum absorbency of the goaf is about 60÷65% of the absorbency of the non-tightened goaf. The amount of fly ash that can be deposited in such conditions, depending on the height of the sealed layer and the loosening coefficient of goaf area for a=0.7 will be from 20 to 557 kg/Mg and for the coefficient a=0.8 it will be from 18 to 514 kg/Mg of the extracted coal.

Considering the above and taking into account coal density at the level of 1.3 Mg/m^3 and the conversion factor of 1Mg of ash as 0.8 m^3 of hydromixture fed into the goaf area, we can estimate that the theoretical porosity of the caving rubble in the first stage after the advance of the longwall and before the acquisition of the subsidence coefficient of a=0.7-0.8 can be estimated at the level of 8 to 15% with respect to the extracted volume. For further considerations of the model of airflow distribution, the porosity of the rubble was assumed to be at the level of 10%.

4 SIMULATION OF AIRFLOW MIGRATION THROUGH THE CAVING GOAF

The application of the sealing of caving goafs in the case of the potential fire hazards is an indispensable element of fire prevention. Hard coal mines often use hydromixtures made on the basis of fine-fraction power generation wastes (fly ash), which are introduced into caving areas from the excavation face of the mining front or from the return airway. This allows to reduce air migration through caving goaf, and thus to reduce the inflow of oxygen to it, which reduces the risk of endogenous fire and extends the incubation time of the fire or even prevents the initiation of this process. The effectiveness of the sealing process of caving goaf can be assessed, among others, by measuring the volume of airflow at the inlet and outlet of the longwall.

In order to assess the impact of caving goaf sealing on the ventilation parameters of the longwall heading, a model of the ventilation area of the excavated longwall ventilated by the "Y" method was developed (Musioł D. 2009), with the refreshment of unmined coal in the return airway and the discharge of the used air along the caving goaf. The diagram of the model is presented in Figure 3.

For the calculations involving airflow distribution in the ventilation network in the vicinity of the longwall area, the VENTGRAPH program (Dziurzyński W. et. al. 2010) was applied – a system of programs for ventilation engineers for the analysis of ventilation network in a coal mine in normal and emergency conditions, which has been developed by the Institute of Rock Mechanics of the Polish Academy of Sciences in Krakow. The said program enables digital modeling of airflow and the flow of other gases through mine headings and offers multi-variant

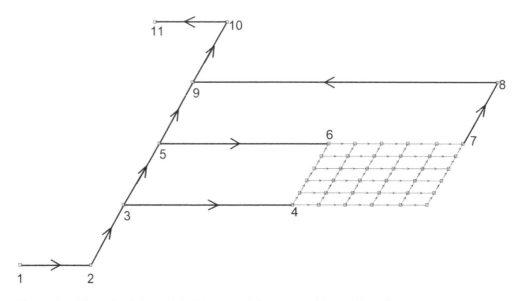

Figure. 3.   Schematic of the model of the area of the excavated longwall ventilated by the "Y" method with the refreshment of unmined coal.

simulations of airflow distribution in mine ventilation networks. The calculation algorithm of the program is based on the one-dimensional mathematical model of airflow in a network and makes use of a system of equations describing the flow in the siding. The model describing a given flow is based on the equations of the conservation of mass and momentum in the following form:

$$G = \rho v a = const, \qquad (11)$$

$$\Delta p = h_w + h_n - W_o - \sum W_i, \qquad (12)$$

where:

G	- mass expenditure in the siding, kg/s,
$\rho = \rho(T,C,p)$	- air density, kg/m^3
$T = T(s)$	- temperature distribution, K
$C = C(s)$	- distribution of gas concentration, %
$p = p(s)$	- pressure distribution along the siding, Pa
$\Delta p = p_2 - p_1$	- pressure difference at the outlet and inlet of the siding, Pa
$v = v(s)$	- flow rate, m/s
$A = A(s)$	- cross-section, m^2
s	- current coordinate of the length of heading, m
$z = z(s)$	- leveling depth of the siding, m
$h_n = -g \int_0^l \rho dz$	- natural air pressure, Pa
h_w	- depression of the ventilator in the siding, Pa
$W_0 = \int_0^l 2\rho v \vee v \vee \frac{1}{2S} ds$	- loss of thrust on distributed resistances, Pa
$\sum W_i$	- sum of depression decreases on local resistances, Pa

The combination of these equations in the network through nodal and mesh equations makes a complete system of equations to be solved with the unknown mass expenditures G. The program adopts a numerical algorithm for solving this system based on the modified Euler method given by Hardy-Cross (Dziurzyński W. et. al. 1997, McPherson M., 1993).

4.1 Preparation of the model for calculations and calculation methodology

The digital model of the network was prepared in the VENTGRAPH program. Geometric parameters of sidings used in the model network were adopted as for real mine headings. The model of caving goaf was prepared in the form of a mesh with the square size of 50x50 m as presented in Figure 3. It was assumed for the numerical calculations of the basic model that the free spaces (porosity) occurring in the caving goaf, as assumed in point 3 of the work, constitute 10% of their volume along the entire length of the caving goaf in the longwall heading, which is equal to the area of 50 m². It was also assumed that the distribution of free spaces along the entire length of the caving goaf in the longwall heading is uniform. For such assumptions, airflow resistance through each "siding" of the goaf was calculated in line with the formula:

$$R = \alpha \frac{LB}{A^3}, \frac{kg}{m^7} \tag{13}$$

where:
R – aerodynamic resistance of the siding, kg/m⁷,
α – aerodynamic drag coefficient, kg/m³,
L – length of the heading, m,
B – perimeter of the heading, m,
A – cross-section of the heading, m².

For the basic model prepared in that way, the airflow distribution in the ventilation network and airflows through the caving goaf were calculated. The results of the airflow distribution are presented in Figure 4.

The rectangles contain volume flow rates of air (m³/min) passing through the sidings, while the values in the nodes represent the values of total pressure losses (Pa). For the sake of comparison, in subsequent numerical calculation steps, the size of free spaces occurring in the caving goaf was reduced by 1% until it reached 3% of its volume, while calculating at the same time the aerodynamic drag of the goaf sidings. Figure 5. and Figure 6. present an exemplary airflow distribution for 6% and 3% of the volume of free spaces in the caving goaf.

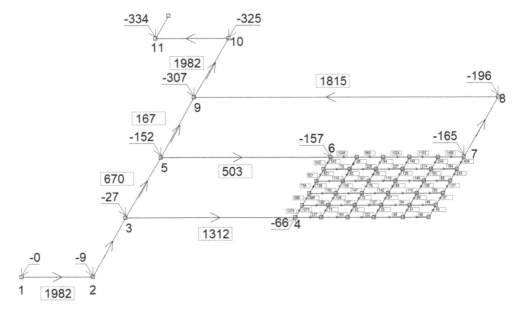

Figure 4. Graphic representation of the results of airflow distribution in the network of the basic model.

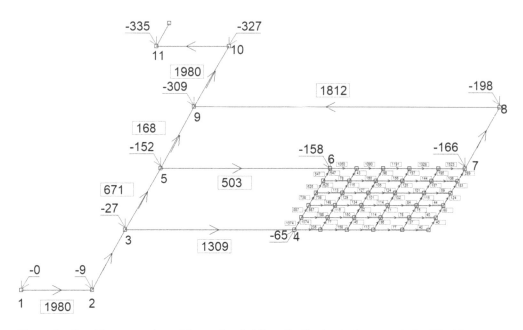

Figure 5. Graphic presentation of the results of airflow distribution in the network for 6% of free spaces in the caving goaf.

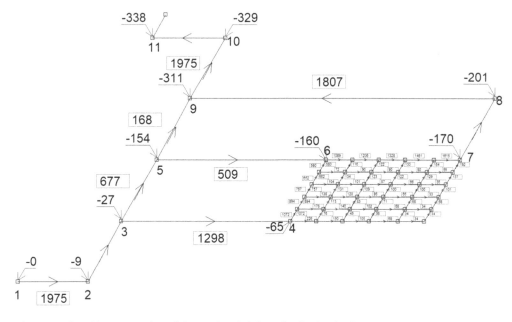

Figure 6. Graphic presentation of the results of airflow distribution in the network for 3% of free spaces in the caving goaf.

The numerical calculations made for the models in which the sealing degree of the caving goaf was being increased indicate that the amount of air migrating in it was reduced. The change in the volume of airflow migrating through the caving goaf in relation to the degree of its sealing is presented in Figure 7.

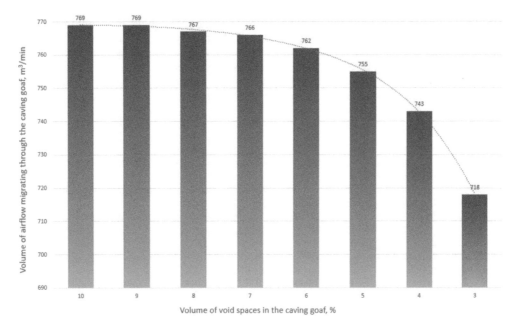

Figure 7.   Impact of the sealing degree of the caving goaf on the volume of airflow migrating through it.

As we can see in Figure 7, along with the increase in the sealing degree of the caving goaf of the exemplary longwall, the porosity of the caving rubble decreases within the range from 10 to 3%. In effect, the migration of air passing through the goaf is reduced, which results in the rise of the airflow volume passing through the longwall heading. We can estimate that with the specified range of sealing, the said rise will be around 7%. Any increase in the airflow volume in the longwall heading contributes to the improvement of safety in terms of a potential hazard in the longwall by effectively reducing the concentration of methane. At the same time, the reduction of air migration through the goaf has a positive effect on the reduction of the risk of endogenous fires.

5 SUMMARY

The paper presents a theoretical foundation for the determination of the absorbency of caving goaf sealed with a fine-fraction (ash and water) hydromixture, which allows with sufficient accuracy in terms of mining practice to determine the amount of fly ash which can be located in caving goaf. As presented in the paper, this amount depends on many factors, among which the most important are: the type and porosity of caving rubble, height of caving, type and properties of roof rockmass, thickness of the extracted bed, tightness degree and accessibility of goaf, the sealing method of goaf, and migration properties of ash-water mixture.

The paper presents the ventilation analysis based on the numerical modeling of an exemplary longwall ventilated by the "Y" method, for which we considered a different sealing degree of the model caving rubble which was created after the longwall advance with a decreasing porosity within the range from 10 to 3%.

As presented in the work, along with the increase in the sealing degree of caving goaf, the porosity of the caving rubble decreases, which results in the reduction of the migration of airflow passing through the goaf, which in turn brings about the rise of airflow volume through the longwall heading. Model calculations demonstrate that the sealing of caving rubble in the presented scope will contribute to effective reduction of the amount of air migrating through the goaf and also to the rise of airflow volume in the longwall to about

7%. We can state that the said rise contributes to the improvement of safety in terms of methane hazard occurring in the longwall due to the reduction of methane concentration in the longwall. At the same time, the reduction of air migration through the goaf contributes to the minimization of the risk of endogenous fires in the caving goaf.

The model calculations presented in the paper enable to forecast ventilation conditions for headings in the vicinity of the longwall, aiming to reduce the natural hazards occurring in them.

BIBLIOGRAPHY

1. Piotrowski Z., Mazurkiewicz M. 2006. *Chłonność doszczelnianych zrobów zawałowych*, Górnictwo i Geoinżynieria, Issue 3 pp. 37–46.
2. Mysłek Z. 1999. *Hydraulic grouting of caving areas with fine-grained waste.* Zeszyty Naukowe. Górnictwo, Politechnika Śląska, Z. 244 pp.107–116.
3. Plewa F. Mysłek Z. 1995. *Teoretyczne podstawy wyznaczania stopnia wypełnienia rumowiska zawałowego mieszaniną odpadów drobnofrakcyjnych z wodą.* Zeszyty Naukowe. Górnictwo, Politechnika Śląska. Z. 225 pp. 203–213.
4. Popczyk M. 2018. *Perspective directions of the development of hydrotransport gravity mixers installations in the light of existing industrial solutions.* 4th Polish Mining Congress-Session: Human and environment facing the challenges of mining 20–22 November 2017, Krakow, Poland, IOP Conf. Series: Earth and Environmental Science 174 (2018)
5. Popczyk M. 2018. *Optimization of the composition of fly ashwater mixture in terms of minimizing seepage water and the possibility of gravitational hydrotransport into the underground workings.* Mineral Resources Management, Volume 34 Issue 3, pp 151–166
6. Popczyk M. 2018. *Wpływ zwiększonego udziału wody zarobowej w hydromieszaninie wykonanej na bazie wybranego odpadu energetycznego na wydajność grawitacyjnej instalacji transportowej.* Konferencja Naukowa Górnictwo Zrównoważonego Rozwoju. Gliwice.
7. Popczyk M. 2019. *The influence of increased proportion of mixing water in a hydromixture made on the basis of selected energy waste on the gravity performance of a transport installation.* IOP Conference Series: Earth and Environmental Science, Mining of Sustainable Development 28 November 2018, Gliwice, Poland, volume 261,
8. Popczyk M., Jendruś R. 2019. *Impact of ash and water mixture density on the process of gob grouting in view of laboratory tests*, Arch. Mining Sci. Volume 64 Issue 3, pp. 625–634.
9. Dziurzyński W., Pałka T., Krawczyk J. 2010. *Ventgraph dla Windows, System programów inżyniera wentylacji do analizy sieci wentylacyjnej kopalni w stanach normalnych i awaryjnych – symulacja nieustalonego przepływu powietrza i gazów pożarowych*, Strata Mechanics Research Institute Polish Academy of Sciences, Kraków 2010, ISSN 1509–2593,
10. Dziurzyński W., Nawrat S., Roszkowski J., Trutwin W. 1997. *Computer simulation of Mine Ventilation Disturbed by Fires and the Use of Fire Extinguishers*, Proceeding of the 6th Int. Mine Ventilation Congress, USA
11. McPherson M. 1993. *Subsurface Ventilation and Environmental Engineering*, Chapman&Hall, London-Glasgow-New York-Melbourne-Madras, First Edition
12. Musioł D. 2007. *Poly-optimization as a metod for calculation o fair distribution In a colliery during restructuring* Gluckauf Volume 10, pp. 491–495.
13. Musioł D. 2009. *Ventilation of longwall region (Y system with reblowing) – considerations based on PC simulation*, Kwartalnik Górnictwo i Geologia Volume 4, Issue 3, Wydawnictwo Politechniki Śląskiej, pp. 75–85.

Case studies

Geomechanical analysis of the rock mass stability in the area of the "Regis" shaft in the "Wieliczka" salt mine

Grzegorz Dyduch, Patrycja Jarczyk & Marek Jendryś
Silesian University of Technology, Gliwice, Poland

SUMMARY: The "Regis" shaft is one of the key workings which ensure the safety and smooth flow of tourist traffic in the historic Wieliczka salt mine. Its fault-free operation requires the monitoring of the technical condition of the supports and the surrounding rock mass. The results of the conducted monitoring and macroscopic observations of the lining in the shaft and of the adjacent workings have exhibited gradual and progressive degradation of the rock mass structure. In order to assess the impact of these processes on the shaft and on the nearby workings, a spatial numerical model of the rock mass has been created using the FLAC 3D software. The model geometry was determined after analysing the geological documentation of the area in the form of maps, cross-sections, and on the basis of visitations in the workings located in the vicinity of the shaft. On this basis, it was determined that the behaviour of the "Regis" shaft is primarily affected by a large post-mining chamber (the "Kloski" chamber) located in its immediate vicinity, on the north-eastern side. The tool used for the simulation (FLAC 3D software) allowed for the mapping of the behaviour of the continuous medium and the simulation of the destruction of the material by assigning to it (after reaching the limit stress state) the features of a plastic medium. On the basis of conducted calculations and simulations, it was determined that leaving the rock mass without any protective works may pose a threat to its stability and the stability of the "Regis" shaft lining. To protect the shaft and the rock mass, it was proposed to fill the "Kloski" chamber and the shaft below level IV. Such solution would stop the propagation of damage zones and minimise the rock mass displacement. Creating a stress and deformation state in the model, achieved as a result of backfilling a part of the shaft and the chamber, is highly likely to ensure the stability and safe operation of the workings located above it, in particular the section of the "Regis" shaft from the surface to level IV. The presented example showing, that backfilling used even many years after excavation of chamber can bring good results in terms of improving rock mass stability.

Keywords: backfill, numerical modelling, rock mass, shaft

1 INTRODUCTION

Safe exploitation and the possibility of making workings in historic underground mines available for tourist traffic require preserving their ongoing stability. This is particularly important in the case of development workings, which not only perform transport and ventilation functions, but also act as key evacuation routes. Any damage to their supports can constitute a serious threat to the operation of an entire mine. Workings of this type include the "Regis" shaft of the Wieliczka salt mine, currently acting as the main transportation shaft for the tourist traffic. Its history dates back to the fourteenth century. As a working providing access to the further parts of the deposit, it was maintained for several hundred years, performing mainly mining and ventilation functions. The exploitation of salt deposits carried out at that time in the immediate

DOI: 10.1201/9781003205906-31

vicinity of the shaft caused irreversible changes to the rock mass structure. Their impact on the shaft tube, rock mass, and terrain surface, despite the cessation of exploitation, is also currently observed in the form of progressive deformations (Kwinta 2012; Bruneau et al. 2003; Tajduś et al. 2003; Majcherczyk and Lubryka 2003). The rheological processes taking place in the rock mass may be associated with a large volume of the deposit exploited in the vicinity of the "Regis" shaft. Post-exploitation voids have only been partially liquidated, which means that they continue to significantly affect the behaviour of the rock mass around the shaft tube. An additional factor adversely affecting the condition of the rock mass in the vicinity of the "Regis" shaft is the presence of unexploited salt blocks in its profile, which, due to rheological processes, increase the load on the lining (Chudek et al. 200;, Kortas 2010).

Due to the complexity of the issue, geomechanical analyses of the rock mass are usually performed using methods based on numerical modelling. Due to the relatively high degree of complexity of the geomechanical conditions in the vicinity of the "Regis" shaft, the FLAC 3D software based on the finite difference method has been used to perform the analysis. The selection of the mode of analysis is dictated by the confirmed effectiveness of this method, which has already been repeatedly employed to assess the condition of the rock mass and workings at the Wieliczka salt mine, as evidenced by the publications on this subject (Cała et al. 2009; Cieślik et al. 2009; Cała et al. 2016).

2 NUMERICAL SIMULATIONS OF THE ROCK MASS BEHAVIOUR IN THE VICINITY OF THE "REGIS" SHAFT

2.1 Model structure

In order to map the stress and deformation states in the rock mass, a spatial numerical model has been created which incorporated the rectangular rock mass with the centrally located "Regis" shaft. The model geometry was determined after analysing the geological documentation of the area in concern in the form of maps, cross-sections, and on the basis of visitations in workings located in the vicinity of the shaft in concern. On this basis, it was determined that the behaviour of the "Regis" shaft is primarily affected by a large post-mining chamber (the "Kloski" chamber) with varying degrees of filling, located in its immediate vicinity, on the north-eastern side.

The performed numerical simulations allowed for the determination of stress and deformation changes in the rock mass surrounding the "Regis" shaft. As a result of the redistribution of primary stresses existing in the rock mass before the shaft and its adjacent chambers were created, a secondary state of equilibrium was achieved, characterised by stress concentration zones, occurrence of stress relief zones, and local zones where the rock mass reached a postcritical state. The tool used for the simulation (FLAC 3D software) allowed for the mapping of the behaviour of the continuous medium and for the simulation of the destruction of the material by assigning to it (after reaching the limit stress state) the features of a plastic medium.

Hence, the main indicator adopted for assessing the state of the rock mass in this paper are displacements and plasticisation zones where the material was damaged. In the real rock mass, this may result in its fracturing, spalling of the mass, and formation of collapses. Such manner of assessing the state of the rock mass was also used in papers concerning shafts (Wang et al. 2003, Kleta and Jendryś 2013).

The state of the rock mass can also be assessed on the basis of the effort indicator (D'Obryn and Wiewiórka 2015) or through the stability indicator, determined using the shear strength reduction method (Cała et al. 2009).

The modelled rock mass in the horizontal plane exhibited dimensions of 400 × 400 m, at a depth of 300 m, as shown in Figure 1. The following rock formations were mapped in the model based on the geological documentation of the analysed area:

– marly claystone covering the largest part of the model, from its bottom wall to a depth of 28 m,
– argillous-gypsum buffer zone with a thickness of 20 m occurring directly above the marly claystone,
– quaternary sediments with a thickness of 8 m, constituting the upper part of the model.

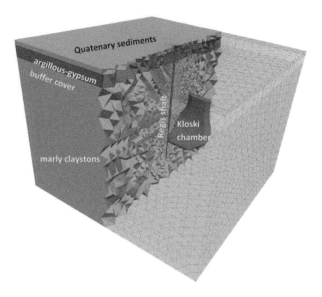

Figure 1. The geometry of the numerical model adopted for calculations.

Figure 2. The model geometry in the shaft bottom zone of level IV.

The shaft was modelled in two sections with varying cross-sections (Figure 2), i.e.:

– from the surface to a depth of 170 m (down to level IV), a circular shaft was modelled with a diameter of 5.0 m and a lining thickness of 0.8 m. For the elements constituting the lining, a material corresponding to C20/25 concrete class was adopted,
– from a depth of 170 m a shaft was modelled with a 4 × 4 m square section from the material corresponding to the parameters of the existing wooden supports.

To the north-east of the "Regis" shaft, the post-mining void, called the "Kloski" chamber, was modelled, with a geometry that was adopted on the basis of a vertical cross-section and maps of individual levels (Figure 3).

In the model, displacement boundary conditions were adopted in the form of supports preventing the movement of the nodes located on the outer walls of the model in perpendicular

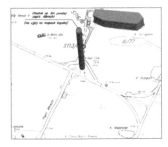 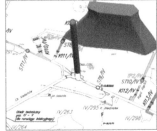 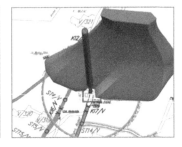

LEVEL III, +114.68 M   LEVEL IV, +72.42 M   LEVEL V, +46.19 M

Figure 3.   The "Kloski" chamber geometry in relation to the maps of individual levels.

Table 1.   Summary of adopted material parameters.

Name of material	elastic shear modulus G [GPa]	elastic bulk modulus K [GPa]	Young's modulus E [GPa]	Cohesion C [MPa]	Internal friction angle φ [°]	Tensile strength Rr [MPa]
marly claystone	0.11	0.25	0.30	2.50	35	0.50
argillous-gypsum buffer zone	0.20	0.27	0.50	0.1	20	0.1
Quaternary	0.04	0.07	0.1	0.02	30	0.01

directions to these walls. These supports were set for the side and bottom walls of the model. The upper wall, reflecting the surface of the terrain, remained free.

For the individual layers constituting the model, a plastic-elastic constitutive model with the Coulomb-Mohr strength criterion was adopted, which was defined by the parameters listed in Table 1. The material properties for the model were adopted on the basis of data provided by the Wieliczka salt mine.

The first stage of the simulation determined the original stress state generated by the volumetric weight of the materials constituting the model. The obtained stress state has generated compressive stress in the vicinity of level IV with the values of 4.17 MPa in the vertical direction and 1.78 MPa in the horizontal plane.

In the second stage, after establishing the original state, the stress and deformation state induced by the creation of the shaft has been mapped. For this purpose, the elements constituting the model inside the shaft contours were assigned a "null" type model, mapping the lack of material, and then balanced, allowing for deformations in the elastic range. The next stage of the simulation involved the introduction of shaft lining defined by a plastic-elastic constitutive model with Coulomb-Mohr strength criterion with the following parameters: shear modulus 0.12 GPa, bulk modulus 0.25 GPa, cohesion 5 MPa, internal friction angle 45°, tensile strength 3 MPa.

After introducing the shaft lining and restarting calculations, the model has been balanced. The stress and deformation state obtained in this way was assumed as the initial one for further analysis of the rock mass behaviour in the vicinity of the "Regis" shaft and the "Kloski" chamber.

At this stage, the modelled rock mass is fully stable and only elements located on the side-walls of the lower part of the shaft, below level IV, have been plasticised (Figure 4).

The main problem in maintaining the stability of the "Regis" shaft and the rock mass that surrounds it is the impact of the post-mining void created after the extraction of a salt block and creation of the "Kloski" chamber. This chamber is one of the largest chambers in terms of cubature in the Wieliczka salt mine and it significantly affects the state of both the rock mass surrounding it and the "Regis" shaft located in its vicinity.

Figure 4.   Plasticisation zones in the area of the modelled shaft before excavating the "Kloski" chamber.

2.2   *The model reflecting the current state*

The fundamental stage of modelling involved obtaining by the model the state that corresponded to the current situation by simulating the exploitation of the "Kloski" chamber and introducing a substitute material, representing its filling (wooden mesh cribs, wooden props, barren rock, and sand). For the material filling the chamber, a linear-elastic material was adopted, characterised by deformation parameters reduced a hundred times in relation to the parameters adopted for marly claystone. For the model prepared in this way, after its stabilisation, the modelling stage was reached, reflecting the current deformation and stress state of the rock mass in the vicinity of the "Regis" shaft and the "Kloski" chamber.

When analysing the plasticisation zones in the surroundings of the "Regis" shaft, at this stage of calculations two areas where the model has plasticised can be distinguished (Figure 5).

The first area includes both the rock mass and the modelled shaft supports and reaches from level IV to a depth of about 133 m. The plasticisation of the material in this area can be

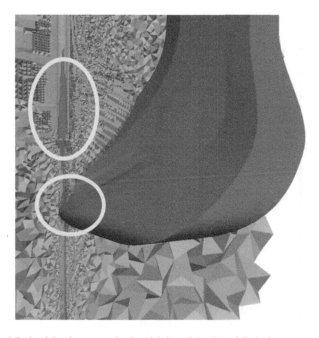

Figure 5.   The model plasticisation zones in the vicinity of the "Regis" shaft.

Figure 6.   Damage to the shaft lining and shaft bottom supports occurring in the vicinity of the level IV.

equated with the damage observed on the supports in the shaft and shaft bottom at level IV (Figure 6).

The second area includes the rock mass in the area where the shaft crosses the "Kloski" chamber at the depth of level V, where the rock mass around the entire shaft has been damaged, over a section of about 10 m.

The occurrence of plasticisation zones in the model can be interpreted in relation to the actual rock mass as its destruction zones manifesting in the fracturing of rocks and excessive load exerted on the supports leading to scratches and fractures.

The plasticisation zones of the model also occur at a greater distance from the shaft, in the ceiling of the lower part of the "Kloski" chamber, and their range usually amounts to about 1 m. Only in the central part of the chamber a plasticisation zone was created, reaching about 15 m from the chamber ceiling.

2.3   The model representing the final state

The purpose of the simulations was to assess the rock mass stability in the vicinity of the "Regis" shaft over a long time horizon and its impact on the behaviour of the working supports. In order to achieve that, it was assumed that with the passage of time and due to the long-term impact of environmental factors, the rock mass mechanical parameters in the plasticisation zones will be subjected to gradual degradation. This phenomenon was modelled by adopting a plastic-elastic constitutive model with weakening, characterised by a total reduction of cohesion and tensile strength in the plasticised elements. This solution has also been used in numerous geomechanical analyses, the results of which have been presented, for example, in Diao et al. 2016, Badr et al. 2003, Szafulera and Jendryś 2012, Reymany and Vakili 2017. It was also assumed that the material filling the "Kloski" chamber will consolidate over time, which was reflected in the model by reducing the stress in the material filling the post-mining void.

The above operations caused further chamber clamping as well as deformations and destruction of the rock mass in the vicinity of the "Regis" shaft. As a result, after the equilibration of the model, its final state was obtained. The results in the form of plasticisation zones and displacement distributions are presented in Figure 7 and Figure 8.

The largest displacements of the rock mass around the "Regis" shaft occurred at the depth of level V (Figure 7). In this location, a large plasticisation zone has formed, resulting in the material undergoing plastic flow, and the displacements would mainly approach a horizontal direction at values exceeding 1.5 m. In the area of level IV, the displacements of the model towards the chamber amounted to 0.53 m.

The obtained results of the plasticisation zones (Figure 8) for the final model indicate a significant increase in the damage zones in relation to the model representing the current

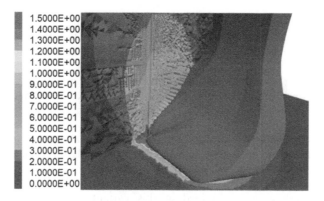

Figure 7. Displacements in the modelled rock mass – final state of the model, vertical cross-section through the shaft in a north-south direction, level V at a depth of 196 m.

Figure 8. Plasticisation zones for the variant representing the final state of the rock mass, vertical section through the shaft in a north-south direction, plasticised elements marked in red.

state. These zones reach to about 15 m above the chamber ceiling and cover the entire section of the shaft between levels V and IV. In this model, at level V depth, the plasticisation zone in the shaft vicinity extends over a distance of over 15 m deep into the rock mass on the side opposite to the "Kloski" chamber.

According to the simulations carried out, a large volume of the rock mass in the area of the level IV was also damaged, and the plasticisation of shaft supports reaches to a depth of 125 m.

In the area of level IV, cribwork was created to ensure the carrying capacity of the shaft section constructed above. The destruction of the rock mass in this area may result in the displacement of rock rubble into the shaft below level IV, the lack of cribwork foundation, and, as a result, a large displacement and destruction of the "Regis" shaft supports, also above level IV.

2.4 The target model taking into account the backfilling of the "Kloski" chamber and sections of the "Regis" shaft

As the presented calculations demonstrated, leaving the rock mass in the area of the "Kloski" chamber without any protective works may pose a threat to its stability and the stability of the "Regis" shaft lining, which currently constitutes the main transportation shaft of the Wieliczka salt mine.

In order to secure the shaft and the rock mass in the vicinity of the "Kloski" chamber, it was proposed to fill the chamber and the shaft below level IV with a liquid self-solidifying mineral injection mixture based on ground blast furnace slag, cement, and brine produced in the Wieliczka mine.

The application of this solution was also analysed in the numerical model by assigning to the elements constituting the model inside the "Kloski" chamber and the "Regis" shaft below level IV the parameters of the elastic medium with a Young's modulus of 3 GPa.

This variant was modelled by introducing the backfill material (marked in green in Figure 9) into the model representing the current state and by adopting plastic-elastic material with weakening for the rock mass with identical parameters as during the simulation leading to obtaining the final state of the rock mass without using the filling material.

The resultant displacements obtained in this variant are shown in Figure 10. The obtained results in the form of displacements indicate only a slight increase in rock mass deformation in the analysed model in relation to the variant representing the current state. The maximum displacements in the model taking into account the filling of parts of the chamber and the shaft are only 5 cm larger than those obtained for the current state and amount to 0.77 m. The displacements in the vicinity of the shaft inlet to level IV amount to about 0.39 m and are only 3 cm larger than those occurring in the reference variant.

The results obtained for the model taking into account the reduction of post-destruction parameters (Figure 11, Figure 12) indicate that filling parts of the "Regis" shaft and the "Kloski" chamber will significantly inhibit the propagation of destruction zones. The ranges of the destruction zones in the chamber ceiling, as well as in the area of the shaft inlet remain practically unchanged. The range of the fracture zone in the shaft lining increased by 1 m, and in the analysed variant it reaches the depth of 132 m.

The increase in the plasticisation zones after backfilling the part of the chamber will occur mainly in its lower part at a depth of about 195 m. However, this phenomenon, due to the filling of the chamber, will not pose a threat to the workings situated above.

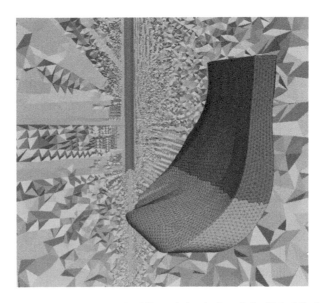

Figure 9.   The model taking into account the filling of the shaft and the "Kloski" chamber below level IV (the backfill material is marked in green).

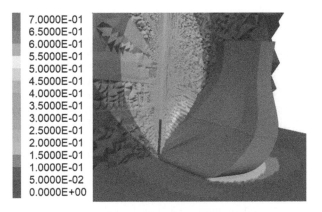

███	7.0000E-01
	6.5000E-01
	6.0000E-01
	5.5000E-01
	5.0000E-01
	4.5000E-01
	4.0000E-01
	3.5000E-01
	3.0000E-01
	2.5000E-01
	2.0000E-01
	1.5000E-01
	1.0000E-01
	5.0000E-02
███	0.0000E+00

Figure 10. Displacements in the modelled rock mass – the model taking into account the filling of the "Regis" shaft and the "Kloski" chamber below level IV, vertical cross-section through the shaft in the north-south direction, horizontal cross-section at a depth of 196 m.

Figure 11. Plasticisation zones for the variant taking into account the filling of the "Regis" shaft and the "Kloski" chamber below level IV, vertical cross-section through the shaft in the north-south direction, plasticised elements marked in red.

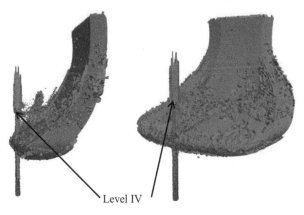

Level IV

Figure 12. Spatial diagram of the plasticisation zones, the model taking into account the filling of the "Regis" shaft the "Kloski" chamber below level IV, plasticised elements marked in red.

3 SUMMARY

The "Regis" shaft is one of the key workings which ensure the safety and smooth flow of tourist traffic in the historic Wieliczka salt mine. Its fault-free operation requires the monitoring of the technical condition of the supports and the behaviour of the rock mass surrounding the shaft. The results of ongoing geodetic monitoring and macroscopic observations of the shaft supports and adjacent workings indicate a gradual and progressive degradation of the rock mass structure. Such phenomena negatively affect the stability of workings located in this area.

In order to assess the impact of the progressive degradation of the rock mass structure on the shaft and its adjacent workings, a spatial numerical model of the rock mass of the area was created, taking into account its current geomechanical situation, using the FLAC 3D software based on the finite difference method. On the basis of the available documentation and the results of observations and analyses carried out, it was established that the greatest impact on such behaviour of the rock mass in the vicinity of the "Regis" shaft is exerted by a large post-mining chamber, called the "Kloski" chamber, located on the north-east side, in the immediate vicinity of the shaft tube.

In order to map the stress and deformation states and to determine the final state of the rock mass, a constitutive model was adopted, assuming the reduction of post-destruction parameters. Consequently, the results indicating a considerable increase in the plasticisation zone were obtained, especially in the shaft section between levels IV and V. The simulations also indicated a significant increase in rock mass deformation, with the maximum displacements resulting from the plastic flow of the modelled material reaching values exceeding 1.5 m, which occurred in the vicinity of the shaft bottom at level V. This behaviour of the model clearly demonstrates the possibility of the formation of collapses and large displacements threatening the stability of the "Regis" shaft lining. Such a situation may pose a threat to the shaft and adjacent workings in the future, if the appropriate protective works are not undertaken.

Based on the results of calculations and analyses, it was recommended to grout the "Kloski" chamber and fill the "Regis" shaft below level IV. Such solution would stop the propagation of damage zones and minimise the rock mass displacement. Creating a stress and deformation state in the model, achieved as a result of backfilling a part of the shaft and the chamber, is likely to ensure stability and safe operation of the workings located above it, in particular the section of the "Regis" shaft from the surface to level IV.

ACKNOWLEDGMENTS

"The author – Patrycja Jarczyk - also expresses gratitude to the National Agency for Academic Exchange of Poland (under the Academic International Partnerships program, grant agreement PPI/APM/2018/1/00004) for financial support of the internship at Montanuniversität in Leoben, Austria which enabled the execution of a part of the work".

BIBLIOGRAPHY

Badr, S., Ozbay, U., Kieffer, S. & Salamon, M. 2003. Three-Dimensional Strain Softening Modeling of Deep Longwall Coal Mine Layouts. In: Brummer et al. (eds.) FLAC and Numerical Modeling in Geomechanics. Swets & Zeitlinger, Lisse": 233–239.

Bruneau, G., Tyler, DB., Hadjigeorgiou, J. & Potvin, Y. 2003. Influence of faulting on a mine shaft—a case study: part I—Background and instrumentation. Int. J. Rock Mech. Min. Sci. 40: 95–111.

Cała, M., Czaja, P., Flisiak, D. & Kowalski, M. 2009. Ocena zagrożenia zapadliskowego wybranych komór KS „Wieliczka" w oparciu o obliczenia numeryczne. Górnictwo i Geoinżynieria Vol. 3/1: 33–44.

Cała, M., Stopkowicz, A., Kowalski, M., Blajer, M., Cyran, K. & D'Obyrn, K. 2016. Stability analysis of underground mining openings with complex geometry. Studia Geotechnica Et Mechanica, Vol. 38, No. 1: 25–32. https://doi.org/10.1515/Sgem-2016-0003.

Chudek, M., Kleta, H., Wojtusiak, A. & Chudek, MD. 2009. Obudowa szybów w warunkach znacznych ciśnień deformacyjnych górotworu. Górnictwo i Geoinżynieria Vol. 3/1: 87–90.

Cieślik, J., Flisiak, J. & Tajduś, A. 2009. Analiza warunków stateczności wybranych komór KS „Wieliczka" na podstawie przestrzennych obliczeń numerycznych. Górnictwo i Geoinżynieria Vol. 3/1: 91–103.

D'Obyrn, K. and Wiewiórka, W. 2015. Renowacja górnicza komory Jezioro Wessel na poziomie III Kopalni Soli,Wieliczka" S.A. Przegląd górniczy No. 7: 69–77.

Diao, XH., Yang, SX. & Wang, K. 2016. Application of strain softening model to numerical analysis of deep tunnel. In: Kim (ed.) Progress in Civil, Architectural and Hydraulic Engineering IV. Taylor & Francis Group, London: 373–378.

Kleta, H. and Jendryś, M. 2013. Przyczyny uszkodzeń obudowy w głębionym szybie w świetle obliczeń numerycznych, Bud. Gór. Tunelowe No. 3: 1–8.

Kortas, G. 2010. Szyb z obudową w górotworze solnym — wstępne badania modelowe, Górnictwo i Geoinżynieria Vol. 2: 395–403.

Kwinta, A. 2012. Prediction of strain in a shaft caused by underground mining. Int. J. Rock Mech. Min. Sci. 55: 28–32.

Majcherczyk, T. and Lubryka, M. 2003. The influence of the depth of an exploited seam on stresses around the shaft, Archives of Mining Sciences 48: 65–80.

Rajmeny, PK. and Vakili, A. 2017. Three-dimensional inelastic numerical back-analysis of observed rock mass response to mining in an Indian mine under high-stress conditions. In Wesseloo (ed.) Deep Mining 2017: Eighth International Conference on Deep and High Stress Mining, Perth: 329-342.

Szafulera, K. and Jendryś, M. 2012. Numeryczna analiza stanu odkształcenia górotworu w sąsiedztwie płytkich wyrobisk porudnych. Górnictwo i Geologia Vol. 7 No. 1: 187–199.

Tajduś, A., Cała, M., Flisiak, J. & Lubryka, M. 2003. Ocena możliwości częściowego naruszania filarów ochronnych szybów na podstawie obliczeń numerycznych stanu naprężenia w otoczeniu rury szybowej, Górnictwo i Geoinżynieria Vol. 3-4: 601–613.

Wanga, JA., Park, HD. & Gaoa, YT. 2003. A new technique for repairing and controlling large-scale collapse in the main transportation shaft, Chengchao iron mine, China, Int. J. Rock Mech. Min. Sci. 40: 553–563.

Minefill 2020-2021 – Hassani et al (eds)
© 2021 Taylor & Francis Group, London, ISBN 978-1-032-07203-6

Case study: Paste plant retrofit

Leslie Correia & Brent Cothill
Paterson & Cooke Canada Inc

Phil Antunes
Simem Underground Solutions Inc

Sean James
Rearden Metals & Mining

SUMMARY: A zinc, copper and lead mine located in Canada has recently restarted operations after several years under care and maintenance. The mine had an existing paste plant that was historically used for surface tailings deposition. The existing paste plant was overhauled and retrofitted to be able to handle full plant tailings and produce cemented pastefill that will achieve the required target strengths underground. In this case study, the various design challenges faced and lessons learned are documented which led to the successful retrofit of the paste plant in 2019.

Keywords: Construction, Design, Retrofit, Paste

1 INTRODUCTION

After an improvement in base metal commodity prices during 2017, a zinc, copper and lead mine located in Canada has restarted operations after a hiatus caused by unfavourable market conditions and high operating costs. Various operating scenarios were considered, but ultimately, a revised mine plan was developed incorporating paste backfill.

Historically, backfill at the mine was produced at an existing hydraulic fill plant using classified tailings. The fines portion of the tailings stream was sent to an existing paste plant used for surface tailings deposition only.

The mine's capability to store tailings on site is limited. One of the requirements for mine restart-up was to maximise the placed density of tailings used as backfill and disposed on the tailings storage facility (TSF) by using full plant tailings. The mine design, therefore, had a longer-term plan that relied on pastefill and phased out hydraulic fill. Using the pastefill has an additional benefit of achieving a more consistent backfill product with better cycle times.

Several different plant design options were considered, each derived from the existing backfill facilities on-site. Analysis of the hydraulic fill plant showed that this plant was not capable of handling full plant tailings required to produce paste. Feasible options identified included repurposing the existing paste plant with a new internal configuration or building a new paste plant using a combination of new and existing equipment. Inspection of the paste plant and subsequent trade-off studies demonstrated that the existing paste plant was in fair condition and held enough value leading in favour of the retrofit upgrade.

This paper documents the design and construction challenges associated with overhauling and retrofitting the existing paste plant providing a summary of lessons learnt.

DOI: 10.1201/9781003205906-32

2 PROCESS DESCRIPTION OF EXISTING PASTE PLANT FACILITY

The existing paste plant was historically used to dewater tailings slimes before deposition on the TSF. The paste plant included the following major pieces of equipment: A cyclone circuit, a high compression thickener, a single vacuum disc filter, a custom-made twin shaft mixer and two positive displacement pumps. A simplified flow sheet of the existing paste plant is provided in Figure 1. Figure 2 provides a 3D model of the existing internal equipment and structure within the paste plant building.

Tailings from the mill reports to a cyclone circuit located inside the paste plant (Figure 3a). The cyclone underflow is pumped to the hydraulic fill plant and the cyclone overflow is transferred to a 25 m diameter high compression thickener (Figure 3b). Thickened tailings (~60%m) is transferred directly to a single vacuum disc filter unit (Figure 3a) to achieve a cake moisture content of 22 to 26%. The filtration unit is a GL&V 3.2 m unit equipped with 12 discs and has a total filtration area of 173 m².

The filtered tailings and trim water were fed to a custom-made continuous twin-shaft mixer (Figure 3c). The discharge from the mixer was via an overflow which allowed paste to be gravity fed to one of two 100 bar rated Schwing KSP 110 V (HD) L (one duty/one standby) piston positive displacement pumps (Figure 3d). The piston pumps distributed the uncemented paste to the TSF.

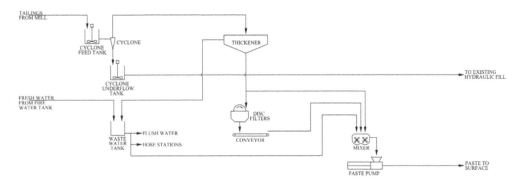

Figure 1. Simplified flow sheet of existing paste plant.

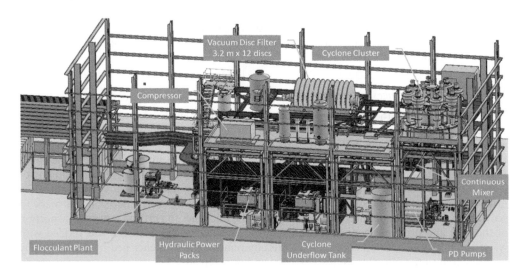

Figure 2. 3D model of existing internal equipment in paste plant building.

Figure 3. Photos of existing paste plant, (a) view of disc filter (left) and cyclones (right); (b) view of thickener adjacent to paste plant; (c) continuous mixer; (d) positive displacement pumps.

3 MECHANICAL UPGRADE/RETROFIT

The existing paste plant was redesigned to produce a cemented paste backfill product to support the mining activities while matching the overall mill utilization of 85 to 92%. A comparison between the existing paste plant 3D model and the as built 3D model complete with the new retrofits is presented in Figure 4.

3.1 *Cyclone circuit*

The paste plant previously processed the fines fraction of tailings only. The retrofitted paste plant would use the total tailings stream and would bypass the cyclone circuit. The cyclone cluster was therefore removed to allow space for a new dust collector. Furthermore, the cyclone underflow tank and pumps were also removed to allow space for the relocation of the existing flocculant plant.

3.2 *Thickener and process water tank*

Remediation of both the thickener and process water tank were relatively well scoped and properly understood, however, difficulties arose during the clean-up of these items. The thickener overflow launder was full to overflowing with tailings fines, while the bottom of the thickener contained a significant bed of very dense material left behind, estimated to be about a metre thick towards the centre. Cleaning of these items required a significant amount of work using a vacuum truck, pressure washers, and labourers manually shovelling tailings in order to remove all of the compacted material. Due to escalating costs a decision was made to only remove enough material from the bottom of the thickener to perform a proper inspection of the rake and drive mechanism from within the tank. Once the thickener was refilled with water the rake was restated and used to carefully remove the thickener bed by incrementally lowering it until the torque readings stabilized and then repeat.

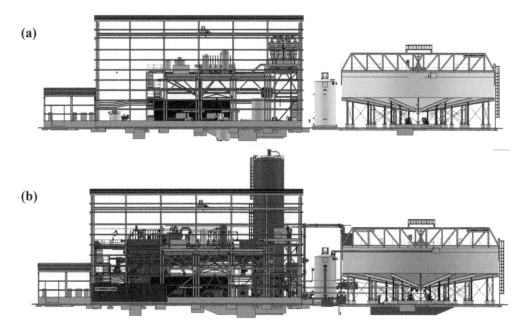

Figure 4. Comparison between (a) Existing 3D model and (b) As built 3D model (new equipment shown as blue).

The process water tank had substantial corrosion and required a more involved remediation or complete replacement. The tank was opened to facilitate sandblasting and painting of the interior. An initial inspection of the tank interior was not done as all the inspection hatches were inoperable due to corrosion. As such, once the process water tank was opened, it was discovered that it was more than half full of tailings. Again, the use of a vacuum truck, pressure washers and labourers were employed to empty the tank.

While these cleaning activities were seemingly small and partly anticipated, the magnitude of the time required to perform the work and the corresponding costs were substantial and noteworthy.

3.3   Thickener feed system

The only major modifications to the thickener were made to the feed system. The existing feed system incorporated a feed tank mounted near the centre of the bridge. Nine feed pipes reported to the tank, some continuous operation, some redundant. The discharge outlet nozzle of the feed tank was vertical and fed through a series of 45° elbows into the feedwell. This arrangement was replaced with a single common feed box mounted on the outside periphery of the thickener tank with one common feed pipe feeding tangentially into the feedwell. This will provide a homogeneous feed that can be injected without air or disturbing the settled bed that lies underneath the feedwell discharge.

The existing feedwell itself was a closed bottom design, 4 m in diameter and extended ~2 m in depth. It had a choked lower section fitted with a deflector cone and cone scraper. The deflector cone gap was fixed and did not allow for any changes in throughput. This configuration often lead to short circuiting of the coarse material within the feedwell creating a beach on the thickener tank floor (usually located either opposite to the feed entry point or further around, depending on the particle size). The beach caused an unbalanced load on the rake mechanism, torque spikes on the drive and intermittent slugs of coarse material to pass through the underflow nozzles. The feedwell was replaced with a correctly sized unit providing more efficient and newer technology that prevents short circuiting. This significantly improved the capture of fines and uniform distribution of solids into the thickener tank.

3.4 Vacuum disc filters

The existing vacuum disc filter was refurbished with a second 3.2 m dia. x 12 disc vacuum disc filter added to provide redundancy and/or boost the solids production capacity if required. The new filter was orientated in a left-hand configuration to compliment the right-hand configuration of the existing filter. This allowed both filtrate receivers and snap air tanks to be orientated in between the two disc filters optimizing the space available.

The support structure was extended to support the second filter (Figure 5 & 6). The existing flocculant plant was moved to where the cyclone underflow tank was located previously while the existing west staircase was moved over against the west side of the building. The new structure allows top access to the filter feed tank and agitator.

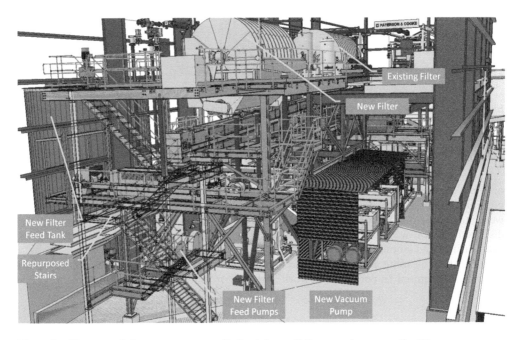

Figure 5.   New extended support structure (Indicated as red) for second vacuum disc filter.

Figure 6.   Comparison between (a) Existing filter level layout and (b) Retrofit filter level layout.

3.5 Filter feed tank and pumps

The existing thickener underflow pumping system fed thickened tailings directly into the existing vacuum disc filter boot. As part of the retrofit, an agitated filter feed tank was introduced to provide a two hour surge capacity for the vacuum disc filters and allow for proper control of the slurry extraction density out of the thickener. The filter feed tank was located on the outside of the paste plant building with the filter feed pumps located directly underneath new vacuum disc filter inside the building. The filter feed pumps maintain a ring main header from which the existing and new filter boots are fed. A bypass on the ring main also allows trim slurry to be introduced directly into the mixer operation, thereby reducing the load on the filters.

3.6 Conveyors

The existing filter cake conveyor was replaced due to its poor condition and increased length required to serve both filters. Due to the limited elevation available between the filters and the new mixer, the belt angle of the main conveyor was set at 5°. The new main conveyor was also equipped with a reversing function allowing filtered tailings to be transported through the side of the plant. This will allow the mill to continue producing tailings in an upset condition where equipment/piping downstream of the mixer fail. In reverse, the main conveyor would transfer filter cake onto a secondary conveyor which in turn transfers the filter cake onto a stockpile allowing trucks to haul the filter cake to the TSF on a short-term basis.

3.7 Continuous mixer

During on-site visits, the paste produced by the existing mixer was witnessed to be full of lumps (filter cake). The filter had a very short residence time (~30 s) and was inadequate to ensure quality backfill production due to low energy input. It was also unlikely that the mixer would be able to adequately disperse the binder and produce a high-quality paste. A low-profile twin-shaft mixer was installed with a residence time of ~2.5 minutes. The new mixer was installed on a new lower steel platform.

3.8 PD pumps

Upon inspection of the two existing piston pumps, it was found that these pumps would need to be replaced due to the poor condition they were in. The associated hydraulic power packs were refurbished and reused. The new PD pumps were moved to align with a new paste hopper.

3.9 Cement system

As the existing paste plant was constructed for surface deposition only, a new cement system was added to the paste plant to allow binder to be fed into the mixer. This included a 168 m³ binder silo, rotary feeder, weigh-belt feeder and screw conveyor.

Initially, it was considered to locate the cement silo on the road side of the plant. However, it was concluded that there may be interference with mine traffic and that the required screw conveyor lengths would be such that it would impede the lifting bay near the mixer. A location behind the paste plant was selected instead that resulted in a short conveyance inside the building to the mixer. To feed the silo, cement trucks park in a lay-by area and transfer cement via tubing across the paste plant to the silo.

Construction of the binder silo was challenging. A bolt up silo was ideal for the limited site access but the tight work envelope made is difficult to take advantage of having a larger crew assembling rings in parallel. This in turn used up more crane stand-by hours waiting on the next ring to be placed.

Another take-away from this is to understand the contractor's capabilities to perform specific specialized work. Construction of a bolt-up silo requires specialized installation expertise in order to get the greatest efficiency out of the construction. The incumbent contractor could rarely dedicate itself to silo construction due to shifting priorities and this resulted in substantial project schedule delays.

3.10 *Piping and valves*

Much of the piping and valving in the plant was installed as new with the exception of some runs of gland water and compressed air piping dedicated to existing equipment.

4 CIVIL/STRUCTURAL UPGRADE/RETROFIT

Initial inspections were conducted to determine the condition of the structure, equipment and tanks that would be reused in the retrofit. While most of the structure was in good condition, any steel that had been in contact with tailings was corroded and required remediation (Figure 7). Fortunately, the equipment and structure with the most significant corrosion and those that had seen resourceful repairs by plant staff were to anyway be replaced by new equipment and structural steel as part of the retrofit. These included the filter cake conveyor, paste mixer and mixer platform structures. The remaining structural members that had minor contact with tailings spillage were sand blasted or wheel abraded in place and coated with a two-part epoxy to protect from future exposure.

4.1 *Building code*

The original paste plant engineering was completed in 2002 using the 1998 British Columbia Building Code while the 2012 British Columbia Building code was required for the plant upgrades. The 2012 code has seen substantial amendments including the seismic design requirements.

The initial structural design approach for the retrofit was to maintain isolation between the original and new equipment structures to minimize upgrades required to the original structures. However, due to the building envelope size, process flow and equipment layout, the

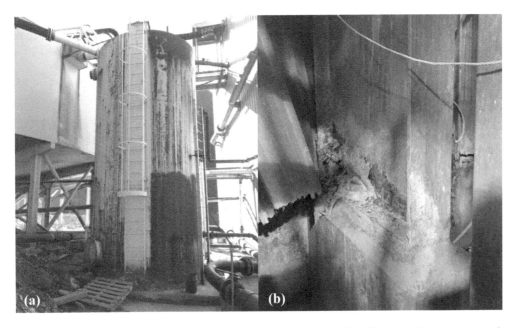

Figure 7. Examples of existing equipment corroded from contact with tailings (a) Process water tank and (b) Building column.

resulting structural load paths required certain upgrades to the existing structure as well as an over design of the new structural components to compensate for the insufficient seismic resistance of the existing.

4.2 Soil/Foundation conditions

Perhaps the most challenging part of the structural design was related to foundation conditions and the resulting impacts to the surrounding building elements. This specifically applied to the binder silo.

The existing structure and layout of the process equipment favoured a design that elevated the silo outlet to allow the binder screw conveyor to feed horizontally into the mixer. The specification required 250 tonnes of live binder storage. The procured silo had a 4.67 m diameter, an outlet height at 12.8 m and a silo eave height of ~22.5 m which resulted in a centre of gravity at 16 m. In addition, through the procurement process, a skirted silo design was selected rather than a custom structural platform for economic and layout reasons.

The silo layout was based on geotechnical assumptions before the commencement of the detailed design and the allowable soil loads could be determined. When past geotechnical data and reports were reviewed, it was found that the existing plant subgrade was largely fill material and the existing building and equipment footings were placed on dynamically compacted subgrade. Furthermore, the densification required for construction was performed only in areas where the footings were constructed in 2002 and the condition of the surrounding grade was unknown. A follow-up geotechnical investigation confirmed that the soils were composed of "*mixed organics, mine waste rock, and other mine waste materials*" to a depth of 3.3 m and limited the allowable soil bearing capacity to 75 kPa.

As per the geotechnical findings and resultant loads, calculations required a raft footing of 10.7 m by 10.2 m to resist to overturning moment of the relatively tall silo and high load. The site layout, however, set the distance from the centreline of the silo to the outside of the existing building to only 3.345 m. This accordingly required that the silo foundation to encroach into the building by over 2 m via concrete and rebar to tie into the structural steel supporting footings on the building's interior. When the need for a large raft footing was determined, a design review was called to determine alternatives. Options reviewed included silo relocation, pile foundations and anchoring to a nearby rock slope. In the end, the raft footing design was retained which resulted in construction sequence modifications.

The results of the geotechnical assessment indicated that a novel approach would need to be taken during excavation in order to avoid undercutting the existing building footings. The process specified by the geotechnical team called for excavating narrow channels, roughly a metre in width, to the specified elevation in order to avoid unloading the footings that the building columns were situated on. Engineered material was then backfilled into the cavity in 300 mm lifts and compacted with a hoe-pack until the final elevation at which the concrete would be formed and poured on was reached. Once an excavation was complete, the excavator would move over and begin the next cut until the full width of the foundation was excavated and backfilled. This was a time consuming and costly exercise that had to be carefully supervised and regularly surveyed. Normally a bulk excavation would be sufficient for such work.

The design challenges that were faced with the binder silo foundation also created additional construction challenges. In order to properly construct the foundation, the stem wall to the building had to be cut out and removed and the floor in the north east corner of the plant was saw-cut and removed. Rebar for the silo foundation was specified as Ø25 mm bar, layered at ~200 mm spacings. To effectively tie the foundation into the interior structural footings, it was necessary to extend the bar into the building and dowel into the existing concrete before pouring new concrete around it (Figure 8).

It was essential that the concrete forming the binder silo foundation was isolated from the building footings while still being fully tied into the upgraded structural steel supporting footings. This was done to avoid complex interactions with the new footings, potential problems due to differential settlement and the need to update to current building and seismic codes.

Figure 8. Silo and filter feed tank raft footings.

An additional complication was that approximately three times as many anchor bolts were required for the silo to comply to the seismic regulations of the code. This configuration was not provided in the original vendor issued for construction (IFC) drawings and, as such, was unknown at the time of the foundation design. Resultantly, no consideration was given to anchor bolt locations in relation to rebar in the foundations. The contractor had an exceptionally difficult time drilling holes for anchor bolts as they continually intercepted rebar due to the tight bar spacing and large bar diameter. In hindsight, a better approach would have been to template and place embeds during the pour.

The filter feed tank foundation was less complex as it did not have to resist the same overturning forces and, with a wider base, had different load characteristics. However, the excavation requirements were largely the same and the same issues with anchor bolt placement was encountered.

4.3 *Structural steel*

A complete 3D model of the building and equipment was created during the design. The principle equipment components and main structural members were well accounted for and accurately depicted in the model. However, some existing ancillary items, such as process piping, equipment components, cable tray routing and building tension bracing, were not all accounted for or correctly modelled. As such, the entire building and contained equipment were photo logged and manually measured to create an accurate model.

There were several cases where items were inaccurately estimated and caused rework onsite due to interferences. While none of these had a direct impact on the critical path of the installation activities, they nonetheless impacted the cost of the work and ultimately caused some delays to the overall schedule.

One example case included a key load-carrying beam on the north side of the plant directly interfered with one of the primary building columns. Similarly, a load carrying beam interference was discovered on the conveyor deck. These example cases required a redesign, additional steel fabrication and additional installation time.

In another case, the interference of the disc filter access deck with the motor for the bay door was discovered. As part of the retrofit design, the existing 3 platform stair tower had to be translated westward by 10 m to allow for the insertion of the new disc filter support structure. The tower also required modification to allow the upper floor to be extended for access and maintenance. The model indicated that the modified deck and tower would be tight up against the west wall but the main overhead door access to the plant would remain clear. However, when the tower was moved and modified, it was discovered that the location of the door motor and track did not match the model even though the main plant overhead door was accounted for. As a result, main load bearing column carrying the upper deck had to be removed and the tower had to be modified, requiring a substantial redesign.

4.4 Cantilever staircase to access binder system

Once the silo foundation design was finalized, the necessary access bridge could be detailed to connect the interior mixer deck to the silo. The initial intention was for a short stair run and platform that would be attached to both ends; one to the silo and the other to the interior mixer deck. It was subsequently found that no provision for loads or attachments of a platform had been made on the skirted silo design. The access platform and staircase therefore needed to be completely supported from the mixer deck structure which required a creative design effort due to limited anchoring members being available. Furthermore, the cantilever platform required a 90° turn which needed to be incorporated into the support design.

The challenges associated with the design of the cantilevered access platform were realized because the precise location of the silo could not be predicted by the design team, partly because of the necessity to situate the base ring wherever the anchor bolts could be drilled around the rebar in the concrete. The resulting stair and platform (Figure 9) are functional although it required site modifications to limit the deflections and minimize the gap between the platform and silo entrance.

4.5 Overland pipe installation

Several challenges were encountered with the overland paste pipeline layout and installation. No as-built survey information was available at the time of design. The design layout of the overland pipeline was therefore indicative and, in order to account for various interferences in the pipeline corridor, specified as "field fit" to the contractor's discretion.

Figure 9.   Cantilever staircase to access binder system.

A "field-fit" slope was indicated on the pipe run between the plant and the primary switch-over station. Sloped pipe supports were designed which could be adjusted to suit the in-situ slope requirements. Due to the way the curve switch-over spools were oriented, this slope could vary considerably without causing any layout issues. The pipe runs both east (to a borehole) and west (to a portal) from the same switch-over station and had no specification for sloped pipe brackets even though both runs were on a slight slope (<2%). This was handled by adjusting the drop spool of the switch-over station and then using grout packs to make up the gaps under the pipe supports.

As a survey laid out the pipe supports required along the retaining wall at the portal entrance, it was noted that the centreline for the pipe would be a little over a metre away from the portal retaining wall. However, the design specified the pipe centreline as being approximately 0.3 metres away from the concrete. This off-set was necessary in order to tie the paste pipeline from surface in with the underground installation which was already in place. An excavation was done adjacent to the wall and it was discovered that the portal retaining wall footing would also be in the way of the planned pipe support. Resultantly, a bespoke pipe support had to be designed that was doweled into the retaining wall footing.

5 ELECTRICAL UPGRADE/RETROFIT

The electrical updates required to support the retrofit required an additional 575 kW to the existing electrical load and approximately 330 control inputs and outputs in addition to various protocoled communication links with field devices.

5.1 *Existing cables and hardware*

The inspection of the cable runs revealed that the cable trays were both overloaded and covered in tailings. Large sections of cable trays were replaced due to poor condition arising from corrosion (Figure 10). Electrical cabling suffered the same fate with large quantities of it being replaced due to cracked sheathing and great difficulties in identifying which cables were in good conditions and which not. The poor condition of the cables and trays in the plant ended up being significant both in terms of cost and having a negative impact on the project schedule.

Figure 10. Cable tray's full and corroded/full of tailings.

5.2 MCC

The Mine specifications required that Motor Control Centre (MCC) were NEMA rated which requires an isolated enclosure, or "bucket" for each motor. Repurposing unused buckets in the existing MCC was problematic as many items were missing or had been modified historically. In addition, the design had to rely on outdated single line diagrams and cabinet layouts. Buckets were, however, repaired and utilized where possible to reduce the required size of the new MCC. Even so, the new equipment increased the plant motor count by over 45 in total and the new MCC required to house all this switch gear was over 6 m long in addition to several subpanels. Several MCC layouts and locations were considered including locating the new MCC in the plant area underneath the relocated stair tower which would separate it from the existing electrical room. This would, however, have complicated the control wiring and communications while requiring an upgrade to an IP65 enclosure. The existing electrical room was therefore reorganized to accommodate all the new switch gear and still meet code clearance requirements.

5.3 Automation hardware and software

The mine had planned to utilize the existing PLC and hardware with the addition of an expansion I/O to accommodate the new equipment. The Wonderware SCADA and HMI system were also to be reworked to minimize impact to the existing automation. After a thorough review it was found that the existing Quantum PLC had been obsolete and while service and repair would be available in the near future, new components were only available through 3rd party vendors who had excess inventory. As the plant life was scheduled until 2028, there was a concern that the existing automation hardware would not support the operation. In addition to the availability and support of the automation hardware, the PLC would also struggle to accomplish the required scan times. Lastly, the communication required to incorporate new vendor PLCs for components like the paste pumps which were supplied with modern PLCs was also of concern.

Once all these concerns were presented, it was agreed that the automation package required an upgrade to modern components, software and operating systems to provide longevity and scalability. Up to this time, the automation and interface programming were impacting both schedule and costs. An alternative solution was made to use the existing I/O's and that only new components added to the system would require new I/O points. The existing PLC was replaced with a communication module and the associated I/O rack would become a slave rack to the new PLC. There was still a risk of having obsolete I/O cards in the existing rack, but these could still be purchased and kept as spares. Importantly, the CPU and communication modules were new and expected to be available and supported well past the life of the plant.

6 BUDGET

The initial budget allocated to the project was sufficient based on reasonable expectations of costs garnered from past projects in conjunction with the scale of the work required. It did, however, suffer from some challenges along the way.

The project budget was based on a pre-feasibility design and fairly recent cost estimates. However, a stifled commodities/trade environment placed pressure on this estimate. During the bidding period, a sudden uptick in available local work caused rising labour costs while ongoing international tariff negotiations caused steel prices to rise, thereby influencing costs of both structural steel and piping. In an effort to control these costs, the project was submitted to tender for a second time to attempt a reduced construction costs through more robust competition. Concessions were made on long lead orders (e.g. piping) to reduce costs in exchange for higher risk on delivery and/or compromises on other commercial terms.

A second item that influenced the budget was the remote location of the project that seemed to be a unique area where potential contractor companies seem reluctant to mobilize. Adding to this, although the project budget was significant, the available construction budget seemed to fall below the threshold for project work that major mining contractors were willing to undertake. Correspondingly, many of the local construction firms were simply not large enough or properly equipped to take on the work. This left very few suitors and created a near monopoly for the very few companies that were both willing to take the work and large enough to be successful on the project. This made getting truly competitive bids very challenging. In hindsight, it would have been wise to make the project more financially attractive by including other parts of the scope in the tender package in an effort to attract some of the larger competitors.

The third key budget item was directly associated with the installation of the paste pipelines and associated infrastructure in the underground mine. Much of the mine was in adequate condition to accept paste line installation with little of the mine rehabilitation performed in advanced. The areas for the paste pipeline installation were beset with a variety of issues including bad ground, redundant or old services, failing services, blockages due to hydraulic fill spills, water, etc. While rehabilitating the mine should not be included in the paste line installation scope, costs should have been allowed for the reconfiguration of the existing services in order to accommodate the paste lines.

Finally, the budget did not make allowance for issues associated with new construction on poor foundation soils in an active seismic zone. Costs associated with upgrading structural foundations, technical excavation of fill materials around building footings, and two sizeable foundation pours in support of the filter feed tank and, the binder silo were categorically out of scope and not typical.

7 SCHEDULE

As with most retrofit projects, the paste plant refit had its fair share of scheduling challenges. The original restart schedule was aggressive as the mine endeavoured to get underway as soon as practicably possible in order to start generating revenues to off-set capital costs.

7.1 Long lead procurement

Initially, accountability for many of the mine's restart projects, including the paste plant retrofit, was with one person. Resultantly, procurement of long lead items languished due to many competing priorities. Due to procurement delays, the principle vendor selection criteria was often the delivery schedule which resulted in price escalation and budget impact. In other cases, the agreed-on delivery date could not actually be met causing construction inefficiencies and associated increase in costs.

Additionally, delays on the procurement side introduced further delays on the detailed engineering side with the most noteworthy delays being to the civil and structural design. A notable example of this, was the binder silo (discussed earlier) as this required significant rework to both the structural members and the footings/foundations.

Allocating sufficient procuremnt time is important to maintain the budget as the owner has time to make informed decisions and to create competition between vendors for new business.

7.2 Design drawings & supply delays

Initially, design drawings were issued for tender with basic assumptions being made in respect to configuration of civils, structural, mechanical, and piping.

Civil and Structural IFC drawings were delayed due to redesign which stemmed primarily from the follow-up site geotechnical assessment. As discussed previously, this incurred significant rework for the concrete footings, the concrete volumes, rebar configuration, and structural steel configuration to access the binder silo.

The changes to the drawings resulted in approximately one month's additional delay for steel fabrication and delivery. The delay closed the window on available floor time with the steel vendor and only part of the original scope was completed. To expedite the remainder of the work, the contractor fabricated a portion of the steel inventory in their own local shop.

While obtaining quotes for the new MCC a review of the drawings and specifications uncovered that the vendor had erroneously used the Issued for Tender drawings and not the IFC drawings when constructing the MCC configuration. While this was the vendor's mismanagement of information, it highlights the importance of carefully managing drawing revisions and the critical importance of reviewing build specifications prior to committing to a purchase. The MCC took close to six months from order to delivery and since it was a critical path item, an incorrect build would have been an enormous setback.

7.3 *Construction tender, bids, and award*

As discussed earlier, the construction scope for this project was ultimately tendered twice. This was largely due to a lack of initial competitive bids but was also deemed necessary because of scope changes resulting from the design updates. The design updates, as has previously been noted, were directly related to both procurement delays and the consequential detailed engineering delays. As a result, the award for the construction contract was delayed by three months.

This schedule impact was further complicated with the addition of the structural steel supply to the construction contract. This was a long lead item and the contractor was forced to begin construction in parallel with structural steel deliveries. While there were activities that preceded the steel installation, it was a tight scheduling window and, as previously identified, there were challenges which caused significant delays.

7.4 *Filter cake production*

Towards the end of construction, the mine decided to run the mill which would require the paste plant to be operational to the extent that it could produce filter cake. In order to accommodate this, temporary electrical connections had to be made to the parts of the circuit required to produce filter cake. This included both conveyors, the original disc filter, process and gland water systems, the thickener, and the thickener underflow pumping system. This change in priorities caused a significant cost escalation and approximately two months delay to the schedule.

7.5 *Underground pours*

At the time that this paper was written, two successful underground paste pours have been completed. The mine is planning to ramp up production shortly to ensure the paste backfill adequately supports the current mine plan.

8   CONCLUSIONS

One of the allures of retrofitting an existing plant is to take advantage of the existing infrastructure. However, there is a risk to this approach and careful consideration should be given to what items can be practically salvaged.

Generally, the mechanical and piping aspects of the plant construction were relatively smooth. There were areas that caused difficulty though, and while the work was anticipated, the magnitude of that work during construction was somewhat underestimated.

Most of the challenges experienced with construction of the plant were associated with the civil and structural packages. Key issues causing difficulties were poor foundation materials surrounding the plant, seismic requirements, design requirements for new binder silo, insufficient as-built information for the interior structural steel and unplanned structural interferences. Ultimately, the mine has been able to successfully conclude the paste plant retrofit job (Figure 11) with cemented paste being poured underground.

Figure 11.   Paste plant building complete with new binder silo and filter feed tank.

Minefill 2020-2021 – Hassani et al (eds)
© 2021 Taylor & Francis Group, London, ISBN 978-1-032-07203-6

Review of sandfill reticulation system at northern Ontario mine

Jacob Landriault & Drew Dewit
Golder Associates Ltd

Jordan Yamine
Anglo American plc

SUMMARY: A mine in northern Ontario Canada has been in operation for several years. The mine produces 1.2 Mtpa of nickel and copper ore. The facility contains a sand plant that has been in operation for 15 years. The current sand plant combines local alluvial sands, classified tailing and reclaim tailings to produce a cemented sandfill that is sent underground via gravity.

The current underground distribution system has been expanded and modified many times throughout the history of the mine and now has a total length of several kilometers. The mine operator has concerns regarding the pressure profile and flow regime of the fill throughout the reticulation system.

This paper will provide an overview and evaluation with regards to the operation of the reticulation system as well as the flow model and layout. It will highlight the pressure trends and indicate where the system can be optimized and modernized.

Keywords: Sandfill, reticulation, pressure, underground

1 INTRODUCTION

A backfill systems reticulation design is crucial to the system's ability to transport the backfill to the desired underground destinations. As such, optimization opportunities exist in most systems to allow more efficient and safer transport underground. High pressure areas in the reticulation system can be of concern to operators and underground personnel and can be minimized through proper design and operation.

The sandfill reticulation system at a mine in northern Ontario Canada is driven by gravity. This requires that, at a minimum, the first borehole operate in slack flow to provide the required head pressure for the remainder of the pipeline. This increases the ability to push material through at the desired flowrate, with some additional height available for upset conditions. The actual reticulation geometry could be improved to allow greater amounts of slack flow from the level piping through the borehole piping.

It is not uncommon to have several boreholes within a reticulation system operating under slack flow conditions, which allows for different areas of the mine to be filled using the available head. For a borehole serving a backfill stope that is relatively close, the borehole may only be a fraction full compared to when the pour site is a further distance away. It is possible that in some instances a borehole may operate in slack flow for some pours and be filled completely for other pours. This is dependent on the geometry of the line, mix design and flowrate contributing to the system pressure.

DOI: 10.1201/9781003205906-33

The intent of this paper is to review the reticulation system and to identify problem areas and balance the geometry of the reticulation system to reduce movement in the line at flexible connections and reduce the overall wear and degradation.

2 UNDERGROUND DISTRIBUTION (RETICULATION) SYSTEM

The underground distribution system (UDS) or reticulation system is a key design consideration for the transportation of cemented backfill underground. Many factors must be taken into consideration when designing a reticulation system such as geometry, pipe size and thickness as well as couplings and instrumentation. A lot of these design consideration are based on the friction loss of the backfill material in the pipeline. To mitigate risk of high-pressure events, robust pressure monitoring is required for any UDS that transports backfill underground. This can be accomplished with pressure instrumentation that can indicate a difference in pressure when it occurs in the line. This will alert the operators that there is a possible leak or a possible blockage. Burst discs or rupture spools should be incorporated into any UDS design to allow the release of pressure from the line. If a line becomes blocked and is not cleared before the cemented material sets, then the line would have to be blasted and replaced. This is time consuming and costly.

The mining method for the northern Ontario mine is cut-and-fill with the use of cemented sandfill in the mined stopes. The mine plan calls for mining underneath the sandfill with the stipulation that a cored sample's strength is at least 1 MPa. The actual cured backfill has a strength much greater than 1 MPa (approaching 2 MPa).

The mine consists of the "upper country", the already mined and filled areas of the mine. Due to the expanse of the system the sandfill travels through these areas before it reaches the operating mining areas. 3370 level (L) is the location of the toe of the boreholes that form the "upper country" and serving the current mining areas. The toe of the boreholes leading to the 3370 L are cased and equipped with a dead leg "Kettle". The majority of the pipe in the distribution system is 5" Schedule 120 induction hardened pipe with external shoulder couplings. The boreholes are DDH holes 3.78 inches in diameter. However, there is frequent use of flexible hoses throughout the distribution network. The use of flex hoses is not ideal for high pressure lines, this was done in the past for cost and layout reasons. Ideally the flex hoses would be replaced with steel pipe.

Free fall in the boreholes is a problem in the upper parts of the mine. The creation of vacuum in the boreholes has been the cause of damage to the reticulation system in the past and results in violent shaking of the line downstream of the borehole toe. In order to remediate this, different apparatii have be used in order to break the vacuum, with more or less success (gate valve, ball valve). Several boreholes are in an advanced stage of degradation, which has led to some big pieces of rock being found in the fill. This will need to be addressed immediately. It is recommended that new cased boreholes be drilled if the current boreholes cannot be repaired to the proper standards.

There are a pressure transducer and a flow meter on each of the two sandfill lines on 3370 L. However, this is the only place where such instruments are located. To better reduce risk the mine should install a greater number of transducers. As well the current pressure transducers could be better positioned at the top of the pipe i.e. 12 o'clock. All of the diverter valves used in the reticulation system are manually operated with no instrumentation present. It is recommended that these valves be modified to be automatic. This would reduce the requirements of the operators.

3 SANDFILL PLANT

The backfill consists of cement, sand and classified tailings that are blended as dry materials by a front-end loader and transported to the mixing tank via a conveyor. The ratio for backfill materials (loader buckets) is two reclaimed tailings, one operation mill classified tailings, and two alluvial sand. These proportions had been used since October 2018. There is no time weighting on the dry material belt and maintaining the correct proportions relies on the loader operator filling the main bin. Once a month, a "belt cut" is manually done in order to validate the feed rate. A tramp screen is in place in order to reject the lumps and oversize material. Typical operations produce 1200-1500 tons per day (tpd) of backfill. This is based on an 8-hour day of pouring. The plant consists of a "twin" system. However, one of them is not operational and has not been used for quite some time.

A three-inch orifice plate is used on top of the boreholes to better control the flow into the reticulation system. The plates must be changed once a week due to rapid wear. There is a "breather" on top of the boreholes, which must be opened more and more as the plate starts to deteriorate.

4 FLOW MODELLING

Three separate iterations of the flow model have been done for the reticulation system. The original geometry data that was provided was remodelled in a standard template with current friction loss estimates. This allowed the model to demonstrate the current state of the system and identify potential problem areas. Two modified flow models were subsequently completed to demonstrate how potential modifications could impact and benefit the system. The two modified flow models use different friction loss data, one from 2012 and the other from 2019.

Table 1 shows the current layout of the reticulation system from surface through the "upper country" to the lower areas of the mine. As a note, the level naming convention at this mine is in feet of elevation and study is metric. The flow model shows that there are several instances

Table 1. Current flow model data.

Node	Vertical distance (m)	Pipe or b/h length (m)	Pressure Loss Start (kPa)	End (kPa)	Available Head Start (kPa)	End (kPa)	Pressure Start (kPa)	End (kPa)
1 (Surface)	209	211	7734	7102	21406	17664	0	0
2	3	9	7102	7088	17664	17615	0	0
3	73	91	7088	6814	17615	16305	free fall	6
4	5	55	6814	6727	16305	16223	6	0
5	50	50	6727	6577	16223	15328	0	0
6	5	51	6577	6495	15328	15246	0	0
7	109	119	6495	6138	15246	13287	free fall	270
8	6	237	6138	5758	13287	13177	270	0
9	130	130	5758	5369	13177	10852	0	0
10	3	20	5369	5337	10852	10803	0	0
11	71	71	5337	5294	10803	9536	0	0
12	8	8	5294	5281	9536	9400	0	0
13	146	146	5281	4842	9400	6780	free fall	106
14	5	118	4842	4654	6780	6698	106	0
15	129	129	4654	4267	6698	4394	free fall	23
16	3	5	4267	4260	4394	4334	23	76
17	56	57	4260	4226	4334	3330	76	1046
18	5	31	4226	4175	3330	3242	1046	1083
19	1	610	4175	3200	3242	3221	1083	129
20	9	180	3200	2913	3221	3062	129	0
21	125	899	2913	1474	3062	830	free fall	644
22	-13	196	1474	1161	830	1064	644	96
23	32	32	1161	1077	1064	486	96	591
24	1	145	1077	643	486	461	591	181
25	16	17	643	632	461	175	181	458
26	10	71	632	421	175	0	458	421
End	0	140	421	0	0	0	421	0
	Total	**3827**						

Source: Review of Sandfill Reticulation System at Northern Ontario Mine.

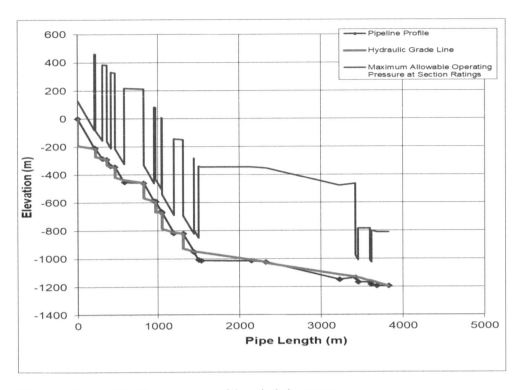

Figure 1.   Flow model of the current state of the reticulation system.

Source: Review of Sandfill Reticulation System at Northern Ontario Mine Golder Associates, Canada 2019.

where the pipeline operates in slack flow for substantial portions of the pipeline. This is also graphically represented in Figure 1. This large portion of slack flow is expected since the system has a substantial amount of elevation change through boreholes with little horizontal piping to balance it by adding backpressure through friction losses in the line.

In short, the flow model can be adjusted several different ways to achieve a more stable process. At this stage in the lifecycle of the operation there is a fixed specific recipe and throughput tonnage which leaves modification of the piping as one of the sole options to remedy the issues.

Table 2 shows the same flow model with several lengths of horizontal level piping elongated (edited values highlighted red). These horizontal lengths correspond to locations following boreholes.

By elongating those sections, changes to pipeline pressure as calculated through the system are realized and the result is a reticulation system that regains control of material flow at the bottom of each borehole where a jog occurs. What is also realized is the modification of the impact zone location from the free fall. The goal is to relocate the impact zone into the borehole, shortening the free fall distance and moving the impact away from the downstream piping which alternatively will require significant mechanical restraint.

These horizontal lengths would be added as control loops to the current system, where the operator installs additional piping that extends past the downstream target borehole and loops back to return and hit the borehole.

Table 2. Modified flow model data.

Node	Vertical distance (m)	Pipe or b/h length (m)	Pressure Loss Start (kPa)	Pressure Loss End (kPa)	Available Head Start (kPa)	Available Head End (kPa)	Pressure Start (kPa)	Pressure End (kPa)
1 (Surface)	209	211	10384	9584	21406	17664	free fall	318
2	3	209	9584	9216	17664	17615	318	0
3	73	91	9216	8870	17615	16305	free fall	191
4	5	155	8870	8597	16305	16223	191	0
5	50	50	8597	8407	16223	15328	free fall	201
6	5	161	8407	8124	15328	15246	201	0
7	109	119	8124	7671	15246	13287	free fall	308
8	6	237	7671	7254	13287	13177	308	0
9	130	130	7254	6761	13177	10852	free fall	197
10	3	140	6761	6515	10852	10803	197	0
11	71	71	6515	6482	10803	9536	0	0
12	8	8	6482	6468	9536	9400	0	0
13	146	146	6468	5912	9400	6780	free fall	214
14	5	168	5912	5617	6780	6698	214	0
15	129	129	5617	5127	6698	4394	free fall	733
16	3	5	5127	5119	4394	4334	733	785
17	56	57	5119	5093	4334	3330	785	1763
18	5	131	5093	4863	3330	3242	1763	1620
19	1	610	4863	3791	3242	3221	1620	570
20	9	180	3791	3475	3221	3062	570	413
21	125	899	3475	1894	3062	830	413	1064
22	-13	196	1894	1550	830	1064	1064	485
23	32	32	1550	1444	1064	486	485	959
24	1	145	1444	859	486	461	959	398
25	16	17	859	852	461	175	398	677
26	10	71	852	567	175	0	677	567
End	0	140	567	0	0	0	567	0
Total		**4507**						

Source: Review of Sandfill Reticulation System at Northern Ontario Mine.

5 RETICULATION SYSTEM DESIGN RECOMMENDATIONS

The original model showed areas in the reticulation system that were in freefall for multiple consecutive boreholes and short sections of 'level pipe' that connected two boreholes.

The redesign focused on eliminating freefall in large sections of the pipeline and minimizing it to select boreholes that would operate in slack flow and have subsequent boreholes filled to maintain order in the system. This results in greatly reducing pressure spikes formed from material exiting completely empty boreholes with no resistance which is hard on the downstream piping and pipe supports to regain control of the mass of backfill. There have been instances, during operation, of flexible piping near borehole bottoms shaking violently during operation due to that phenomenon, this addition of piping would be one solution to that problem.

Other recommendations include locating pressure transmitters away from the bottom of the borehole as the system currently operates. The best practice to ensure the system works at optimum efficiency is to ensure the pipeline is full on the levels when backfilling. This eliminates shocks delivered to instrumentation as material rockets through horizontal piping which can damage the diaphragm.

It is also recommended that two transmitters be located a known distance apart on a straight length of pipe (approx. 300ft) to capture friction loss data of their backfill material

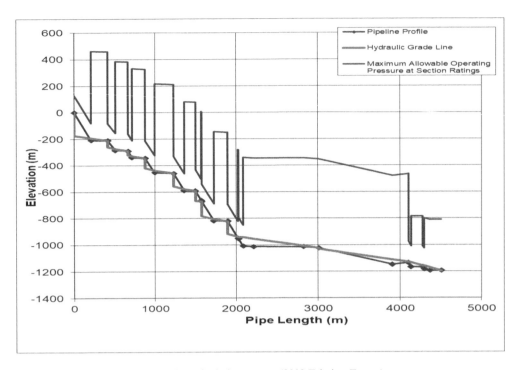

Figure 2.   Modified Flow model the reticulation system (2012 Friction Factor).

Source: Review of Sandfill Reticulation System at Northern Ontario Mine Golder Associates, Canada 2019.

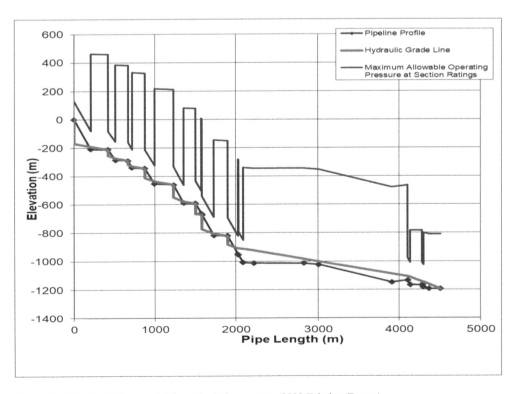

Figure 3.   Modified Flow model the reticulation system (2019 Friction Factor).

Source: Review of Sandfill Reticulation System at Northern Ontario Mine Golder Associates, Canada 2019.

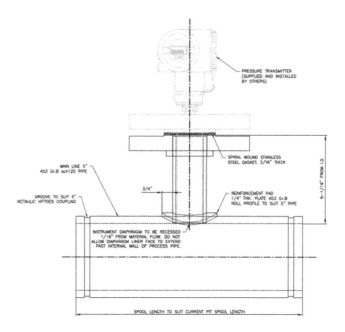

Figure 4. Typical Pressure Monitoring Arrangement.

Source: Review of Sandfill Reticulation System at Northern Ontario Mine Golder Associates, Canada 2019.

which will aid in further tuning of the system to attenuate issues, by inputting real time data into the model.

The use rubber flex hose in the reticulation system can be acceptable, so long as the borehole is sufficiently full, allowing the backfill to exit the borehole in a controlled manner. Operating with a nearly empty borehole causes a lot of movement in the hose and is a potential operation and health and safety risk

Ultimately it is recommended that preliminary efforts be undertaken to elongate the horizontal piping at specific chronic problem areas to demonstrate that the symptoms are attenuated and validate the model and the theory.

BIBLIOGRAPHY

Hatch 2014. *12820-01-Coleman Mine Sandfill System Audit*:
Paterson and Cooke 2012. *Hydraulic analysis Rev 5. (Coleman)*.

Minefill 2020-2021 – Hassani et al (eds)
© 2021 Taylor & Francis Group, London, ISBN 978-1-032-07203-6

Paste-waste design and implementation at Newmont Goldcorp's Tanami Operation

Ryan L. Veenstra & Johannes J. Grobler
Newmont Goldcorp Australia

SUMMARY: Newmont Goldcorp's Tanami Operations consists of the Dead Bullock Soak underground mine (DBS) and the Granites Processing Plant. DBS is currently undergoing an expansion which has created a surplus of underground (UG) waste rock. In order to deal with this surplus of waste rock a paste-waste project was initiated. The goal of this project was to develop and implement a design methodology of maximizing waste rock disposal by depositing the waste rock in primary stopes (usually only filled with 100% cemented paste backfill [CPB]). This paper presents the details of how the design methodology was developed. This includes an analysis of rock chute placement, waste rock versus CPB fill rates, and backfill strength design. Finally, the paper also presents a case study of a paste-waste stope. This study includes an economic analysis of paste-waste, highlighting the importance of paste-waste to DBS in the future.

Keywords: Paste Waste, Cemented Paste Backfill, Design, Stability

1 INTRODUCTION

The Dead Bullock Soak (DBS) underground (UG) mine at Newmont Goldcorp's Tanami Operation (NGT) is expanding to be able to mine deeper and more laterally distance stopes. In order to excavate these stopes, the amount of UG development (i.e. access tunnels to get to the stopes) will also increase, causing a subsequent increase in waste rock (WR).

The DBS mine currently operates in a primary-secondary sequence, with the primaries being filled with cemented paste backfill (CPB) and the secondaries with WR. Due to the expansion, there is not enough secondary stope void to accommodate the WR being generated. Figure 1 shows the life-of-mine (LOM) WR profile. This figure compares the amount of WR that is generated (green column) and the amount of WR void available (red column). The difference (blue columns) between these values is the amount of WR that will need to be trucked to surface. This figure shows that there will be a surplus of WR generated over the next 7 years. In order to reduce the amount of WR transported to surface, a project was initiated to increase the amount of WR being deposited UG. Part of this project was building on DBS's current paste-waste (PW) processes. PW is defined a process that backfills a stope with both CPB and WR.

2 PASTE-WASTE METHODOLOGY

PW backfill masses have been placed at DBS for several years in the form of CPB plugs or caps. However, these were generally secondary stopes that would be undercut or required tight filling. In order to increase the amount of PW, a different methodology was required.

DOI: 10.1201/9781003205906-34

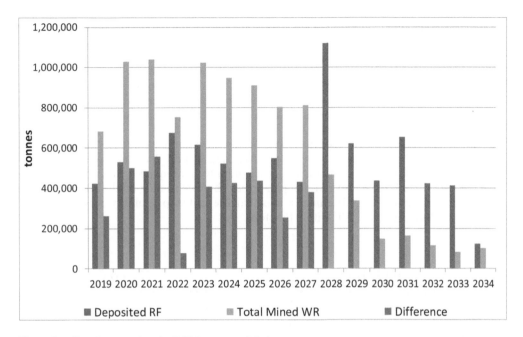

Figure 1. Graph presenting the LOM waste rock balance.

The PW method needed to rely on WR being dumped into the stope via the UG trucking fleet. The majority of the run-of-mine waste rock is too large to reticulate. Figure 2 contains the particle size distributions (PSD) for the current tailings blend being used for CPB production, the run-of-mine WR, and the raise bore fines. The dashed vertical line is a rule-of-thumb for the maximum particle size that could be transported by the DBS reticulation system. Any screening or crushing process to reduce the WR to the required size was unfeasible.

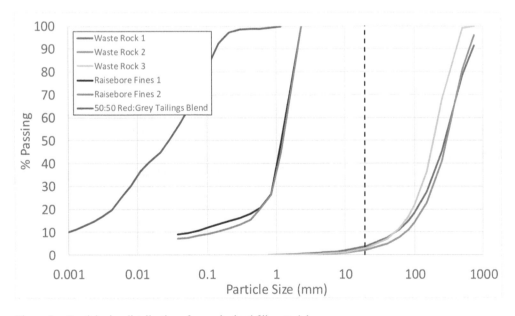

Figure 2. Particle size distributions for onsite backfill materials.

Given the truck dumping requirement, it was decided to pursue an 'encapsulation' type of filling method. In this method, the waste is isolated from an exposure surface by being encapsulated completely by the CPB. The amount of WR that can be deposited using this method varies considerably on the number of vertical exposures and the dumping location.

Figure 3 highlights the difference in the amount of WR that can be deposited into the same sized stope with four vertical exposures versus two touching vertical exposures. Note that if the two exposure surfaces were opposite to each other the amount of deposited WR would be the same as the four-exposure stope.

Figure 4 highlights the difference that the deposition location makes to the amount of WR that can be placed into the same sized stope given. The central diagram represents an end-dump type of deposition while the right-hand diagram would require a rock chute. The left-hand diagram could be either an end-dump or a rock chute depending on the access drive and stope geometries. In order to determine a mass balance and stope stability, it is necessary to determine the location of the WR within the stope.

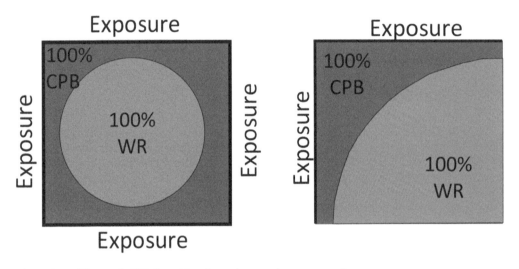

Figure 3.   Difference in WR deposition due to increased exposure surfaces.

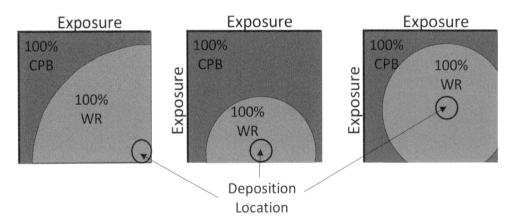

Figure 4.   Difference in WR deposition due to different deposition locations.

2.1 *Segregation modelling*

Segregation modelling refers to determining the position of the WR within the stope. Segregation modelling has two parts. The first part was determining where the WR would be deposited in the stope while the second was to determine what the shape of the WR would be within the stope.

2.1.1 *Deposition location of waste rock*

Determining the location of the WR profile was accomplished by using Trajec3D (Basrock, 2019). This freeware program was originally designed for rock fall analysis for open pit walls but has also been used to show material flow within a stope (Basson et. al., 2015).

Figure 5 a) and b) show two examples of a calibration exercise conducted comparing the surfaces of deposited WR, as determined by cavity monitoring surveys (CMS), with the flow trajectory predicted by Trajec3D. The CMS surveys agree well with the modelled locations. Figure 5 c) shows a modelled flow trajectory for an angled rock chute.

2.1.2 *Waste rock shape determination*

The shape of the WR depends on the deposition rates of the CPB and WR within the stope, the shape of the stope, and deposition location. If both deposition rates are kept constant and

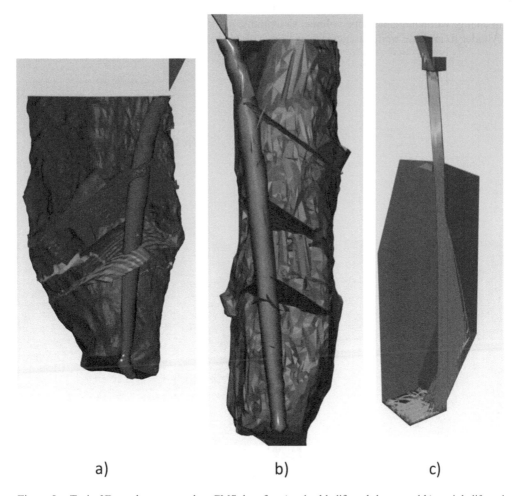

a)   b)   c)

Figure 5.   Trajec3D results compared to CMS data for a) a double lift end-dump and b) a triple lift end dump, and c) a Trajec3D model single lift angled rock chute.

are deposited into a uniform shaped stope then the backfill rise-rate will eventually stabilize to a continuous rate. The backfill rise-rate is the height of the CPB level including the WR within it.

This is shown schematically in Figure 6, where the initial WR cone volume is small. This means that the CPB has a large volume to fill, causing the small combined backfill rise-rate. With subsequent WR deposition, the rate-of-change of the WR volume becomes negligible and the backfill rise-rate stabilizes. Figure 6 also introduces the concept of a WR halo. Essentially this halo is a transition zone between the 100% WR section inside stope and the 100% CPB encapsulating it. Currently, the size of the halo (or even if a halo exists) is unknown. Conservatively, the size of the halo was determined to be the radius of the WR cylinder plus twice the length of the isolated section of each WR cone (as shown in Figure 6).

However, the above deposition model is unrealistic as stopes are not uniformly shaped and the deposition rates, particularly the WR rates, are not constant. The inconstant nature of the WR deposition is due to when the WR trucks are available. In order to deal with this inconsistency, the design was initially bookended by a 'staggered' or an 'instantaneous' design. The staggered approach mean that each truck was deposited at equal-time intervals over a shift whereas the instantaneous approach meant that the entire amount of deposited WR for the shift was deposited instantaneously once during the shift.

An algorithm was developed on-site accounting for the volume changes within the stope as well as the CPB and WR deposition rates. This allowed the size of the WR cylinder and halo to change with changing stope volume. Essentially, a decrease in stope volume caused the WR cylinder to increase while an increase in stope volume caused a decrease. There are limitations

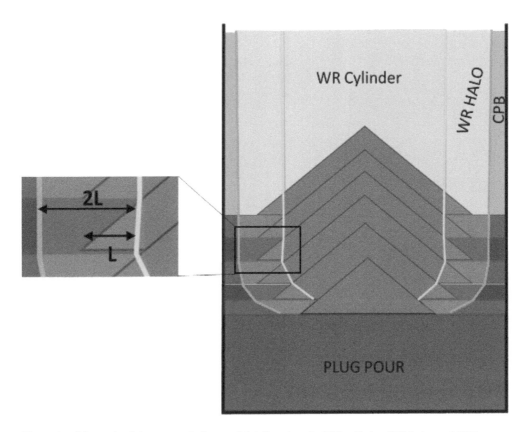

Figure 6.   Schematic of the encapsulation model delineating the WR cylinder, WR halo, and CPB areas.

to the algorithm as it works best when the entire WR volume is encapsulated within the stope volume (e.g. the 4-sided exposure schematic in Figure 3). Algorithm variations have been trialled using half or quarter volume cones but these require stopes with relatively straight sides and/or a 90° corner.

An example of the results of this algorithm are shown in Figure 7. This graph shows the results of a staggered versus instantaneous design analysis by plotting WR radius versus back-fill height. The volume of the stope with height is also shown. A comparison of the WR radius to stope volume line confirms the stope volume to WR radius relationship discussed above.

Figure 7 also shows the staggered model tends to have smaller cylinder and halo sizes but more irregular curves. This is due to resolution issues within the algorithm. Therefore, the instantaneous values were used for stability analysis as this was considered a worst-case scenario.

Figure 8 shows a comparison of the algorithm's results to those of Deswik.CAD's backfill simulator for an instantaneous condition (Deswik, 2019). There was excellent correlation between the two. Unfortunately, the Deswik.CAD analysis took over 5 hours to run which was substantially slower than the site-developed algorithm. However, for more complex

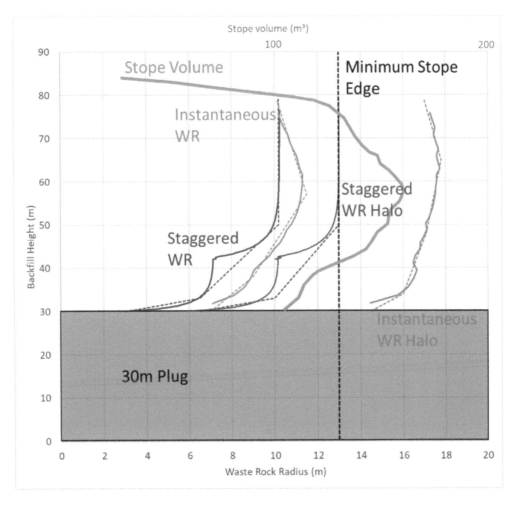

Figure 7.   WR size algorithm results comparing a staggered and instantaneous design. The stope volume is also shown.

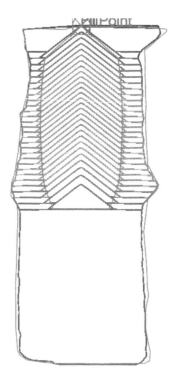

Figure 8. Comparison of WR volume determined by site algorithm (magenta) and Deswik.CAD.

volume relationships (i.e. the WR volume is not enclosed and cannot be approximated using half or a quarter of cone volume) then using Deswik.CAD would be required.

2.2 *Stability modelling*

The stability modelling was completed using Itasca's Flac3D software (Itasca, 2019). This software has a built-in factor of safety (FOS) solver. This solver uses a strength reduction method that modifies the input strength parameters (primarily cohesion and friction angle but can include other parameters such as tensile strength, dilation angle, etc.) by using a safety factor. The equations for the cohesion and friction angle reductions are below, where ϕ is the friction angle and F^{trial} is the safety factor:

$$cohesion^{trial} = \frac{cohesion}{F^{trial}} \tag{1}$$

$$\phi^{trial} = arctan\left(\frac{\tan \phi}{F^{Trial}}\right) \tag{2}$$

The program brackets the minimum safety factor by running a series of models at different safety factors. The solver stops when the minimum safety factor has been determined.

2.2.1 *Modelling inputs*

The basic parameters for any Flac3D model are density, elastic modulus, and Poisson's ratio. The use of an additional failure criterion requires further strength-related parameters. In this methodology, the Mohr-Coulomb failure criterion was used. The cohesive strength, friction angle, and tensile strength were incorporated in that criterion.

Most CPB testing programs involve determining the unconfined compressive strength (UCS) of the CPB with time. For this work, inputs were required for the WR and the WR halo zones. The WR inputs were determined from shear box tests but the WR halo inputs were more difficult to determine.

In order to obtain a better understanding of the strength parameters, above what UCS testing can provide, multiple consolidated undrained (CU) triaxial tests were undertaken of both the 100% CPB as well as various mixes of CPB and raise bore fines (RBF). These RBF mixes were intended to represent the mixing between the CPB and WR. RBF were used as they were small enough to be added to casting cylinders (PSDs of the RBF are shown in Figure 2).

Figure 9 shows the cohesion, UCS, elastic modulus, and friction angle curves for the RBF mixes, which were normalized to 100% CPB control tests. Multiple curing ages and binder contents were included in this plot and, while there is some scatter, there are some general trends. Essentially, the friction angle increases with increased addition of RBF, while the cohesion, UCS, and elastic modulus decrease. It was decided to use a 50:50 CPB to RBF ratio for determining the strength of the halo material as this was assumed to conservative. The tensile strength of the CPB or halo was assumed to be ten percent of the material's UCS.

2.2.2 Modelling results

Figure 10 shows an example of the analysis completed using the FLac3D FOS solver. Figure 10 a) shows the results of a model with no encapsulated WR, generating a relatively small surface 'bowl' failure with a FOS of 2.1. Figure 10 b) shows the configuration of the WR and halo within the stope. Figure 10 c) shows the results from a model of the second section, generating a much larger, full body failure surface with a FOS of 1.87. Note that the CPB strength was not modified between the models.

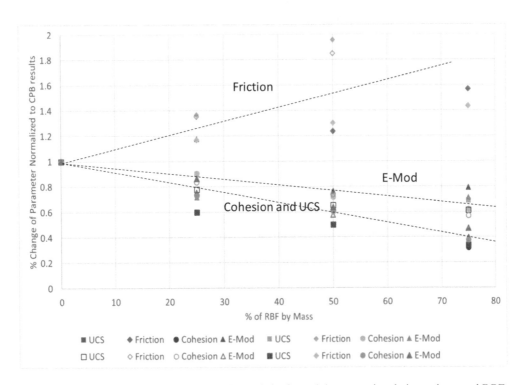

Figure 9. Normalized friction, cohesion, UCS, and elastic modulus curves in relation to increased RBF.

389

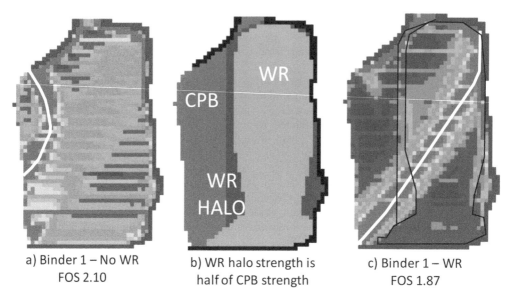

a) Binder 1 – No WR
FOS 2.10

b) WR halo strength is
half of CPB strength

c) Binder 1 – WR
FOS 1.87

Figure 10. Stope sections showing the results for a model without the WR cylinder and halo, the locations of the added WR cylinder and halo, and the results of the model with the WR cylinder and halo.

2.3 *Methodology conclusions*

The methodology shown here provides framework by which an encapsulation paste-waste design can be completed and highlights the optimization that is allowed by this approach. It allows flexibility to the amount of WR that can be deposited in a stope by taking into account the operational or geometrical constraints. Additionally, the amount of binder required to produce a stable backfill mass can be optimized. Several key assumptions have been made in order to do this analysis:

1. There is a transition zone between the areas of 100% CPB and 100% WR
2. The strength of this transition zone can be approximated by the RBF strength testing
3. That using the 50:50 CPB/WR strength ratios are a worst-case scenario

Future work is planned to determine the validity of these assumptions.

3 CASE STUDY – 085A STOPE

The 085A was the first PW stope to be filled using the methodology discussed in Section 2. The stope was located approximately 800 m below surface in the Upper Auron portion of the DBS mine. It was a double-lift primary stope that was scheduled to be completely filled with CPB. The stope will be exposed vertically on its east and west sides and has a horizontal 'ledge' exposure. These are shown in the right-hand drawing in Figure 11. The exposures are numbered in the sequence that they were to occur.

The upper level access drive was undercut which required the WR to be deposited through a rock chute from the level above the stope. However, the CPB was able to be deposited into the stope from the upper level access drive. The rock chute was designed to be vertically inclined.

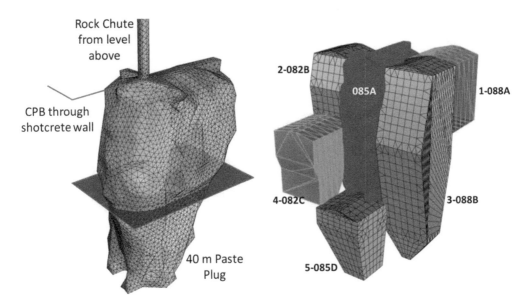

Figure 11. Drawings showing a) the 085A filling locations and the shape of WR cylinder, and b) the backfill exposures.

3.1 *Design*

The initial design had the PW process starting once an initial 15 m plug was installed. The plug was designed to deal with the 'ledge' exposure. However, due to operational demands, the first lift was completely filled with CPB.

The WR deposition rate targeted was 12 trucks per shift, which works out to approximately 1,400 tonnes per day, an amount that accounts for approximately 70% of DBS's total daily WR deposition. The onsite algorithm was used to determine the shape of the WR cylinder based on the location of the rock chute (green cylinder in Figure 11). Note that a Trajec3D model (Section 2.1.1.) was not completed for this stope as part of the initial design.

The design was further refined to determine the total of amount of WR and CPB that would be placed into the stope and the expected filling times. These plots are show in Figure 12. The total stope void (above the initial plug) was approximately 62,000 m³, of which the WR was going to account for approximately 18% (or 11,000 m³). There was also an expected 4-day decrease in filling time due to PW of this stope.

The stability analysis indicated that the third exposure (088B stope) was the critical exposure and the binder content was modified to achieve a FOS of 2. This minimum FOS was determined from a previous calibration exercise undertaken by DBS backfill personnel (Veenstra, 2019). Figure 13 shows a graph of the binder contents with height. The blue line represents the binder content determined from using NGT's old analytical calculator for a stope with 100% CPB, the purple line is the binder content determined using Flac3D for a stope with 100% CPB (FOS of 2), and the green line is the binder content determined using Flac3D for the above PW design (FOS of ~2). The approximate failure surface plot is also shown in Figure 13, with the failure surface initiating at the top of the original plug, running through the bottom cylinder, and then along the interface of the WR cylinder and the CPB.

3.2 *As-Built*

PW operations in the 085A stope started on September 28th, 2019. There was a 1.5 day lag in WR deposition to account for the blasted rock chute material. The first truck used the rock

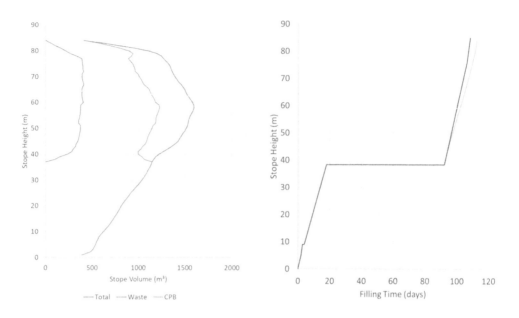

Figure 12. Graphs showing designed volume of CPB and WR within the stope and the estimated minimal stope filling time.

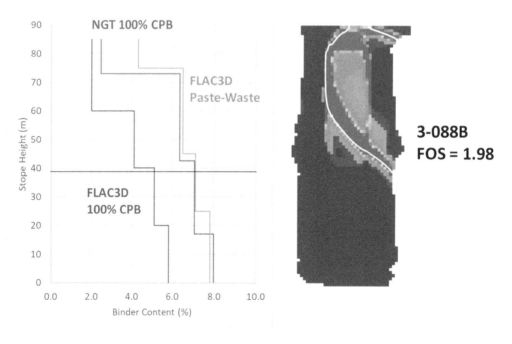

Figure 13. Graph comparing the design PW to 100% CPB binder contents and the modelled failure surface of the critical exposure.

chute on the September 30th, 2019. Figure 14 shows the cumulative gain of the PW material during the 085A pour. There are two distinct filling areas separated by a multi-day paste plant (PP) maintenance shutdown. There was a period just after the shutdown where limited WR

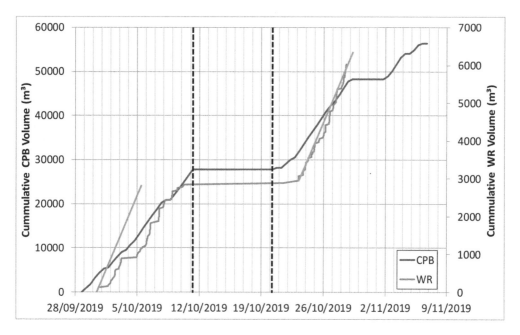

Figure 14. Cumulative depositional rates for CPB and WR into the 085A. The green lines are the maximum WR deposition rate allowed (12 trucks per shift).

was deposited. This was due to the rock chute's stop block being damaged and requiring repairs. The green lines show the design PW rate of 12 trucks per shift. A comparison of these lines to the actual deposition show that the WR placement rates, prior to the shutdown, were jagged and significantly below the design rate. Some of these were due to miscommunications between the paste plant operator and Mine Control resulting in the rock chute being closed even though the PP was operating. The WR deposition rates after the PP shutdown were much improved. This improvement was driven by the experience derived from the previous deposition and improved direction and communication between the stakeholders.

There were some interesting results that came out of the deposition rate monitoring. At DBS, truck movements are the responsibility of Mine Control (or Pitram). Prior to filling the 085A stope, it was anticipated that, given conversation with Mine Control, it would be easier to space the trucks over the shift. However, as observed in Figure 14, it rapidly became apparent that it was easier to dump several trucks over a short time period. In this way, the actual deposition patterns were closer to the 'instantaneous' model (discussed in Section 2.1.2) than the expected behaviour.

In order to monitor the pour, CMS were taken every couple of days while the stope was filling (Figure 15). The 'start of paste-waste' surface was taken after the plug was finished. The CMS taken on the September 30th was to determine if the rock chute material had been sufficiently covered prior to open the rock chute for WR deposition. There was no WR deposited between the 21st and the 23rd and this is shown in how the surfaces' shape change between CMSs taken on the 19th and the 23rd. The surfaces also show some interesting 'ponding' effects where flow of the CPB was modified by the WR cone.

Also shown in Figure 15 is the design and as-built WR cylinders. Note that the as-built was generated using the CMS surfaces and WR deposition log, and is intended to be a worst-case scenario. A spatial comparison shows good alignment between the design and the as-built in the north-south direction but a poor alignment in the east-west direction. For this particular stope, the alignment in the north-south direction was most important as this kept the WR material away from the exposed surfaces.

393

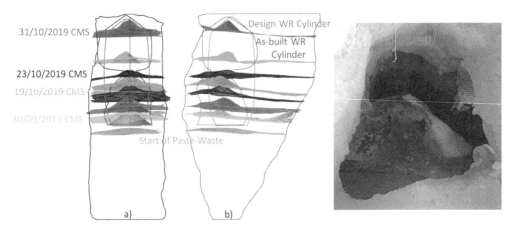

Figure 15.   Cavity monitoring surveys taken during the pour with a) looking north, and b) looking west. The initial WR cylinder design shape (magenta) and the 'worst-case' as-built (dark blue) are also shown. Also shown is a photograph of the top of the stope taken on 31/10/2019.

However, it would be beneficial to know if the east-west alignment issues could be explained. To this end, a Trajec3D model was completed using the CMS of the rock chute as an approximate guide. The actual rock chute was not vertical but was slightly inclined towards the south. The results are shown in Figure 16, and show better correlation to WR cone locations than the original WR design shape (shown in magenta). However, some of the surfaces, particularly the 19/10/2019 CMS, do not match the Trajec3D analysis. This difference highlights the importance of tracking the backfill surface over the duration of the pour.

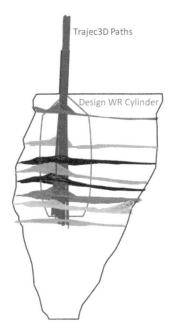

Figure 16.   Traject3D model compared to the CMS surfaces obtained while filling the 085A and a photograph of the top of the stope after WR deposition finished.

Stability modelling is not usually redone unless there is an issue during pouring. In this case an erroneous calculation on the stope run sheet after the PP shutdown meant that the PP operator started the plant at a lower binder content. This mistake was eventually rectified but it meant there was a weak layer in the backfill. Therefore, a stability model was completed to identify if there were stability issues and to determine if additional binder was required in the upper portions of the stope. The results of this model are shown in Figure 17. Using the as-built WR cylinder shape, the FOS for the critical exposure models without a weak layer and with a weak layer were 2.09 and 2.00 respectively, meaning there was no stability issue.

There are two points of interest that come out of these stability models. The first was that the FOS of the non-PW model was higher than the design model. This was due to the decreased WR deposition during the first deposition phase. The other was that the failure surface, in the PS model, was shallower compared to the surface generated in Figure 13, due to the smaller size of the lower WR cylinder.

3.3 *Reconciliation*

A comparison of the original design and the as-built highlights the following:

- The original design called for a deposition of approximately 20,000 tonnes of WR. The actual amount of deposited WR was approximately 12,000 tonnes. There are several reasons for this, mainly the decreased WR deposition rates obtained prior to the shutdown and the time spent repairing the stop block.
- It was anticipated that PW would decrease the filling time by around 4 days. The actual decrease in time was about 3 days.
- The original design binder cost was approximately $345,000 more than the non-PW design (purple line versus green line in Figure 13). Normalizing this cost to the amount of WR deposited, the unit cost is around 12 $ binder per deposited WR tonne. The as-built results give a unit cost of 37 $/tonne.

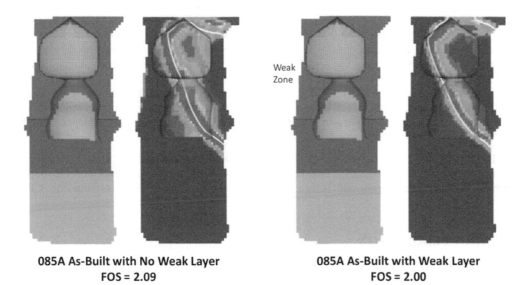

085A As-Built with No Weak Layer
FOS = 2.09

085A As-Built with Weak Layer
FOS = 2.00

Figure 17.   Stability model results comparing models with and without the weak CPB layer.

3.4 *Future work*

The 085A stope will continue to be monitored during the subsequent exposures. There are also plans to mine into the backfill along the mid-level drive in order to get an idea of the size of the halo and what it looks like. An attempt to obtain drill core from the fill mass will also be attempted dependent on drill availability.

4 CONCLUSIONS AND RECOMMENDATIONS

Due to the expansion of the DBS UG mine, there will be a WR surplus for the next 7 years, requiring this material to be transported and disposed of on surface. One of the mitigation measures, for reducing this surplus, was to increase the amount of PW being deposited UG by increasing the amount of in-stope WR deposition.

The first part of this paper presented a design methodology for designing a PW backfill mass that utilizes an encapsulation approach. Included in this methodology were discussions related to determining how the WR enters the stope, the approximate volume and shape of the encapsulated WR within the stope, and the stability of the backfill mass. The stability analysis also looked at how the modelling inputs were developed and utilized.

The second part of this paper was a case study dealing with the 085A stope, which was the first backfill mass designed using this methodology. Included in this case study was a monitoring program that allowed the design to be compared to how the stope was actually filled.

The design was relatively complicated due to multiple, large exposures on two sides and limited access to the stope (due to the southern wall overbreak). This limited the amount of WR that could be deposited.

The CMS taken of the stope and the deposited material logs maintained while it was filled allowed for an as-built of the stope to be created. There were several notable deviations from the design:

- The depositional surfaces generated from the CMS showed that the WR surfaces migrated southwards during filling.
- The first was that there was significantly less WR deposited. The main reasons for this were the rock chute closure while the stop block was repaired and the slower initial WR deposition rate. The deposition rate improved during the later stages of the stope filling.
- The second was that a weak CPB layer was poured due to a mistake made at the paste plant. The impact of this weak layer was analysed by using the stability modelling methodology presented in this paper and it was found that the weak layer did not reduce the stability FOS below the site standard.

In general, the response to this initial trial has been positive and additional stopes are being identified and ranked as PW targets. Future work is also planned for the 085A which includes monitoring the future exposures and obtaining samples from the CPB and WR halo zones.

ACKNOWLEDGEMENTS

The Authors would like to acknowledge the help of the DBS Pitram Operators, Adam Zajac, and Ceinwen Mackay in collection the data used in this paper. The Authors also acknowledge the Newmont Goldcorp Tanami Management Team for their support in the writing of this paper.

BIBLIOGRAPHY

BasRock. 2019, '*Trajec3D. Version 1.7.2.3*'. BasRock, Perth, WA, Australia.

Basson et. al. 2015 – Basson, F.R.P., Dalton, N.J, Barsanti, B.J, and Flemmer, A.L. 2015. Simulate waste rock flow during co-disposal for dilution control. [In:] *Proceedings of Design Methods in Underground Mining*. Perth: ACG.

Deswik. 2019, '*Deswik.CAD. Version 2017.1*'. Deswik, Brisbane, QLD, Australia.

Itasca Consulting Group, Inc. 2019, '*FLAC3D: Fast Lagrangian Analysis of Continua in 3 Dimensions. Version 6.0*'. ICG, Minneapolis MN, USA.

Veenstra, R.L. 2019. *Flac3D FOS Solver Calibration to Tanami Stopes*. Newmont Goldcorp Australia: Internal Communication.

Minefill 2020-2021 – Hassani et al (eds)
© *2021 Taylor & Francis Group, London, ISBN 978-1-032-07203-6*

Summary of improvements to the backfill system at DBS operations

Adam Zajac & Ryan L. Veenstra
Newmont Goldcorp Australia

SUMMARY: Newmont Goldcorp Australia's Tanami Operations consists of the Dead Bullock Soak underground mine (DBS) and the Granites Processing Plant. The Backfill Group at DBS has recently introduced a number of improvements to benefit the UG reticulation system. These improvements focused on three areas: reduction of major blockage downtime, continuous monitoring of tight filling operations, and paste plant automation. The first part of the paper will cover the installation of dump/diversion valves throughout the DBS reticulation system. This paper will present why these valves were installed, the design behind determining where they were placed, and several case studies showing how their installation has benefited DBS. The second part of the paper will detail the changes that allow for continuous monitoring of tight filling and present case studies on how this has benefited DBS. The third part of the paper present the results of a trial paste plant automation project which was run in conjunction with NGA's Process Control Group. This section will detail how this automation works and present case studies on how this automation has benefited DBS.

Keywords: Continuous Improvement, Valves, Cemented Paste Backfill, Automatic Pilot, Downtime

1 INTRODUCTION

Newmont Tanami Operations is in the Northern Territory in the Tanami Desert, Situated 540 km north-west of Alice Spring on Aboriginal freehold land. The camp is fully equipped 1000-bed village with the underground mine located 42kms from the village. Modern mining has been ongoing since 1985 in the Tanami region but the history of gold discovery dates back to 1898 explorer and prospector Allan Davidson, along with his team of four men and nine camels, set off to explore the Tanami Desert. The underground mine is currently 1.4km deep with a trucking fleet of 18 trucks.

The DBS cemented paste backfill plant (PP) was commissioned and begun production in March 2012. Until recently, the tailings used at the PP were harvested from decommissioned tailings ponds near the mill, approximately 45 km away from the DBS. For the first 4 years of production the PP ran these harvested oxide tailings. A dewatering system (cyclone bank and filter belt) was installed at the start of 2018, which utilises the tails stream directly off the mill circuit and allows the blending of the harvested oxide and dewatered tailings. Various blends have been trailed since the dewatering system was brought online with 50/50 Blend currently being used.

DBS is currently expanding and as mining operations start excavating stopes that are both deeper and laterally more distant from the PP, the reticulation network will also need to be expanded. The required placement rate of CPB also increases from approximately 550,000 m³ per year to 775,000 m³ per year over the next few years. This expansion will require improved

DOI: 10.1201/9781003205906-35

control over the backfill system as well as a reduction in downtime. In order to improve the operating capabilities of the DBS backfill system a series of improvements were implemented. Three of these improvements are summarized in this paper:

1. reduction of major blockage downtime,
2. continuous monitoring of tight filling operations and
3. paste plant automation.

2 REDUCTION OF MAJOR BLOCKAGE DOWNTIME

During the early years of backfill production, the PP had minimal down time due to major blockages. A major blockage is classified as a blockage that causes more than a half-a-shift of downtime. This lack of downtime is mainly attributed to new infrastructure and the relatively shallow stopes being filled resulting in short reticulation runs that were easy to unblock. As the mining front moved deeper, clearing blockages became more problematic. During 2016, the PP was down for over two months as a result of multiple major blockages (Figure 1).

This escalating trend of downtime highlighted an area for concern and improvement opportunity. At the start of 2018 NTO started the installation of underground dump and diversion valves.

These valves were chosen because of their ease of use, rapid execution, and improved safety due to reduced manual handling and operator exposure. Dump valves work by decreasing the amount flow resistance while leaving the driving force the same. This allows the CPB to drain through the valve until it reaches equilibrium; flush water is then used to clear the remaining paste upstream of the valve. When valves are used in cascades it allows for rapid clearing of the reticulation system.

Previously, clearing blockages were a slow and labor-intensive process. This would involve independent firing of blast-T caps to release the CPB from the line. If everything went well, it usually took around 1.5 hours to fire a blast-T, drain the paste, flush the line, and re-instate the blast-T. The blasting and manual handling accounted for the majority of this time. The current valves, dependent of the type of valve, take between 15-45 seconds to rotate.

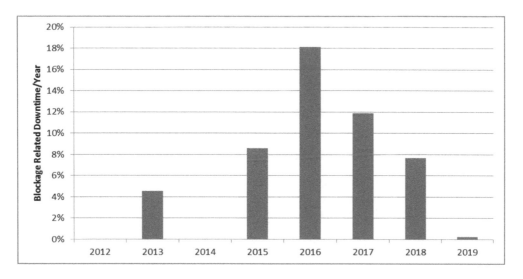

Figure 1. Blockage related downtime per year.

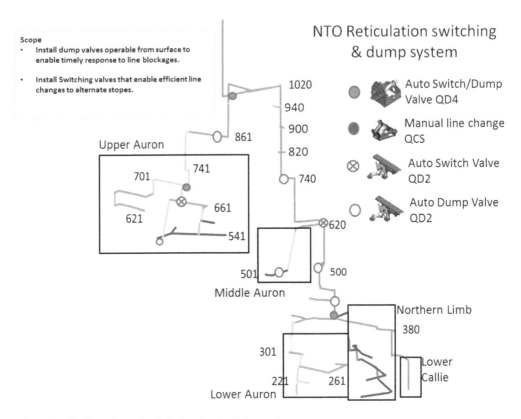

Scope
· Install dump valves operable from surface to enable timely response to line blockages.
· Install Switching valves that enable efficient line changes to alternate stopes.

NTO Reticulation switching & dump system

1020
940
900
820
740
861

Upper Auron

701   741
621   661
541   501   500
620

Middle Auron

⬤ 🛠 Auto Switch/Dump Valve QD4

⬤ 🛠 Manual line change QCS

⊗ 🛠 Auto Switch Valve QD2

○ 🛠 Auto Dump Valve QD2

Northern Limb
380

301
221   261
Lower Auron

Lower Callie

Figure 2. Design schematic of the DBS reticulation system.

Three different types of vales were selected and installed. Automatic hydraulically-driven three-way valves were selected for the main branches to allow flow in three directions (the two branch directions and to dump), automatic pneumatically-driven two-way valves were selected for either dumping or diversion, and manual quick-change valves were selected to allow rapid changes between reticulation lines (i.e. between two stopes). Note that the manual valves are not used to dump the line but only for dry changeovers. An additional benefit to these valves is the reduction of manual handling in order to enact a reticulation line change.

Figure 2 is an example design schematic of the reticulation system, showing the locations of the valves within the system. The schematic also highlights what each valve is being used for (diversion, dumping, diversion/dumping, etc.).

2.1 Dump valve case study – 136D blockage (no dump valves) versus 127H blockage (dump valves)

Figure 3 shows the locations of the 136D and 127H stopes. Both stopes were filled using the same line and were at approximately the same location in the mine (the 136D stope was one level deeper and required a much longer horizontal line to fill).

The 136D stope started filling in October 2017. At around 0700 on the 12th of October the PP recorded a decrease in tailings moisture from approximately 10.5% to 9% over a 5-minute period (Figure 4) and the PP responded accordingly. However, subsequent studies indicated that the installed moisture sensor was uncalibrated and inaccurate, and that the actual decrease seen by the PP was closer to 7%. This dry material caused a slug of thick CPB to enter the reticulation system. It took approximately 5 minutes to travel through the 1020 level pressure drop measurement loop. This is shown by the rapid rise in friction factor (kPa/m)

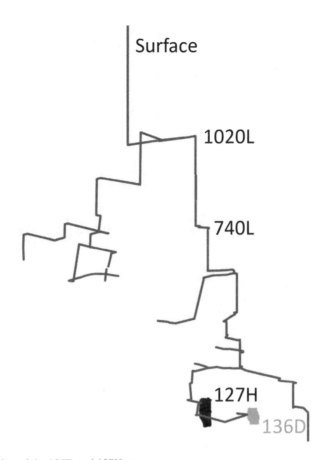

Figure 3.   Location of the 136D and 127H stopes.

from 5.5 to 11.5 kPa/m over approximately 10 minutes. During this time the PP operator did respond and dropped the PP's percent solids setpoint but then brought the percent solids up again very quickly. Approximately 30 minutes after the moisture changed the flow decreased to zero. The PP operator was unable to unblock the line using the PP and the underground (UG) operators started UG unblocking operations (firing blast T's). It took 12 hours to clear the reticulation system from surface to the 740L. After this the UG operators started cracking pipes and a high-pressure washer and diamond drill contractor were brought in clean out the pipes downstream that could not be unblocked. This blockage took 2 months to clean from the 740L to the 133G stope.

The first dump valve was installed at the 1020 Level in January 2018 with the 740L valve being installed soon after. These valves were fired for the first time on February 25, 2018.

The causes for this blockage were similar to what caused the 136D blockage. In this case there was a rapid increase in moisture content. However, the PP had not been well tuned to the new moisture analyzer and the PP took too much water away from the mix, causing a slug of thick tailings to enter the reticulation system. This dense paste resulted in a friction factor increase from ~7 to ~11 kPa/m. The PP operator responded and adjusted their solids, and then shut the plant down and tried to flush the lines using water. The UG operators reached the 1020L dump valve first and fired that dump valve manually (the dump valve did not have telemetry with the PP control room yet and needed to be fired manually). This allowed the reticulation system to be cleared of paste to the 1020 level. The paste plant then started filling the reticulation system with water. Once the UG operators reached the 740L valve it was also fired manually, clearing the reticulation system to the 740L. The paste plant was then able to clear the remaining paste through to the stope.

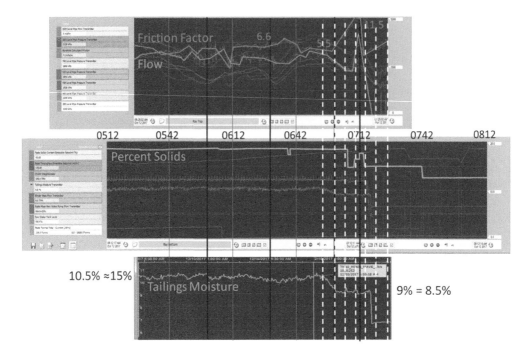

Figure 4.　Citect trends from the 136D blockage.

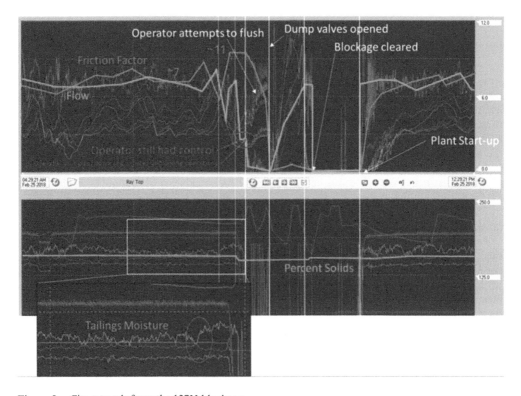

Figure 5.　Citect trends from the 127H blockage.

The 127H blockage was cleared within 45 minutes. It required approximately 15 minutes for the underground crew to arrive in position and activate the 1020L and 740L valve, and another 30 minutes to clear the blockage. Once the blockage was cleared the PP restarted and continued to fill the stope.

This case study highlights how useful the dump valves were in reducing the possible downtime for similar blockage situations. The major advantage to dump valves is the speed at which they can be activated and reset. Additionally, now that telemetry has been established between the UG valves and the PP control room the entire operation can be conducted without requiring UG operators to be present. Dumping points are located in isolated areas where operators are not exposed when the valves are activated.

Since the dump valves were installed there has only been one major blockage, caused by an elbow blowout a level below the last dump valve. The blockage was cleared quickly to the dump valve, but the PP operator was unable to flush the lines once the elbow had been replaced. This example highlights that dump valves are an excellent tool, but they will not clear all blockages.

3 CONTINUOUS MONITORING OF TIGHT FILLING

A number of improvement steps were taken to reduce down time during the tight fill process. Tight filling occurs when the last few vertical meters of the stope are being filled which, as this is the case at DBS, generally occurs at the upper access point to the stope. This means that there is a shotcrete wall located at this level which the CPB will be filling against.

Historically, an UG operator was required to act as a spotter; monitoring exclusion areas and visually inspecting the shotcrete wall, and then relaying any relevant information to the PP operator. As the spotter was not available over shift change, normal operating guidelines dictated that the PP shut down once the spotter left and would not start up again until a replacement spotter was in place. This time delay could vary from 4 to 12 hours depending on crew or vehicle availability, and blast fume clearing.

In order to limit this downtime, DBS introduced live feed camera network that along key points of the reticulation system using the PP dedicated fiber network (Figure 6). Fixed cameras were installed on dump/diversion valve locations (allowing visual confirmation that the

Figure 6. Paste plant operator screen showing the live feed camera network.

valve had rotated and that the dump zone was clear) as well as problematic areas (areas with high vibration potential etc.). The system also allowed mobile cameras to be placed at filling points. This allowed the PP operator to have a continuous, visible reference that CPB was entering the stope as well as helping eliminate down time over shift change. An additional benefit was that the spotter was no longer required and could be assigned other duties.

However, it was found that the current 25mm breather pipes were not sufficient enough to allow the PP to observe if the stope was full. To better aid in the tight fill process the bulkheads were re-designed. An 8-inch (200 mm NB) HDPE drainpipe was installed at the top of the bulkhead. The drain pipe was installed at a decline from the crown of the stope allowing paste to flow into the containment area when tight filling was completed (Figure 7 Figure 8).

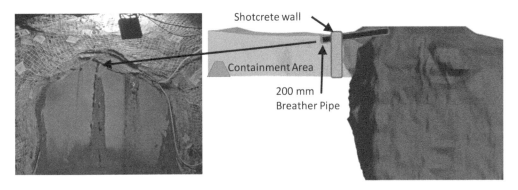

Figure 7. Photograph showing placement of large diameter drainpipe and a schematic showing the ideal placement of the drainpipe through the shotcrete wall.

Figure 8. Exposed CPB once wall removed showing 200mm HDPE drain pipe.

This further removed error when tight filling by allowing either the UG operator via camera or the UG spotter a better reference that the stope was full.

Initially there were concerns that the addition of this large diameter pipe would prevent the tight filling of the stope due to a 'short circuit' between the delivery pipe and the drainpipe. There is evidence that this type of 'short circuit' does occur. However, when both pipes are installed properly there has been no evidence that the large diameter pipe installation prevents the tight filling of the stope. Figure 8 is a photograph showing a CPB exposure from behind a shotcrete wall. The buried end of the large breather pipe is visible, and the stope has been tight filled to the top of the drive. Additionally, this 'short circuit' behavior also protects the shotcrete wall from over-pressurization.

3.1 *Large diameter pipe case study – tight filling with no continuous monitoring of the 145U stope*

The PP was filling the AUR_145U stope during the early morning and was within 200 m³ of starting the normal tight filling procedure as per the CMS run sheet. This run sheet determines the height of the CPB level within the stope based on dead reckoning using the amount of CPB volume placed versus the total volume of the stope. The CPB containment area bunds were in place. A large diameter drainpipe had been installed into the shotcrete wall. However, no camera had been installed as the fiber network could not be expanded in time.

DBS requires a reticulation check (visually inspect the shotcrete wall, containment area, and the reticulation system etc.) every 6 hours. However, no line checks had been completed on night shift. The CPB level was higher in the stope than what was determined on the CMS run sheet and, subsequently, CPB started filling the containment area at ~04h45. This was not rectified until the first dayshift lined checks approximately 5 hours later. A total of 510 m³ were overfilled in total: ~440 m³ within the containment area and ~50 m³ was spilled over the exclusion bund. Figure 9 shows a level plan with the approximate locations of the bunds and the overflow as well as two pictures from the drive after the CPB was removed.

This case study highlights importance of the entire system being in place. The large diameter drainpipe worked as it was supposed to and gave a visual reference that the stope was full. The unfortunate part was that nobody was watching.

Additionally, the installation of the drainpipe protected the shotcrete wall from over-pressurization to the point that the PP operators did not observe any changes in the pipeline pressures. The shotcrete wall showed no signs of stress-related damage when inspected. This single control eliminated the potential for a bulkhead failure and a sudden inrush or a pipe blockage.

4 AUTOMATIC PASTE DENSITY SETPOINT CONTROL

Another improvement introduced into the DBS backfill system was a PP control system referred to as the 'Automatic Paste Density Setpoint Controller' or OTTO. This controller is not being developed to replace the operator but to allow the operator an 'autopilot' freeing the operator for other PP related tasks. The controller was developed between DBS site personel and Newmont Goldcorp Australia's Process Control Group.

OTTO uses the 1020L pressure value as input to a standard PID (proportional integral derivative) controller that manipulates the paste mixer feed density setpoint. This controller manipulates the same setpoint edited by the control room operator via the Citect popup. The PID controller makes small adjustments based on the rate of change and comparing the actual value of the 1020L pressure value to its setpoint (which varies given the configuration of the reticulation system). The paste density setpoint is only allowed to vary within the following operator-specified limits. Ideally, OTTO would first aim to control a 1020L flow in order to then control the 1020L pressure. However, issues with inconsistent tailings feed often result in a highly unstable 1020L flow, and therefore not possible to control flow automatically. Instead, to further emulate typical control room operator actions some additions applied to the basic pressure controller described below:

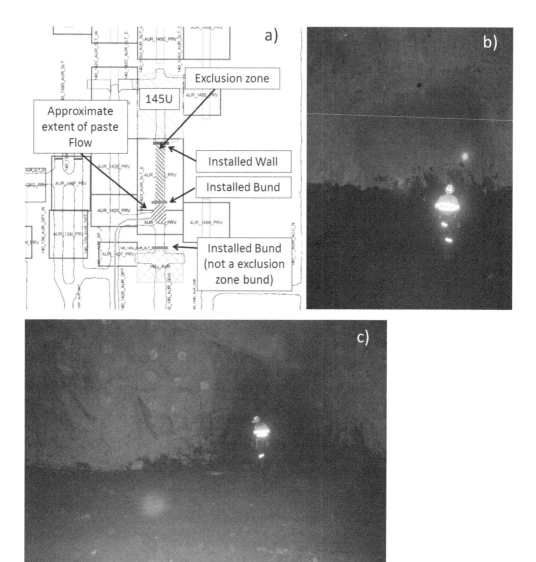

Figure 9. Paste 145U overfill showing a) level plan with approximation of location of bunds, b) paste staining at the shotcrete wall location, and c) paste staining along the drive.

- If 1020L pressure increasing and/or 1020L pressure above setpoint -> paste mixer feed density reduced
- If 1020L pressure decreasing and/or 1020L pressure below setpoint -> paste mixer feed density increased

While the controller is primarily a pressure controller, 1020L flow can optionally be included as a feed-forward input to the pressure controller. This will immediately make small adjustments to the paste density setpoint (when enabled) based on the changes in 1020L flow as per below:

- If 1020L flow increases, slightly increase paste mixer feed density.
- If 1020L flow decreases, slightly decrease paste mixer feed density.

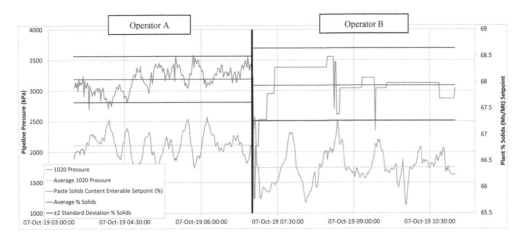

Figure 10.   Citect trends showing the difference between the performance of OTTO and a PP operator.

The assumption is that a change in flow will result in a subsequent change in pressure.

Finally, as an option, the borehole vacuum pressure can be included as a feed-forward input to the pressure controller. This emulates a typical operator response to a loss of bore-hole pressure. This control is more aggressive than the 2 above, as it must assume the loss of flow and vacuum is real, and therefore allows for radical changes of the paste density setpoint

In practice the PP operators generally run OTTO only using the basic 1020L control. Occa-sionally the operators will run with the flow control if the tailings feed is consistent. An upgraded flow meter has recently been installed on the reticulation system and this has improved the flow measurement. None of the operators utilize the borehole vacuum pressure option.

Further improvements to OTTO are planned. These primarily deal with utilizing the lag time between the mixer and the flow and pressure measurements at the 1020L to optimize OTTO's performance. However, this has not been fine-tuned as the trial of the basic system is ongoing.

Figure 10 shows Citect trends for the percent solids setpoint and the 1020L pressure over a 24-hour period. During this time there where two different operators running the plant: Operator A was running OTTO, while Operator B controlled the PP with manual density changes. OTTO and Operator B maintained average percent solids setpoints of 68.1% and 67.9% respectively. However, the average 1020L pressures, over this time, were 2111 kPa for OTTO and 1756 kPa for Operator B. OTTO was set to 2100 kPa.

Relatively speaking the percent solids averages are similar. However, the percent solids vari-ability between the two operators is large and is highlighted by the purple lines denoting ±2 standard deviation of the percent solids. The tighter percent solids control allowed OTTO to maintain a line pressure that was approximately 350 kPa higher over the time period monitored.

5   CONCLUSIONS

The expansion of the DBS mine requires the subsequent expansion of the reticulation system as well as an increase in the CPB placement rate and requires improved control over the back-fill system as well as a reduction in downtime. Multiple improvements have been initiated at DBS, with three of these improvements being summarized in this paper.

The installation of dump/diversion valves have decreased the occurrence of major blockage downtime from over 10% in 2016 and 2017 to less than 1% in 2019. It has also reduced reticu-lation change downtime and the amount of manual handling required to enact these changes.

This control system was designed and installed to improve productivity, but the safety impact by reducing manual handling of pipes when blockages occur, almost outweighs the operational improvements.

The continuous monitoring improvement has decreased downtime due to spotter constraints during tight filling operations by allowing the PP operator to visual observe the shotcrete wall in real-time. This improvement also allows the PP operator to have a visual reference to how the reticulation system is performing at key areas as well as allowing the PP operator to observe CPB entering the stope. Part of this was the installation of large diameter drainage pipes that provides a better visual reference that the stope is filled. An additional benefit of this larger diameter pipe is that it prevents the shotcrete wall from being over-pressurized but still allows the stope to be tight filled.

The last improvement was the introduction of OTTO or the PP 'autopilot'. In OTTO's basic form it modifies the percent solids setpoint in order to maintain a set pressure at the 1020L pressure sensor. It does this by observing the rate of change and the value of this pressure sensor. OTTO is still being trialled, but initial results are encouraging. Additional modifications and improvements are ongoing.

ACKNOWLEDGEMENTS

The Authors would like to acknowledge the Newmont Goldcorp Australia Process Control Group, particularly Craig Sempf and Aidan Hill, for their ongoing help and support with OTTO. The Authors also acknowledge the Newmont Goldcorp Tanami Management Team for their support in the writing of this paper.

Minefill 2020-2021 – Hassani et al (eds)
© *2021 Taylor & Francis Group, London, ISBN 978-1-032-07203-6*

Case study – design and implementation of a high density cemented hydraulic fill system at OZ Mineral's Prominent Hill mine

Matthew Helinski
Outotec Pty Ltd

James Shaw
Oz Minerals Ltd

SUMMARY: The combination of recent advancements in paste fill technology and some unfortunate incidents with poorly managed hydraulic fill systems has seen the implementation of Cemented Hydraulic Fill systems in Australian mines reduce significantly in recent years. This paper describes a high density cemented hydraulic fill system that has performed very well in recent years at OZ Mineral's, Prominent Hill Operations, Malu underground Mine. The paper details the design methodology and specific features of the system that were developed to address the underground mining demands. This is followed by a description of how the fill system was implemented and the innovative management systems used on an ongoing basis to maximise production while managing safety risks. Based on data gathered during this period the success and failure of various design considerations are discussed.

Keywords: High Density Hydraulic Fill, Design Principles, Operational Experience

1 INTRODUCTION

OZ Minerals owns and operates the Prominent Hill underground copper gold mine in South Australia. Underground mining at Prominent Hill commenced in 2010 with the development of the Ankata Mine, and then subsequently the Malu underground mine. The Ankata mine voids are filled with pastefill, while Cemented Hydraulic Fill (CHF) is adopted to meet initial backfilling requirements in the Malu underground mine. This paper describes the design and operational experience for the Malu CHF system.

Stoping operations in the Malu underground mine commenced in 2014. 2019 annual mine physicals included 12.5 kilometres of mine development, and production of 3.3 – 3.5 million tonnes of ore hauled from the underground mine using trucks. In the Malu mine sublevels have been at 25 and 50 m intervals with 75 m high open stopes with a nominal footprint dimensions of 20 m by 20 m.

Due to the ongoing resource definition work at this time, a backfill solution was sought, that satisfied the mining requirements with relatively low capital investment. A study was initiated to address this and, giving consideration to the proposed mining geometries and schedule, CHF was considered a feasible solution.

This paper presents the results of detailed design, analysis and testwork carried out to develop a very high density cemented hydraulic fill system for the Malu mine. In addition, this paper also presents the system implementation as well as the challenges and triumphs incurred throughout operation of this system.

DOI: 10.1201/9781003205906-36

This paper illustrates that, with the correct approach to design and operation, CHF can provide a cost-effective solution to satisfying the needs of many underground bulk mining operations.

2 SYSTEM DESIGN

CHF systems are typically very simple, consisting of cyclone desliming/dewatering of mine tailings, binder addition (to the underflow product) and turbulent flows through the underground reticulation system to the stope.

In addition to a simple lower capex solution, for filling the Malu underground mine it was also the intention that the proposed hydraulic fill plant would later be relocated to the Ankata Mine to facilitate filling of secondary stopes with uncemented high density hydraulic fill. To address this requirement, Outotec originally proposed a compact "skid mounted" cemented hydraulic fill plant, similar to Outotec's HF-Compact product, presented in Figure 1.

2.1 *Process design testwork and analysis*

While the hydraulic fill manufacturing process is very simple, as noted by Liston (2011), subtle errors in the system design can create a legacy of negative issues during operation. To ensure this wasn't the case an extensive laboratory scale testwork program was undertaken.

It is well documented (Cowling *et at.* 1988, Kuganathan, 2002) that the key to safe CHF operations is careful pore water management. This is typically achieved through campaigns of "filling" and "resting", where the stope is allowed to drain pore water (without addition of

Figure 1.   Schematic and photograph showing the originally proposed skid mounted fill plant model.

more fill water) during the "the resting periods". The duration of filling periods relative to resting periods directly impacts the ability to fill a given stope (and in-turn the mining cycle). The quantity of water to be drained is controlled by the placed fill density, while the rate that free water drains is controlled by the fill permeability. As these characteristics are controlled by the cyclone underflow characteristics, and the cyclone underflow characteristics are controlled by the cyclone configuration, critical to success of a hydraulic fill system is careful design and operation of the cyclone desliming circuit.

Design of the desliming circuit was initially completed using cyclone simulations and laboratory scale "gravity decanting" to remove fines. Results from this testwork are presented in Figure 2, which shows the decanted PSD and permeability for the resulting material.

The results presented in Figure 2 show the samples created from "decanting" to be "well graded". The nature of this grading curve means relatively low recoveries, but high fines content, which leads to the low permeability presented in Figure 2. Interestingly, even though the "1 min x 5 pours" grading curve has significantly less fines than the industry accepted maximum fines criteria (less than 10% finer than 10 μm), due to adverse mineralogy, the resulting permeability is very low, at 20 mm/hr.

To assess the implications of this relatively low permeability on the filling/resting schedule a typical Malu stope was modelled using the Outotec Water Balance Model (Helinski & Grice, 2007). Output from this model is presented in Figure 3. This figure shows the fill surface, water table surface (saturated zone) and stope-drawpoint intersection pore water pressure plotted against time throughout filling. The "zig-zag" steps in the model represent discrete periods of filling and resting, where the fill schedule was set to maintain the water surface (blue line) below the fill surface (black line) throughout filling.

The results presented in Figure 3 show that, with a fill permeability of 20 mm/hr (maximum measured for the decanted samples) it would be expected to require 60 days to fill a typical stope. This was too long for the proposed mining schedule.

Given the unfavourable "bench scale" underflow properties it was decided to complete full scale cyclone simulations. The cyclone test rig is fitted with a complete cyclone and is configured to allow the cyclone setup, operating pressures and feed properties to be adjusted to allow the underflow product to be optimised. A photograph showing the test apparatus is presented in Figure 4. Also presented in Figure 4 is a comparison between the PSD curve of the sample created using the decanting method and that from the cyclone trial.

Figure 4 shows that, relative to the decanting method, separation through full scale cyclone classification provides a far more uniformly graded cyclone underflow. Optimisation testing showed that dilution of the cyclone feed resulted in increased uniformity, resulting in an

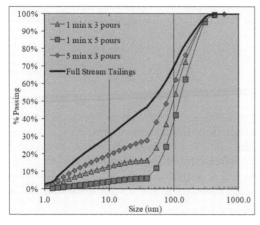
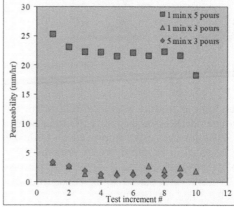

Figure 2. PSD and permeability of samples made using the laboratory decant method.

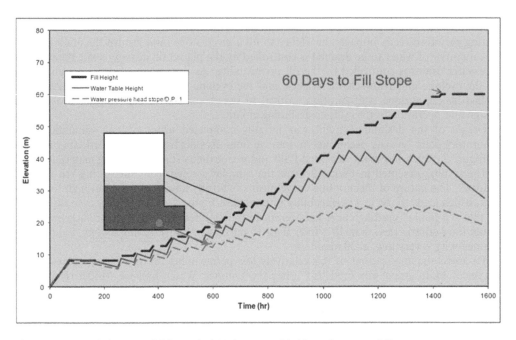

Figure 3. Water balance model for typical Malu stope with 17 mm/hr permeability.

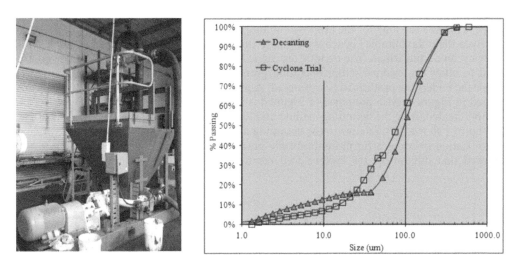

Figure 4. Cyclone test rig and cyclone trial results.

underflow product with increased permeability and solids content. Figure 5 presents the cyclone underflow permeability plotted against the cyclone feed solids concentration.

The impact of the improved underflow product on the underground mining operation is illustrated in Figure 6, which presents the water balance model output for the design underflow product. Figure 6 shows the permissible filling sequence with a fill permeability of 32 mm/hr and placed solids concentration of 77%, illustrating that a typical stope could be safely filled within 30 days.

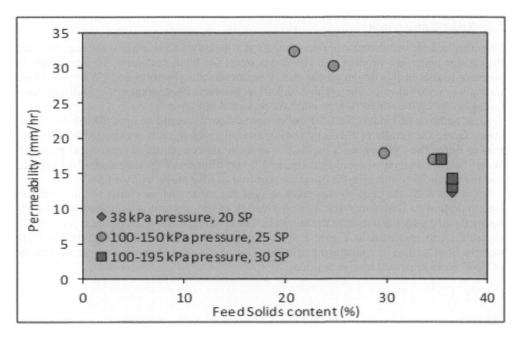

Figure 5. Cyclone feed density versus cyclone underflow permeability.

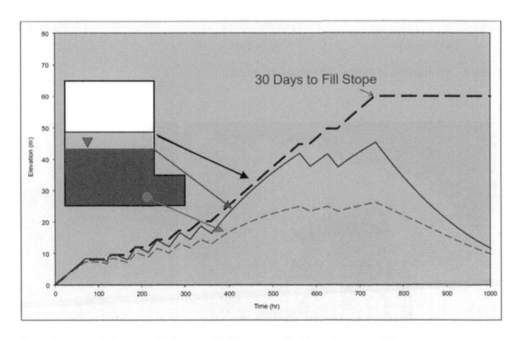

Figure 6. Water balance model for typical Malu stope with 32 mm/hr permeability.

While the test results show that a reduction in cyclone feed density (from 38 to 20%) delivers a significant increase in permeability and underflow solids content, the drawback is a 3-fold increase in flows. This results in the need for a significant increase in pumping and cyclone cluster capacity.

2.2 Binder dosage system

The other notable compromise of the originally skid mounted CHF plant is the need to pump the cyclone underflow product into the vortex mixer for binder addition. To investigate the potential impact of this design on the overall performance, cylindrical and "Warman" slump testing was undertaken on the cyclone underflow product. Photographs of cylindrical slump testing on the cyclone underflow product are presented in Figure 7.

After addition of binder, the CHF solids content (w/w) would be upto 80% during operation. Given the results of rheology testing and pump analysis it is expected to be necessary to dilute the CHF product to 75-76% to facilitate conventional centrifugal pumping. Water balance analysis showed that this additional dilution of cyclone underflow product had the potential to increase the duration required to fill a stope by over 65%.

Given this delay the decision was taken to raise the entire cyclone cluster to allow the cyclone underflow to flow under gravity into a vortex mixer, where it is combined with binder, before flowing under gravity into the borehole. The resulting CHF plant layout is presented in Figure 8. Also presented in Figure 8 is a photograph of the produced high-density CHF material after settling in a measuring cylinder. This photograph shows the very small amount of bleed water generated after deposition.

| 79% Solids | 77% Solids | 75% Solids |

Figure 7.   Photographs of CHF at different solids contents.

Figure 8.   Photograph of Malu CHF plant and Malu CHF material.

414

3 OPERATING EXPERIENCE

3.1 *Stope drainage*

As detailed by Kuganathan (2002) and Helinski & Grice (2007), fundamental to safe and efficient hydraulic filling operations is careful stope water balance management. At Prominent Hill the stope water balance is managed through the combination of water balance modelling and *in situ* pore water pressure monitoring. Water balance modelling combines the actual placed fill characteristics (solids content and flowrate from plant quality control testing), with stope geometry data to represent the stope performance. The model is calibrated against *in situ* fill height and water pressure measurements in accordance with the method described in Helinski & Grice (2007).

This approach proved to be largely successful with back calculated permeability and density measurements typically being very similar to values measured in the quality control testing. However, on a number of occasions increased and reduced stope drainage rates were identified through the *in situ* pore pressure monitoring.

Examples of different stope drainage conditions are presented in Figures 9 and 10, which show *in situ* piezometer measurements superimposed over water balance model output from stopes that drained and did not drain effectively. Fill with favourable drainage conditions showed a much slower increase in water pressure during filling and a much faster drop in water pressure during drainage periods.

In cases where improved drainage was observed filling rates were increased to accelerate stope turn-around times. Examples of cases where stope drainage was shown to be problematic include:

- A case where rouge groundwater was entering the stope leading to increased stope water addition and consequently higher than expected pressures.
- A case where an error occurred while calculating fill permeabilities. Following higher than expected water pressures, further analysis of quality control testing showed lower than expected fill permeability.

On these occasions filling and drainage periods were adjusted to manage the adverse stope drainage conditions.

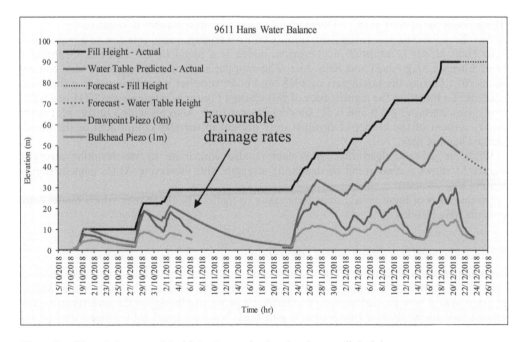

Figure 9.   Water balance model with *in situ* monitoring showing a well-draining stope.

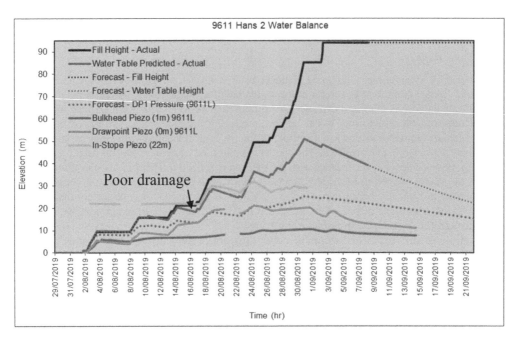

Figure 10.   Water balance model with *in situ* monitoring showing a poorly draining stope.

3.2 *Quality control testing*

The quality control testwork regime included unconfined compressive strength testing after hydration periods of 3, 28 and 240 days hydration. Each of the 3-day tests was multiplied by an amplification factor to extrapolate the value to a longer-term hydration point. Where extrapolated 3-day strengths indicated a trend of lower longer-term strengths, the fill binder content was increased to a level necessary to, not only achieve the target strength, but to a level that adequately compensated for the lower strength layer to ensure exposure stability. Throughout the operation this adjustment was necessary on a number of occasions.

Another interesting learning from quality control testing was in relation to binder types. Low Heat binder is adopted in operation, which is a blend of GP Cement, Granulated blast Furnace Slag (slag) and Kiln dust. The supplier's specification notes a Slag content of 50-70%. During the latter part of 2018 the binder supplier steadily increasing their slag content. To illustrate the significance of this change Figure 11 presents the CHF strengths (for mixes containing 6.0 and 6.5% binder) against time. Also presented in this plot is the Al_2O_3 content of the binder (a product only present in the slag component of the blend) plotted against time.

While its acknowledged that other factors also contribute to the resulting strength, Figure 11 shows a clear trend of increasing strength with increasing Al_2O_3 content (representing slag content).

On this basis of these results steps were taken to tighten the range over which binder components were blended.

3.3 *Underground reticulation system pipework*

To maintain the high-density CHF in suspension it is necessary to transport the material at velocities in excess of 2.1 m/s. At such high velocity pipe wear can be problematic and to manage wear it was deemed necessary to include a protective lining for the underground steel reticulation pipework. For this project 6 mm thick hot vulcanised rubber lined steel piping was adopted. Flanged couplings were adopted to ensure the rubber lining properly wrapped

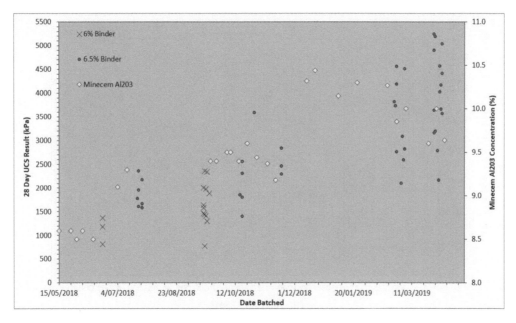

Figure 11. CHF strength and Al₂O₃ content plotted against time.

Figure 12. CHF reticulation system pipework.

around the pipe ends and minimise the likelihood of delamination. A photograph showing this pipe is presented in Figure 12.

The experience with this pipeline solution was favourable with limited issues relating to wear or rubber "lifting" during the first 3 years, or approximately 500,000 m³ of operation (design life). After 3 years some wear and rubber lifting issues were encountered. This manifested in rubber flaps lifting and blocking the reticulation system.

Figure 13 presents the cumulative system downtime, attributed to rubber lining issues verses fill quantities placed.

Figure 13 shows that the rubber pipe lining performed well for the first 500,000 m³, but after this the associated downtime increases exponentially. This suggests that, in this case, the design life of the solution is approximately 500,000-750,000 m³.

Due to concerns about damaging the rubber lining during pipe cleaning all cleaning was undertaken using high pressure water sprays (rather than drill strings). This method was largely successful for clearing cured CHF and post cleaning inspections indicated that this method did not appear to damage the lining.

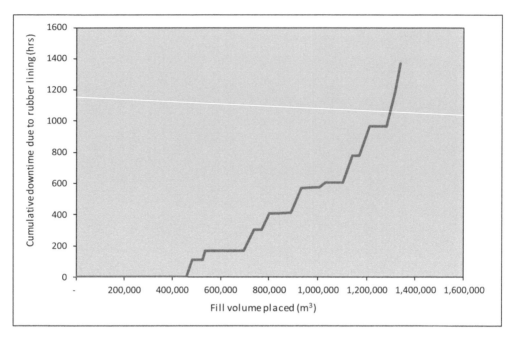

Figure 13.   Reticulation rubber lining downtime verses fill placed.

Surface boreholes were lined with 200NB Schedule 40 steel with 10 mm alumina ceramic sleeve inner lining. Some problems were encountered during installation due to the accumulation of minor casing length discrepancies and the lesson taken from here is to stipulate the ceramic lining length, rather than the total pipe length (i.e. including threading). While we understand that some issues have been encountered with this product at other sites, this casing solution proved largely successful in our application.

Due to the favourable ground conditions, internal boreholes are typically left unlined. Over the life of the operation this has proven to be a good decision with no reported cases of borehole collapse.

3.4   Internal horizontal borehole

One particularly unique component of the reticulation system is 110 m long horizontal boreholes. These are used to avoid reticulation pipework being installed in major travel ways, such as the decline, to prevent any disruptions to ore haulage in the event of reticulation issues.

A schematic showing the geometry of the boreholes is presented in Figure 14. Also presented in Figure 14 is a photograph showing installation of the casing in a horizontal borehole.

Initial attempts to drill long horizontal holes using an in the hole hammer drill rig were not successful, with holes drifting well off target. This issue was then resolved through the introduction of a raisebore drill. The raisebore drill was oriented horizontally, with holes targeting a design incline 1 degree above horizontal, which was primarily to assist with flushing of drill cuttings. In total 2 operational holes were successfully drilled. The holes are cased with an outer 200NB schedule 80 steel casing with threaded joints between lengths and then an inner casing of 125NB schedule 120 steel pipe with threaded joint. The horizontal holes are lined to facilitate periodic replacement of the casing. The outer casing provides protection against the 125NB pipe becoming grouted in should a failure occur. In the event of a blockage, the casing could be easily removed and replaced rather than re-drilling with a raisebore.

Figure 14.   Horizontal internal borehole (a) design and (b) being drilled.

3.5   *Reticulation system blowback*

The surface borehole collar area was designed in a manner that the hydraulic fill is permitted to flow directly into the borehole with a breather hole located at the borehole entry point. After operating for a short period of time (approximately 15 minutes) significant borehole "blowback" occurred, which effectively stopped CHF flowing. The blowback is expected to be a result of air entering the system at the vortex mixer, an expansion of the borehole (from 8" to 12" diameter, see Figure 15) midway along the surface borehole leg as well as the low viscosity of the CHF (which prevented release of this pressure).

To resolve the issue, a series of "breather holes" were drilled into the borehole on the intermediate level. These holes act to vent any air pressure accumulation. A photograph showing the mid-height level with breather holes venting water, during a borehole flush, is presented in Figure 15.

This experience demonstrated that, while borehole air entrainment is not typically problematic with (more dilute) hydraulic fills, when producing high-density CHF care should be taken to manage borehole air entrainment using similar design strategies to those commonly adopted on paste fill systems.

3.6   *Horizontal exposures*

The mining sequence at Malu requires fill masses to be exposed over very large horizontal expanses (up to 30 m x 30 m) when extracting underlying ore. Our experience has shown that

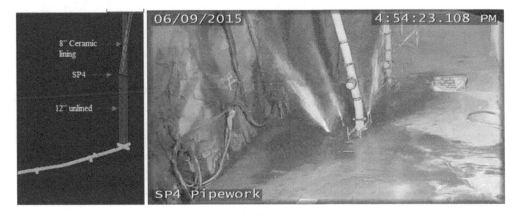

Figure 15.   Surface borehole sketch and photograph showing blowback at mid borehole level.

during deposition the meandering "beaching" process tends to create stratified layers with horizontally oriented cold joints. An example of this is presented in Figure 16, which shows a photograph of an exposed paste fill mass that was placed with a continuous fill run. Due to these "cold joints" it was not considered appropriate to adopt the analytical solutions proposed by Mitchell (1991) (which assume a continuum "massive" fill mass) for the proposed large exposures. Rather, analysis was undertaken using a "dis-continuum" modelling methodology, where horizontally oriented cohesionless interface elements were inserted to represent the cold jointing. This model is presented in Figure 16, which shows the deformed mesh with interface elements represented by pink lines.

With the presence of the cold joints extreemly high strengths are required to totally eliminate dilution, therefore rather than providing a deterministic solution, the results were presented in a series of design charts that relate the expected dilution to prescribed fill layer strength. This design chart is presented in Figure 17.

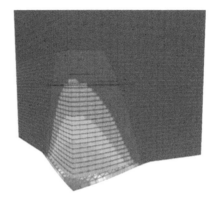

Figure 16. Photograph of cold joints and model used for horizontal exposure analysis.

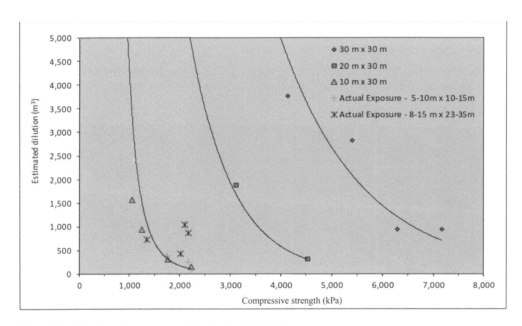

Figure 17. Horizontal exposure design charts with field data.

420

Superimposed over Figure 17 are observed dilution levels against actual fill strength (from Malu). The faviourable comparison btween the modelling and acutal dilution levels indicate that the modelling approach provides a reasonble representation of the performance.

4 CONCLUSION

This paper presents a case study of the cemented hydraulic fill system at OZ Mineral's Malu underground mine. The major learnings from the design and operational experience include:

- The importance of full-scale cyclone trials during design to:
  - ○ ensure that the cyclone underflow geotechnical characteristics can be properly represented.
  - ○ ensure that the process design, most notably the desliming circuit, can be properly sized to suit the needs of the mining operation.

- With suitable process design focus high density hydraulic fill systems can be developed. However, subtle design shortcuts can compromise the ability to produce high density hydraulic fill, which have a significant impact on the rate at which underground stopes can be filled.
- Water balance modelling combined with calibration against *in situ* pore pressure monitoring can provide an effective means of accelerating fill rates where appropriate and managing safety risks under adverse conditions.
- Minor variations in binder constituents can have a significant impact on resulting fill strengths. When establishing fill operations care should be taken to establish tight control over any binder blending.
- Hot vulcanised rubber lining can form an effective pipe wear protection lining system, but in our experience the lifespan of this system was realistically less than 750,000 m^3 of fill.
- With high density hydraulic fill systems, specific consideration should be given to minimising air entry into the reticulation system as the lower viscosity of the slurry can prevent air release leading to significant borehole "blow back" events.
- When simulating large horizontal exposures, the presence of horizontally oriented "cold jointing" makes it difficult to totally eliminate dilution. Continuum analysis doesn't provide a reasonable representation of unravelling during horizontal exposures and rather discontinuum modelling appears to provide a rational approach for estimating dilution.

BIBLIOGRAPHY

Cowling R, Grice A.G. and Isaacs L.T. 1988. Simulation of hydraulic filling of large underground mining excavations. Proc 6th Conf on Numerical Methods in Geomechanics, Innsbrück, Balkema, Rotterdam, pp 1869–1876.

Helinski, M. Grice, A.G. 2007. Water management in hydraulic fill stopes, Minefill '07 Innovations and experience in minefill design, Montreal April 29-May 3. # 2480.

Kuganathan, K. 2002. 'A method to design efficient mine backfill drainage systems to improve safety and stability of backfill bulkheads and fills', Proceedings of the 8th AusIMM Underground Operators' Conference"growing our underground operations", Australasian Institute of Mining and Metallurgy, Carlton, pp. 181.

Mitchell, R.J. 1991. Sill Mat Evaluation Using Centrifuge Models. Mining Science and Technology, Elsevier Science Publisher B.V., 13: pp. 301–313.

Minefill 2020-2021 – Hassani et al (eds)
© *2021 Taylor & Francis Group, London, ISBN 978-1-032-07203-6*

Green mining – use of hydraulic backfill in the Velenje Coal Mine

Jože Kortnik
Faculty of Natural Sciences and Engineering, University of Ljubljana, Ljubljana, Slovenia

SUMMARY: The paper presents the use of hydraulic backfill in the Velenje Coal Mine and the development of some procedures and devices for optimizing the basic properties of backfill mixtures. The technical, economic and environmental aspects of the hydraulic transportation of thick backfill mixtures (pastes) may be estimated only if the rheological properties are previously determined. For this purpose we developed the ball-pull test device, the tube viscosimeter and the pressure leaching test (PLT) equipment. A brief outline of tests of plastic and leaching characteristics and the method for determining the pressure gradient required for the specific flow of a time-independent non-Newtonian backfill mixtures and its application in the Velenje Coal Mine will be also presented.

Keywords: circle economy, green mining, longwall mining, hydraulic backfill

1 INTRODUCTION

Mining in the future need to be ecologically sustainable and economically effective. By establishing of closed ecological and technological circle of extraction, processing or energy transformation and returning mining waste materials into excavated open space it is possible to organize environment friendly, sustainable metal/non-metal/coal mining or circle economy in mining industry. By depositing mining waste into open underground spaces, earth surface can be disburdened. Suitable underground locations should take existing natural barriers into account otherwise some technical measures have to be taken to protect surrounding formation against adverse impact. The good results of this research should assure by selection of appropriate waste materials for suspensions preparation, transportation and deposition by proper technology. The perspective of coal mining in the Slovenia and in the world is mainly related to the introduction of environmentally friendly technologies into extraction and conversion into electricity. Given the fact that the content of ash and free sulfur in Slovenian coal is relatively high, it is necessary to take appropriate care of the products of combustion and flue gas cleaning during energy conversion in the Šoštanj Thermal Power plant. Use or return of large quantities of waste materials generated during the extraction and conversion of coal into electricity in excavated cave spaces significantly reduces the impact of excavation on the surface, improves the stress-strain state in the rock around the excavation and at the same time eliminates the load on landfills.

The perspective of coal mining in Slovenia and around the world is mainly related to the introduction of environmentally friendly technologies for coal extraction and conversion into electricity with the least possible impact on the environment. Given the fact that the content of ash and free sulfur in Slovenian coal is relatively high, it is necessary to take appropriate care of the products of combustion and flue gas cleaning during energy conversion in the Šoštanj thermal power plant. Useful use or return of large quantities of waste materials generated during the extraction and conversion of coal into electricity in excavated cave spaces significantly

DOI: 10.1201/9781003205906-37

reduces the impact of excavation on the surface, improves the stress-strain state in the rock around the excavation and at the same time eliminates the load on landfills. We are talking about the quantities of several millions m^3 of secondary waste materials that still need to be transported from the place of origin or the thermal power plant to the place of disposal or the excavated underground spaces (Bajželj, 1994).

The connected operation of a coal mine and a thermal power plant, with the return of secondary waste materials (Directive 2006/21/EC) to the underground premises of a coal mine, is called a closed ecological-technological cycle of electricity production or circular management of electricity production (EU Directive 2018/851). more friendly extraction of coal energy and its conversion into electricity. In the process of circular management in electricity production, the following negative impacts occur on the environment (Bajželj, 1994, 1995):

– In the phase of extraction:
   - Change in the primary stress state in the rock,
   - Deformations as a result of demolition processes,
   - Deposition of pit tailings on the surface.

– In the phase of mineral processing:
   - Disposal of separation tailings on the surface,
   - Disposal of separation sludge on the surface,
   - Discharge of technological water into the environment.

– In the phase of conversion of coal into electricity:
   - Disposal of slag, gypsum, ash and fly ash,
   - Emissions of flue gases into the environment (SO_2, NO_x, CO_2).

Most of the negative impacts mentioned before can be eliminated by the introduction of Ecologically-technologically closed cycle of electricity production or circular economy or green mining as shown on Figure 1 below.

Taking into account the principle of waste management, that each waste producer must ensure the proper treatment of waste alone or with an officially recognized contractor (Directive 2008/98/EC or EU Directive 2018/851), the mining (coal extraction) industry has the advantage, that it can use the excavated open spaces for the disposal of practically all secondary waste materials generated in the process of extraction and energy conversion of coal into electricity, as well as secondary waste materials of other industries (see Figure1). In this respect, the connection of the coal mine and the thermal power plant into an ecologically-technologically closed cycle of electricity production or circular management in the production of electricity is an optimal solution.

With the introduction of ecologically-technologically closed cycle of electricity production or circular management in the production of electricity, we gain many advantages, such as (Bajželj, 1995, 2001):

– reduction of coal excavation losses,
– higher quality or calorific value of extracted coal,
– increasing productivity at excavations and in the preparation of preparatory lines,
– reduction of maintenance costs of cave facilities,
– reduction of the possibility of the collapse of hangingwall and the intrusion of water, sludge or liquid sands,
– smaller settlements on the surface,
– Avoid the construction of surface landfills for cave and separation tailings, ash, EF ash and gypsum, flue gas cleaning products.

These advantages can be achieved that the excavated open spaces (goaf areaa) are filled with backfill mixtures of appropriate physical and mechanical properties. The technology of preparation, transport and installation of backfill mixtures must meet the technological requirements of coal extraction.

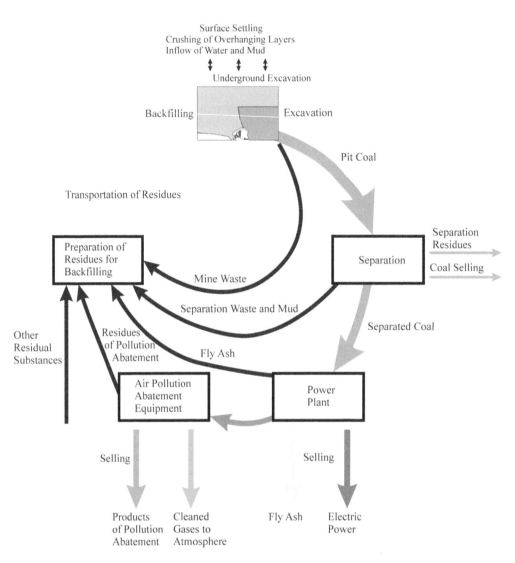

Figure 1. Ecologically-technologically closed cycle of electricity production or circular economy in the production of electricity from coal (Bajželj, 1994).

2 DETERMINATION OF THE BASIC BACKFILL MIXTURES PROPERTIES

Backfill mixtures suitable for transport over long distances of several kilometers and then for injection into the stope goaf area must have well-defined physical, chemical, transport and mechanical properties. By using various secondary waste materials of industry, we can positively influence the required properties of backfill mixtures, but we must take into account the principle without influence or environmental protection. Transporting large quantities of backfill mixtures over long distances requires the preparation of stable suspensions in rheological, chemical and mechanical terms, especially to prevent binding and settling of solid particles during transport and to ensure adequate strength of injected backfill mixtures into the collapsed old excavation work.

Backfill mixtures must achieve the appropriate properties in terms of (Kortnik, 2020):

– Stability (thixotropy) and consistency (ball pull test, measurement of the paste spread on the glass plate)
– Transportability or pumpability (tube viscometer),
– Mechanical properties of injected backfill mixtures (high consolidometer),
– Leaching properties of injected backfill mixtures (pressure leaching test).

The stability (thixotropy) of the backfill mixture is determined by the ball-pull test. The self-made device is shown in Figure 2 (right). The time stability of the backfill mixture is determined by measuring the force required to pull the steel ball. The ball is located at the bottom of the measuring vessel. Measurements are performed at various time intervals after the preparation of the backfill mixture. The measured resistance force of the steel ball at different aging times of the backfill mixture is an assessment of the stability (thixotropy) as a consequence of the settling of solid phase particles in the backfill mixture mainly behave as time-independent non-Newtonian media. This can be expressed by the equation (1);

$$F_{\max} = F_{\max}(t) \tag{1}$$

Where it means,

F_{max} the maximum value of the pull-out force of the steel ball which the suspension has at the bottom of the measuring vessel. Since the measured resistance force is the largest at the bottom of the measuring vessel, it was taken as relevant in estimating the dynamic viscosity coefficient.

t aging time or measurement time after preparation of the backfill mixture.

For easy, fairly accurate and fast determination of the consistency (resistance to shape change) of the backfill mixture, we used the measurement of the paste flow on the glass plate or paste flow value measurement. The measurement is taken from civil engineering standards (flow table test or flow test; ASTM C230, DIN EN 12350-5) and is performed at various time

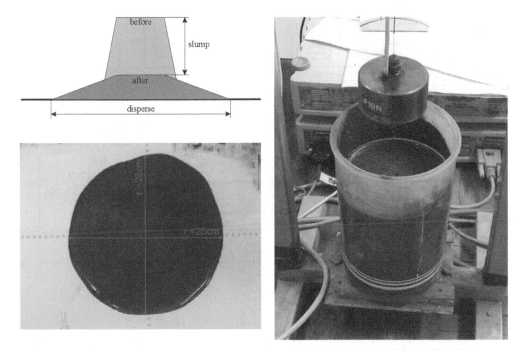

Figure 2. Ball-pull test device for the backfill mixture time stability determination by pulling the steel ball (right) and the slum test for determining the consistency of the backfill mixture by measuring the flow value (left).

intervals. Place a metal cone with a volume of 0.34 dm^3 in the middle of a dry glass plate (1 x 1 m) and fill it with the backfill mixture. Raise the cone quickly and measure the diameter resulting flow of the backfill mixture in two perpendicular directions, as shown in Figure 2 (left).

The backfill mixture flow value at different aging time is equal to the arithmetic mean of both readings; see equation (2);

$$R_{\max} = \frac{r_1 + r_2}{2}(t) \tag{2}$$

Where it means,

R_{max} the maximum flow value of the backfill mixture at different time intervals.
r_1, r_2 diameter resulting flow of the backfill mixture in two perpendicular directions.
t aging time or measurement time after preparation of the backfilling mixture.

The transportability (pumpability) of the backfill mixture, as non-Newtonian suspensions, is determined by a laboratory test of pumping in a tube viscosimeter of our own production, which is schematically shown in Figure 3. The tube viscosimeter has two measuring sections, both 6 m long. The first has a tube diameter of 25.4 mm and the second of 50.8 mm *(Bajželj, 1995)*. The velocity or flow of suspensions, density and temperature are measured in the part of the pipeline which is common to both sections. The purpose of the experiment is to measure the pressure losses when pumping the backfill mixture over long distances, to measure the starting pressures for different aging times of the backfill mixture in the pipeline in case of different pumping failures/interruptions and also to determine the pumping capacity of the backfill mixture.

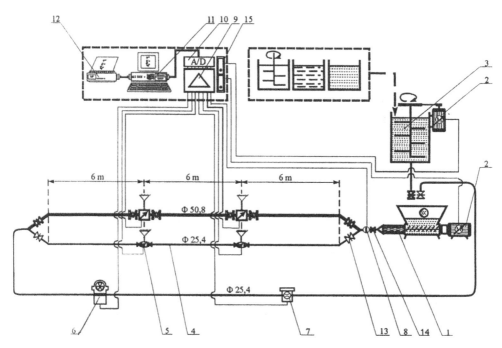

Figure 3.   Shematic representation of the tube viscosimeter (Bajželj, 1995).
LEGEND: 1. Moineau pump; 2. squirrel cage motor; 3. mixer; 4. pipeline; 5. pressure gauge; 6. densimeter; 7. flowmeter; 8. thermometer; 9. carrier amplifier; 10. AD converter; 11. PC computer (486/66MHz): MS-DOS, Windows 95, DOS-ADC, Mathematica; 12. printer; 13. manual valve; 14. compensator; 15. inverter;

The starting pressures for different aging times of the backfill mixture in the pipeline can be calculated by following equation (3);

$$\Delta p_{start} = \frac{p_2 - p_1}{l}(t) \qquad (3)$$

Where it means,
Δp_{start} starting hydraulic pump pressure per meter of the pipeline.
p_1, p_2 measured pressure with pressure gauges (25.4 or 50.8 mm) in tube viscosimeter.
l length between two pressure gauges (6 m) in tube viscosimeter.
t aging time or measurement time after preparation of the backfilling mixture.

Leaching properties of injected backfill mixtures into the goaf area were determined using standard leaching tests (BS EN 14405:2017, etc.) and pressure leaching test (PLT) of own production, which is schematically shown in Figure 4. The use of a PLT enabled to establish similar conditions to those which are present during and after consolidation of injected backfill mixture into the goaf area behind the stope, and simulates leaching after the penetration of groundwater into the old excavated area consolidated with a backfill. The PLT was developed to study the leaching of partially bound granulated and solidly bound or monolithic samples. During measurement of the leaching parameters, PLTs can also be used to measure the initial hydraulic gradient, the coefficient of permeability, and the porosity of injected backfill mixtures.

The mechanical properties of injected backfill mixtures in the old excavation work are determined in laboratory in a high consolidometer equipped with an electronic displacement and compression force meter. Samples of crushed coal/hanging-wall material with injected backfill mixture from a high consolidator are aged for up to 60 days and measurements of basic geomechanical characteristics are performed in the laboratory. The measured values are shown in Table 1.

The measured values of the mechanical properties of the backfill were used to make mathematical geomechanical models of excavation with different hydraulic backfilling of the stope goaf area over entire width or only in areas of delivery and transport stope roads.

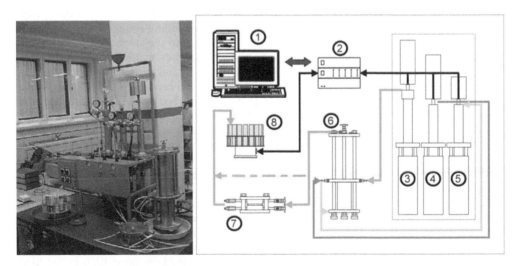

Figure 4.   Schematic representation of pressure leaching test (PLT) equipment (Kortnik, 2000, 2003).
LEGEND: 1. PC computer (486/66MHz): MS-DOS, Windows 95, DOS-ADC, Mathematica; 2. AD converter; 3-5. Hydraulic drive unit; 4. vertical water injection unit into sample; 5. Sample axial pressure; 6. Pressure leaching cell; 7. Open diffusion cell; 8. Automate scale for separate eluate sampling (7x70ml);

Table 1. Geomechanical characteristics of geological and backfilled hardened materials (*Likar, 1994*).

Geotechnical description	Elastic modulus E (MPa)	Poisson's ratio v	Shear module G (MPa)	UCS N (MPa)	Tensile strength T (kPa)	Shear strength R (kPa)	Specific weight γ (kN/m³)
Hanging-wall	300	0.20	125	2.0	250	370	18.7
Coal seam	200	0.35	75	4.0	400	730	13.0
Crushed coal above the stope	3	0.35	1.1	0.3	30	55	11.0
Crushed coal with injected backfill mixture RIL-1 (paste)	80	0.25	32	1.0	120	200	14.0
Crushed coal with injected backfill mixture RIL-1 (paste)	40	0.28	16	0.7	100	155	13.0

UCS . . . Uniaxial compressive strength

3 BACKFILLING SYSTEM IN VELENJE COAL MINE

At the Velenje Coal Mine is excavated the coal seam, with a heat value of 8.4-10.5 MJ/kg, in the form of an elongated lens, 8.3 km long, 2.5 km wide and with a maximum thickness of 170 m at a depth of 200 to 500 m. Coal has been mined here since 1875 (146 years of operation) in several mine-fields as Škale mine, Pesje mine, Preloge mine, Gabrke mine, etc. The excavation technology was developed through simple excavation methods to the high-productive Velenje long-wall mining method (average production 4 million t/year) with fully mechanized equipment (hydraulic shield support with electro-hydraulic control, two-drum cutting extraction machine, front chain conveyor frequency and directional, etc.), the possibility of vertical and horizontal concentration of excavations and the possibility of backfilling of the excavated space behind the stope (VCM, 2020).

Backfilling was in Velenje Coal Mine carried out intensively mainly in the period of years 1992-2004 at stopes k.+157, B/5b, B/6, B/6b in the Škale mine, stope D k.+25 in the Pesje mine, at stope L2 and stopes G1/A, G1/B and G1/C in the north-western part of the Preloge mine (Gaberke mine). A schematic representation of the backfilling system in the Škale mine is shown on Figure 5. and the backfilling mixture injection into the stope goaf area is shown on Figure 6. It was found that the construction of new level excavation roads under areas backfilled with backfill mixtures (pastes) is significantly safer, and with the use of lighter support measures also faster, mainly due to improved geotechnical conditions by backfilling hardened rock. Furthermore, it was found also that the disturbed coal seam hardened with embossed backfill mixture can represent a technical solution to improve geotechnical conditions in continuing coal mining on lower levels towards the seam floor-wall and also in preventing oxidation processes of coal in stope safety pillars during excavations on the level (Lajlar, 2004).

The backfill mixtures receipes were adapted mainly to the better transport conditions (lowest pumping pressure), the requirements for instalation into the old excavation work (easy flow of the backfill mixture), the achievement of the required final geomechanical characteristics and the lowest impact of the eluates on the environment (solidification/consolidation of backfill mixture), especially the hydrosphere. The backfill mixtures were made of various materials such as fly ash, cement, lime, bentonite and various additives to inhibit or accelerate the binding.

The backfilling technology (receipe RIL-1; see Table 2.) used on stope B/5b in the Škale mine first consisted mainly of three main phases (see Figure 5): the phase of backfilling mixtures preparation in the surface mixing station, the phase of hydraulic transport of the backfill mixture by pipeline to the stope and the phase of backfill mixture injection into the goaf area (*Bajželj, 1994*). The achieved results of backfilling the goaf areas of stopes B5/b and B6/b are

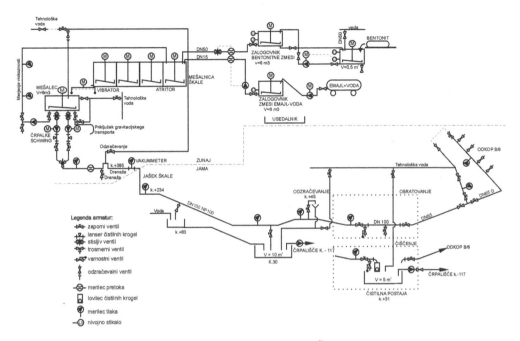

Figure 5.　Schematic representation of backfilling system in the Škale mine (*Bajželj, 1995*).

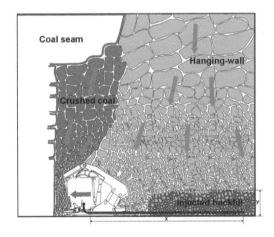

Figure 6.　Schematic representation of backfilling the stope goaf area with the backfill mixture.

collected in Table 3 and Figure 7 graphically shows the heights of the injected backfill mixtures into the stope goaf area B5/b (average 19 cm/m^2).

The backfill technology (RIM-UNI recipe; see Table 2) used on stope D k.+25 in Pesje mine, on stope L2 and stopes G1/A, G1/B and G1/C in the north-western part of Preloge mine was consists of five main phases such as the first phase of RIM-UNI/A liquid backfill mixture preparation at the surface mixing station, the phase of liquid backfill hydraulic transportation by pipeline to the underground mixing station, the phase of preparation of the RIM-UNI/B thick backfill mixture in the underground mixing station, the phase of hydraulic transport of the thick backfill mixture by pipeline to the stope and and the phase of thick backfill mixture injection into the goaf area of the stope. The achieved results of backfilling

Table 2. Recipes for backfill mixtures used at the Velenje coal mine.

Backfill mixture/receipe	RIL-1	RIM-UNI/A	RIM-UNI/B
Flow value	26 cm	40 cm	25 cm
Volume weight	1.55 t/m^3	0.82 t/m^3	1.40 t/m^3
Ratio water:solid	37:63	45:55	40:60
Fly ash	81.8 %	87.6 %	69.9 %
Red-mud Kidričevo	7.3 %	-	-
Hydrated lime Zagorje	7.3 %	8.0 %	6.3 %
Bentonite IBECO S-80	3.6 %	-	-
Portland Cement	-	4.0 %	3.2 %
R-2 retarder TKK	-	0.4 %	0.3 %
Mortar M50	-	-	8.3 %
Bentonite gel (solid:water=1:9)	-	-	11.9 %
Thickener L TKK	-	-	0.1 %

Table 3. Achieved parameters of backfilling with backfill mixtures (pastes) on stopes B5/b, B6/b, G1/A and G1/B.

Stope No.		B5/b	B6/b	G1/A	G1/B
Backfilling operation	(days)	31	40	95	48
Injected paste quantity	(m^3)	947	2,389	3,606	2,097
Max. paste injected heigh	(cm/m^2)	24	65	70	51
Averg. paste injected height	(cm/m^2)	19	25	38	30

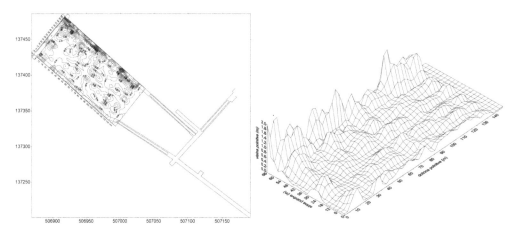

Figure 7. Graphic representation of backfilling results over entire width on the stope B5/b (Bajželj, 1995).

the goaf areas of stopes G1/A and G1/B are collected in Table 3 and Figure 8 graphically shows the heights of the injected backfill mixtures into the stope goaf area G1/A (average 30 cm/m^2).

The main purpose of the backfill introduction in the Velenje Coal Mine were:

– consolidation of the excavated old area behind the stope and improvement of geotechnical conditions for the construction of lower-lying floor excavation road,

430

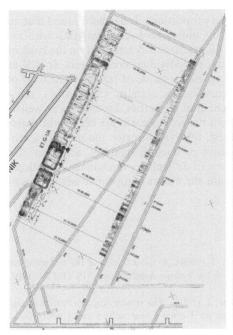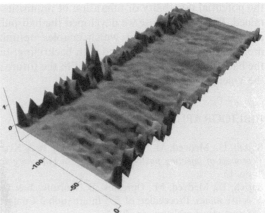

Figure 8. Graphic representation of backfilling results on stope G1/A only in close areas of delivery and transport stope roads (Kortnik, 2020).

- – prevention of oxidation processes in coal in safety columns during the stope excavation,
- – construction of an indicative backfill to control the excavation height on the lower-lying stopes,
- – return of secondary materials (fly ash, gypsum, secondary materials of industry) to the underground openings and solidification of backfilled areas.

Despite many established advantages, the backfill method in the Velenje Coal Mine was intensively used until the end of 2004, but then due to relatively good geomechanical conditions and due to the reduction of excavation costs in the north-western part of Preloge mine, the backfilling method was gradually abandoned.

5 SUMMARY

Mining methods with backfill enable the introduction of Circular economy or closed Eco-logical-Technological circle and thus Environmentally friendly mining or Green mining of mineral resources. This means that various waste materials generated during the extraction and processing of mineral raw materials are used for the preparation of backfill mixtures for filling and consolidation of open underground spaces, and at the same time we reduce the quantities of these substances deposited in surface landfills. In addition, some secondary industrial waste materials can be also used to achieve the required properties of backfill mixtures, in doing so, we are obliged to observe the principles of environmental protection. The technical, economic, environment and other aspects of the thick backfill mixtures (pastes) may be estimated only if their rheological hydraulic transportation properties are previously determined.

The basis of all backfill mixtures used in Velenje Coal Mine is fly ash from the nearby Šoš-tanj Thermal Power Plant, to which we add various additives and secondary materials of industry to ensure stability. The stability of the backfill mixture is greatly influenced by the average particle size, their granulation distribution and proportions. It has been found that an appropriate time-stable backfill mixture is achieved by adding secondary materials with the smallest possible particle size. The suitability of backfill mixtures is determined on the basis of investigations of the physical and chemical properties of individual components as well as whole backfill mixes. For backfill mixtures, depending on the mode of transport, method of filling and required geotechnical properties after installation, stability (thixotropy), rheological properties, pressure gradient of pumping, mechanical properties, leachability as well as the potential possibility of migration of impurities into the environment are need to be determined. For this purpose we developed the ball-pull test, tube viscosimeter and pressure leaching test (PLT) equipment which enables optimization of backfill mixtures and thus also modification of mining method with backfilling. In case of difficult geomechanical conditions, the backfilling method can be reused in the future again and ensure competitive advantages of the Velenje Coal Mine.

BIBLIOGRAPHY

Bajželj, U., Medved, M., Oprešnik, B., Jenko, B. & Križan, J. (1994). *Avtomatization of the technological process of injecting paste made of fly ash into the goafs in the Velenje lignite mine*, APCOM'94, Bled, pp. 13.
Bajželj, U., Medved, M., Oprešnik, B., Kortnik, J. & Likar, J. (1994). *The return of waste materials into active mines.* Proceeding of 2nd International Conference on Tunnel Construction and Underground Structures, Ljubljana, pp. 20.
Bajželj, U., Likar, J., Medved, M., Oprešnik, B., Kortnik, J. (1995). *Priprava in namen uporabe pepelnih past ter tehnologija vtiskanja v razrušene odkopane prostore.* – Rudarsko posvetovanje ob 34. skoku čez kožo, Ljubljana, p. 73–97 (in Slovene).
Bajželj, U., Žerdin, F., Medved, M., Veselič, M., Šubelj, A., Fece, V., Dervarič, E. & Veber, I. (1995). *Ecological Aspects of Thick Coal Mining* - Report on accomplishment of the Slovenian-American Research Project. IRGO, Ljubljana, pp. 29.
Bajželj, U., Kortnik, J. (2001). *Tehnični aspekti uporabe odpadnih snovi v rudarstvu.* Zbornik 37. skoka čez kožo, Ljubljana, pp. 13 (in Slovene).
Kortnik, J. (2000). *Use of the pressure leachate control test for the selection of multi-barrier disposal systems.* Proceedings of 5th International symposium on Environmental Geotechnology and Global Sustainable Development, 17.-23. August 2000, Belo Horizonte, Minas Gerais, Brazil, pp. 6.
Kortnik, J. (2003). *Backfilling waste material composites environmental impact assessment.* Journal of the South African Institute of Mining and Metallurgy, July/August 2003, p. 391–396.
Kortnik, J. (2020). *Circle Economy in mining industry – Underground mine hydraulic backfill use in an ecologically and technologically closed circle.* Proceedings of 2nd International Multidisciplinary Geo-Science Conference – IMGC2020, 8.-9. October 2020, p. 11–21.
Lajlar, B. (2004). *Zapolnjevanje in utrjevanje s pepelnimi mešanicami pri odkopavanju severozahodnega predela jame Pesje* – seminarska naloga. UL-NTF, Ljubljana, pp. 30 (in Slovene).
Likar, J. (1994). *Rezultati geotehničnih raziskav možnosti utrjevanja starega dela v Rudniku lignita Velenje* – poročilo. IRGO, Ljubljana, junij 1994, pp. 30 (in Slovene).
VCM - Velenje Coal mine, [visit on: 18.09.2020]. Retrieved from: http://www.rlv.si/en/.

Author Index

Antunes, P. 360

Bawden, W. 127
Bertrand, V. 244
Brown, D. 244

Correia, L. 360
Cothill, B. 360
Creber, K. 53
Czaban, S. 90, 143
Costa e Silva, M. 169

Dennis, B. 67
Dewit, D. 375
Dyduch, G. 349

Erismann, F. 3
Evans, R. 185

Fabien, B. 44

Grabinsky, M. 108, 118, 127
Grobler, J.J. 382
Grodzicka, A. 317
Gruszczyński, M. 90, 143
Guo, L. 80, 227

Hahn, J. 194
Haley, B. 53
Hansson, M. 3, 169
Hassani, F.P. 234
He, M. 234
Helinski, M. 67, 217, 409

Jacobs, J. 53
Jafari, M. 118
James, S. 360
Jarczyk, P. 349
Jendryś, M. 349

Kępys, W. 102
Kermani, M.F. 234
Kortnik, J. 422

Landriault, J. 375
Lee, C. 204, 244
Liu, G. 80
Lutz, T. 22

Mcgrath, S. 217
Mcguinness, M. 53
Mertuszka, P. 288
Mierzejowska, A. 272
Moraczewska-Majkut, K. 317
Musioł, D. 337

Niu, H. 234

Palac-Walko, B. 288
Pan, A. 108
Peschken, P. 22
Pierzyna, P. 327
Pomykała, R. 102
Poniewiera, M. 272
Popczyk, M. 337
Pratkowiecki, R. 143
Pytel, W. 288

Ruptash, K. 44
Rzepecki, S. 317

Shaw, J. 409
Silva, M. 169
Skrzypczak, Z. 143
Smolnik, G. 261
Sokoła-Szewioła, V. 272
Stefanek, P. 143
Stewart, M. 244
Stone, D. 35, 44
Strozik, G. 307

Thompson, B. 127
Timmis, B. 244
Ting, B. 44
Trinker, G. 185

Van Der Spuy, B. 153
Veenstra, R.L. 382, 398

Wei, X. 227
Wickens, J.D.V. 13
Wilson, S. 13
Wu, W. 80

Yamine, J. 375
Yang, X. 80

Zajac, A. 398
Zhang, Z. 80
Zieliński, S. 90